Multiscale Simulations for Electrochemical Devices

JENNY STANFORD PUBLISHING

Multiscale Simulations for Electrochemical Devices

edited by

Ryoji Asahi

Published by

Jenny Stanford Publishing Pte. Ltd.
Level 34, Centennial Tower
3 Temasek Avenue
Singapore 039190

Email: editorial@jennystanford.com
Web: www.jennystanford.com

British Library Cataloguing-in-Publication Data
A catalogue record for this book is available from the British Library.

Multiscale Simulations for Electrochemical Devices

ISBN 978-981-4800-71-6 (Hardcover)
ISBN 978-0-429-29545-4 (eBook)

Contents

Preface

In recent years, environmental protection and sustainability have become major concerns in society as the industry and economy have grown substantially to meet the energy demand. Therefore, a reduction in CO_2 emission and the implementation of clean energy are inevitable challenges for scientists and engineers today. The development of electrochemical devices, with a particular focus on fuel cells, Li-ion batteries, and artificial photosynthesis in this book, is vital for solving environmental problems, and this is where a lot of research in science and industry has focused. The development of a practical device requires multiple steps, such as materials development, device design, and operational system design. These steps are interdependent, and maximizing a single characteristic does not significantly improve the overall performance. In most cases, the optimization among variables such as performance, cost, and reliability involves serious trade-offs. Therefore, developing a device with commodity value and competitiveness is a process that takes substantial time and is costly. A powerful tool to significantly increase efficiency of such research and development is computational simulations that enable analysis, understanding, and prediction and thus lead to the design and optimization of a device through suitable computational modeling. The central issue we aim to cover in this book is such computational simulations specifically used for electrochemical devices.

Computational modeling and simulation has been well recognized as a useful means of understanding phenomena. After confirming that the method and model of choice sufficiently reproduce experimental results, deep physical insights can be gained by analyzing the output quantities that the simulation can access. Taking this procedure further, a systematic study by changing materials, input parameters, and environmental factors gives an opportunity for more efficient optimization and improvement of the device than can be gained by only performing experiments. This so-called computational design is enabled by the rapid development of computers and computational algorithms in recent years, and already many successful examples can be seen in literature.

Among computational simulations, the first-principles calculation based on the density functional theory (DFT), where only atomic arrangement is required as an input to predict the physical properties and stability of arbitrary crystals and molecules, has been recognized as an accurate and robust tool for materials design. However, even with the current technology, the scale that can be handled by the first-principles calculation is typically within 1000 atoms or a few nanometers and the timescale is limited to approximately 100 ps. When dealing with actual materials, it is often required to simulate microscale texture and diffusion phenomena in more than nanosecond time span. Several techniques have been proposed to enable simulation on a large scale, for example, force-field molecular dynamics, which simulates atomic-level dynamics with approximate force-field potential; coarse-grained modeling, which can handle a polymer or protein by coarsening it; and phase-field modeling and finite-element methods, which treat material structure as a continuum to simulate thermodynamics and multiphysics quantities.

The packaging or linkage of these different time and space simulations is called multiscale simulation. When designing a whole device, it is important to predict the macroscopic behavior that can be observed in experiments and often appears as the performance and reliability of the device. On the other hand, dominant factors attributed to such a macroscopic behavior often originate in microscopic properties. This is a situation in which there is a substantial need for multiscale simulation. In practice, however, it is not easy to transfer inputs and outputs between different scales of simulations to predict multiscale physical properties because each simulation is governed by physical and chemical equations in their approximated forms valid only for the space and timescale of the simulated object. Therefore, one should take careful steps, understanding the limits and precision of the equations, to set an appropriate model for each simulation using data from the prior steps. The multiscale simulation then should be a powerful tool for comprehensively understanding and solving the problem.

In this book, we examine a variety of electrochemical devices related to energy and environmental issues. The authors have been involved in the development of actual energy devices and materials in the industry for many years. In each chapter, after reviewing the calculation methods commonly used in the field, we focus on a specific computational approach that is applied to a realistic problem crucial for the improvement of the device. We introduce the simulation technique not only as an analysis tool to explain experimental results but also as a design tool in the scale of interest. At the end of each chapter, a future perspective is added as a guide for the extension of research.

The computational approach and the device application for each chapter is listed as follows: in Chapter 1, DFT calculations are employed for materials design of hydrogen storage; in Chapter 2, DFT with a global reaction route mapping method is employed to understand electrochemical reactions of electrolytes in a Li-ion battery; in Chapter 3, DFT is extended to treat the electrochemical interface often appearing in fuel cells; in Chapter 4, DFT and nonadiabatic molecular dynamics are employed to understand key factors in artificial photosynthesis; in Chapter 5, a hybrid quantum-classical simulation is employed to treat a large-scale atomistic simulation in a Li-ion battery; in Chapter 6, force-field molecular dynamics and coarse-grained simulation are employed to understand dynamical behavior in the polymer electrolyte membranes of a fuel cell; in Chapter 7, the phase-field model is employed to understand structural and thermodynamical behaviors in microstructures of electrode materials; in Chapter 8, thermal abuse modeling is employed for the device simulation of a Li-ion battery; and in Chapter 9, macrohomogeneous device modeling is employed to understand the performance loss accompanying the catalyst loading reduction needed in commercializing a fuel cell.

It is not possible to encompass a wide range of simulation techniques, and that is not the purpose of this book. Instead, we emphasize the selection and construction of methods to create essential physical models and link them to tackle complex real problems. If this book is useful in guiding not only researchers working now on energy materials and devices but also students and learners who are about to work on them, it would be our great pleasure.

We appreciate the staff in the editorial office of Jenny Stanford Publishing, who gave us this opportunity and patiently helped us to realize this book. In addition, we thank our collaborators in Toyota Central R&D Laboratories, Inc., Toyota Motor Corporation, Tohoku University, Nagoya Institute of Technology, and University at Buffalo SUNY, who have worked together with us in devices and materials development for years and contributed considerable scientific and technological achievements, which are in part introduced in this book. We would like to give special thanks to Atsuo Fukumoto and Shi-Aki Hyodo, who initiated computational methods as leaders in the TCRDL. Finally, I would like to express my gratitude toward Arthur J. Freeman, who believed in the power of theory and has guided me as a mentor.

Ryoji Asahi

2020

Chapter 1

Computational Materials Design for Hydrogen Storage

Kazutoshi Miwa
Toyota Central R&D Laboratories, Inc., Nagakute, Aichi 480-1192, Japan
miwa@cmp.tytlabs.co.jp

1.1 Background

Hydrogen has attracted growing interest as a secondary energy source for a sustainable society. Since hydrogen is in gas phase at ambient conditions, efficient storage methods are required to reduce its enormous volume. There are several options for hydrogen storage [54], for example, compressed hydrogen, liquefied hydrogen, physisorption on material surfaces, hydrogen intercalation in metals, and complex hydrides. The last two classes of materials are the main subject of this chapter.

Metal hydrides are able to store hydrogen in a solid form. Some alloys and intermetallic compounds can reversibly absorb and desorb hydrogen under moderate operating conditions [47], which

Multiscale Simulations for Electrochemical Devices
Edited by Ryoji Asahi

ISBN 978-981-4800-71-6 (Hardcover), 978-0-429-29545-4 (eBook)
www.jennystanford.com

offers a storage means with high safety and volume efficiency. The most popular materials of this type are $LaNi_5$ and related alloys. They show reversible hydriding/dehydriding reactions near room temperature with a volumetric hydrogen capacity of about 120 kg H_2/m^3, which is almost the double of liquid hydrogen. A $LaNi_5$-based alloy has been recently utilized in a commercialized large-scale energy supply system [26]. Other metallic compounds for hydrogen storage are also available, such as TiFe and V-based alloys. Currently, all the available reversible hydrides working under ambient conditions are composed of transition-metal elements. Because of their large atomic masses, the reversible gravimetric hydrogen density is limited to about 2 mass percent H_2. Although this limitation is not so severe for the stationary system, it becomes a serious drawback when considering mobile and portable applications.

Complex hydrides are also potential candidates for hydrogen storage due to their high gravimetric densities of hydrogen [42]. Examples include alanates and borohydrides such as $NaAlH_4$ and $LiBH_4$, which involve 7.5 and 18.5 mass percent H_2, respectively. Unlike conventional hydriding alloys, complex hydrides show dehydriding reactions accompanied by decomposition to some products. Because of this nature, no one ever thought that complex hydrides can show reversible hydriding/dehydriding reactions from the gas phase point of view. In this concern, it was first reported in 1997 [5] that the gas phase reaction for $NaAlH_4$ can be made reversible by the addition of Ti catalysts. Motivated by this discovery, a lot of attention has been paid to complex hydrides as hydrogen storage materials. However, they are, in general, thermodynamically rather stable and desorb hydrogen only at elevated temperatures.

In this chapter, theoretical studies on two types of hydrogen storage materials, transition-metal hydrides and borohydrides, are presented. For transition-metal hydrides, the thermodynamics for hydrides is briefly explained and then the energetics of the CaF_2-type dihydrides is discussed. For borohydrides, the fundamental properties of these compounds are investigated. From the obtained results, it is shown that the electronegativity of cation elements is a good indicator to measure the stability of borohydrides.

1.2 Methodology

First-principles calculations based on density functional theory [21, 23, 25] have been successfully applied for a wide variety of materials and phenomena. The calculations stand on universal approximations and essentially do not require any experimental data and/or empirical parameters to be fitted. This universality is beneficial for investigating hydrogen storage materials, since hydrogen is a highly reactive element and can form hydrides with many other elements, forming various types of bonds, such as metallic, ionic, and covalent.

All the density functional theory calculations presented in this chapter have been performed using the ultrasoft pseudopotential (USPP) method [52] with the plane wave basis set. The USPP approach provides rapid convergence of the plane wave expansion regardless of the elements, keeping good transferabilities. The analytic evaluation of atomic forces and stress tensor—which are essential for efficient structural optimization, in particular for low-symmetry systems—is easily implemented [27].

In addition to the conventional total-energy calculation, the linear response method [3, 4, 8, 14, 15] has been utilized to compute the total-energy derivatives efficiently. On the basis of the $2n + 1$ theorem [12, 13], we are able to access the total-energy derivatives up to the $(2n + 1)$th order when the nth-order wave functions are available. For example, the atomic forces and stress tensor mentioned above, which correspond to the first derivative of the total energy, can be calculated from the ground-state wave functions (namely, $n = 0$). Thus, computing the first-order wave functions by the linear response calculation, the second and/or third derivatives of the total energy are obtained. These derivatives are closely related to various properties of a material, such as dielectric susceptibility tensors, Born effective charges, force-constant dynamical matrices [14, 15, 41], elastic constants [18], piezoelectric constants [7], nuclear-magnetic-resonance chemical shifts [45], phonon anharmonic couplings [43], nonlinear optical constants, and Raman intensities [38, 53].

By diagonalizing the force-constant dynamical matrix, the phonon eigenmode frequencies are obtained, from which the zero-point energy (ZPE) is evaluated as described in the next section. The Born effective charge tensor as well as the dielectric susceptibility tensor are required to describe the long-range dipole–dipole interaction in polar crystals. The Born effective charges are also useful to characterize the bonding nature of insulators as seen in Section 1.4.1, which provide the unique charge partitioning in a physically transparent way.

1.3 Transition-Metal Hydrides

1.3.1 Thermodynamics for Hydrides

The main purpose of this chapter is to investigate the thermodynamical stability of hydrides. Assuming the equilibrium condition for the hydrogenation reaction, the relation between dissociation pressure p and temperature T is represented by a well-known van't Hoff equation,

$$\ln\frac{p}{p_0}=\frac{\Delta H}{RT}-\frac{\Delta S}{R}, \qquad (1.1)$$

where ΔH and ΔS are the enthalpy and entropy changes, respectively, R is the gas constant, and p_0 (= 0.1 MPa) is the standard pressure. Since ΔS is mostly due to the entropy loss of molecular hydrogen gas, it can be approximated in most cases by the standard entropy of hydrogen with the negative sign, $\Delta S = -130$ J/(mol H_2·K). Therefore, it is crucial to control ΔH for tuning the operating condition of hydrogen storage materials. To obtain an equilibrium pressure range $p = 0.1$–35 MPa at $T = 300$ K, the enthalpy charge should be $\Delta H = -39$ to -24 kJ/mol H_2.

The enthalpy change ΔH for the reaction is usually calculated from the total-energy difference between reactants and products. In the first-principles total-energy calculation, the atomic nuclei are treated as classical particles and the quantum mechanical effect for them is ignored. Because of the light mass of hydrogen, however, the contribution of the ZPE to ΔH becomes considerable. This contribution can be estimated with the harmonic approximation: using the phonon eigenmode frequencies ω_k, ZPE is given by $E_{zero} = \Sigma_k \hbar\omega_k/2$ [24].

Table 1.1 shows the calculated ΔH for the CaF_2-type dihydrides, TiH_2, and VH_2 according to the following reaction [33]:

$$M + H_2 \rightarrow MH_2. \quad (1.2)$$

In this case, ΔH corresponds to the heats of formation of hydrides. To clarify the ZPE contribution to ΔH, the values both with and without ZPE are given. The amount of ZPE correction is about 20 kJ/mol H_2 for both hydrides, and this correction improves the agreement with the experimental data significantly. Because the ZPE correction is comparable in size to the target range of ΔH, it is important to take it into account for a quantitative prediction.

Table 1.1 Heat of formation for CaF_2-type dihydrides, ΔH (kJ/mol H_2)

	Theory		Experiment[a]
	without ZPE	with ZPE	
TiH_2	−145	−125	−124
VH_2	−65	−43	−40

[a]Experimental data are taken from Ref. [11].

1.3.2 Energetics of Transition-Metal Dihydrides

In this section, the energetics of 3d transition-metal dihydrides in the CaF_2-type structure is investigated using first-principles calculations [33]. The CaF_2-type dihydride is commonly observed for the early transition metals, and their alloys, such as a Ti-V-Cr alloy, are an important class of hydrogen storage materials [1].

To examine the chemical trends along the 3d series in the periodic table, five hydrides, TiH_2, VH_2, CrH_2, FeH_2, and NiH_2, are considered, where the latter two hydrides are hypothetical. Though CrH_2 has been reported in literature [49], no thermodynamic data are available.

Hydrogenation reactions are often accompanied by the structural transformation and lattice expansion of host metals. To discuss these effects separately, the hydride formation process is divided into three steps, as shown in Fig. 1.1. Figure 1.1a shows the structural transformation of the host metal to a nonmagnetic fcc structure, Fig. 1.1b the lattice expansion of the fcc host metal from its equilibrium spacing to that of dihydride, and Fig. 1.1c the insertion of hydrogen atoms into the expanded fcc metal lattice. The energetics of the former two steps is mainly determined from the nature of the host metal. The hydrogen-metal interaction is strongly reflected in the last step. For simplicity, the ZPE correction is not applied in this section. It is expected that this treatment gives only a minor effect for the present analysis, since the size of the ZPE correction probably is not so sensitive to the metal elements.

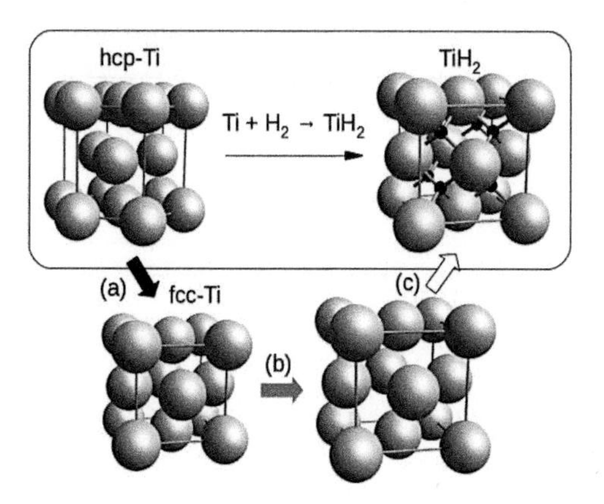

Figure 1.1 Hypothetical hydrogenation process for Ti. The large white and small black spheres indicate Ti and H atoms, respectively. (a) Structural transformation, (b) lattice expansion, and (c) hydrogen insertion.

The results of the energy decomposition are given in Fig. 1.2, together with the heats of formation for the dihydrides. The structural transformation energy of Ti is small because of the similarity between the hcp and fcc structures. For Ni, this energy cost is due to the elimination of the ferromagnetic structure and also small. The other metals have a bcc structure, for which the structural transformation energy is moderate.

The lattice expansion energy monotonically increases with increasing an atomic number until Fe and then decreases for Ni. This trend can be understood from the d-band filling model proposed by Friedel [10]. Transition metals have the maximum in cohesion at the center of the transition-metal series, where the d bands are half filled. Hence, the energy cost for lattice expansion also becomes maximum in the middle of the 3d series. The convex behavior of this energy contribution may correspond to Miedema's reverse stability rule [32]. Owing to the deformation of the metal lattice, the above two contributions always give the energy cost.

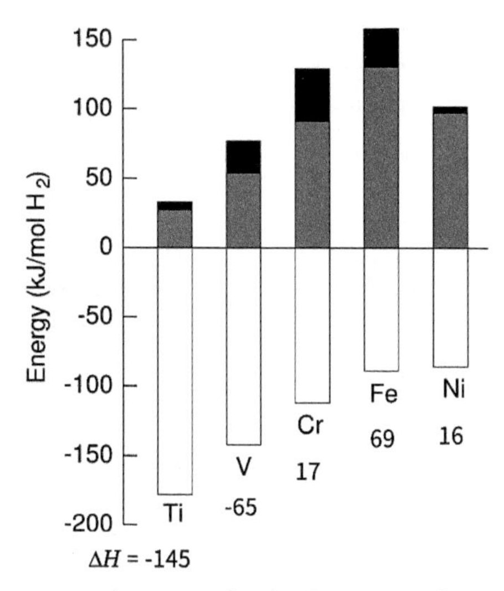

Figure 1.2 Decomposed energies for the formation of CaF_2-type transition-metal dihydrides. The black, gray, and white parts indicate the structural transformation energy, the lattice expansion energy, and the hydrogen insertion energy, respectively. The heat of formation (the sum of three energy components) for each hydride is given at the bottom, with the unit kJ/mol H_2. Reprinted (figure) with permission from Ref. [33]. Copyright (2002) by the American Physical Society.

The remaining step is hydrogen insertion. It has been found that this contribution can be approximated by a simple geometric model. Figure 1.3 shows the hydrogen insertion energy E_I as a function of the hydrogen-metal interatomic distance R. The points indicate the calculated values, where the left and right end points for each element correspond to the equilibrium lattice spacing of the fcc metal and the dihydride, respectively. The relation between E_I and R is approximately represented by a single curve regardless of the host metal elements. This suggests that the hydrogen-metal interatomic distance, or the interstitial hole size, is a dominant factor for the hydrogen insertion energy. According to Ref. [17], the calculated data are fitted to the analytic function

$$E_I = \alpha R^{-n} - \beta. \tag{1.3}$$

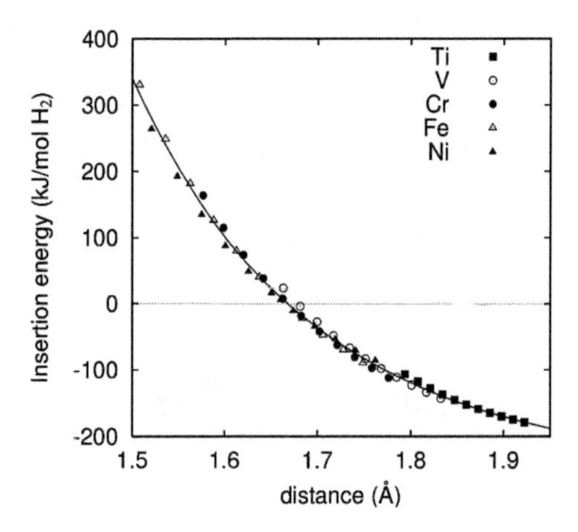

Figure 1.3 Hydrogen insertion energy for CaF_2-type transition-metal dihydrides as a function of hydrogen-metal interatomic distance. Reprinted (figure) with permission from Ref. [33]. Copyright (2002) by the American Physical Society.

When E_I is expressed in the unit of kJ/mol H_2 and R in Å, the least-squares fitting gives the parameters as $\alpha = 1.393 \times 10^4$, $\beta = 2.680 \times 10^2$, and $n = 7.723$, with a root-mean-square error of 7.4 kJ/mol H_2. The extension of this relation to alloy systems is also available [33].

The hydrogen insertion energies are negative for all metals considered here. The heats of formation are determined from the cancellation between this energy gain and the cost due to the metal lattice deformation. From Ti to Fe, the energy gain decreases and the cost increases, and so the heats of formation vary monotonically from strongly negative to positive. For Ni, the hydrogen insertion energy is almost the same as that for Fe, but the deformation cost is reduced because of the small lattice expansion energy. This is responsible for the small positive ΔH of NiH_2. A similar mechanism is expected for the exothermic reaction between Pd and H, which is a noble exception in the late transition metals.

1.4 Borohydrides

1.4.1 Fundamental Properties of $LiBH_4$

Among complex hydrides, lithium borohydride has an extremely high gravimetric hydrogen density of 18.5 mass percent H_2. This material is, however, thermodynamically too stable and so desorbs hydrogen only at elevated temperatures (>650 K). At ambient conditions, the crystal structure of $LiBH_4$ has been reported to be orthorhombic, with the space group *Pnma* [50]. Figure 1.4 depicts the crystal structure of orthorhombic $LiBH_4$, in which a boron atom forms a tetrahedral complex with the surrounding four hydrogen atoms.

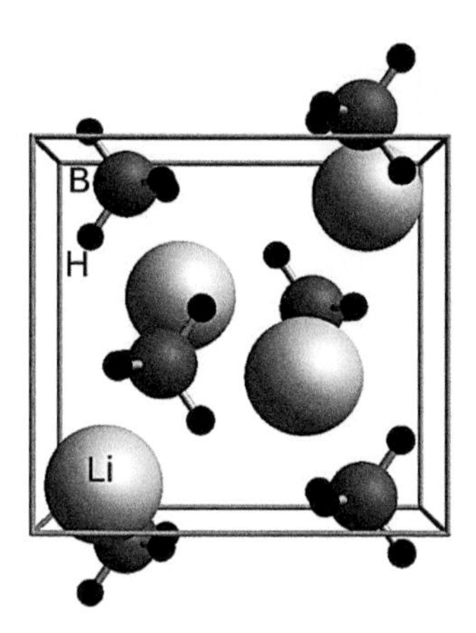

Figure 1.4 Crystal structure of $LiBH_4$ in the orthorhombic phase. Large white, middle gray, and small black spheres indicate Li, B, and H atoms, respectively.

At high temperatures (~381 K), the compound undergoes a phase transition to a hexagonal structure with the space group $P6_3mc$, where tetrahedral BH_4 complexes are also held. In this section, the first-principles calculations are carried out for orthorhombic $LiBH_4$ to investigate its fundamental properties [34].

Table 1.2 shows the result of structural optimization, together with that of the synchrotron X-ray powder diffraction experiment [50]. Although the agreement between them is fairly good for Li and B positions, some discrepancies are found for H. The calculation gives a nearly ideal tetrahedral geometry for BH_4 complexes: B–H bond lengths are almost constant (d_{B-H} = 1.23–1.24 Å) and bond angles are θ_{H-B-H} = 108°–110°, which are close to the ideal value of 109.5°. On the other hand, the experiment suggests a strongly distorted shape for BH_4: d_{B-H} = 1.04–1.28 Å and θ_{H-B-H} = 85°–120°. These discrepancies are probably due to the experimental difficulties in identifying hydrogen positions because of their weak X-ray scattering power. The recent single-crystal synchrotron diffraction study [9] indicates that the BH_4 group in $LiBH_4$ has a nearly ideal tetrahedral geometry, as predicted in our calculation.

Table 1.2 Crystallographic parameters of orthorhombic $LiBH_4$.

			Theory			Experiment[a]	
Atom	**Site**	*x*	*y*	*z*	*x*	*y*	*z*
Li	4c	0.1552	1/4	0.1137	0.1568	1/4	0.1015
B	4c	0.3141	1/4	0.4229	0.3040	1/4	0.4305
H1	4c	0.9131	1/4	0.9263	0.900	1/4	0.956
H2	4c	0.4061	1/4	0.2656	0.404	1/4	0.280
H3	8d	0.2145	0.0246	0.4224	0.172	0.054	0.428

[a]Experimental data are taken from Ref. [50].

Source: Reprinted (table) with permission from Ref. [34]. Copyright (2004) by the American Physical Society.

Note: Space group *Pnma* (No. 62). The lattice constants are fixed at the experimental values $a = 7.179$ Å, $b = 4.437$ Å, and $c = 6.803$ Å.

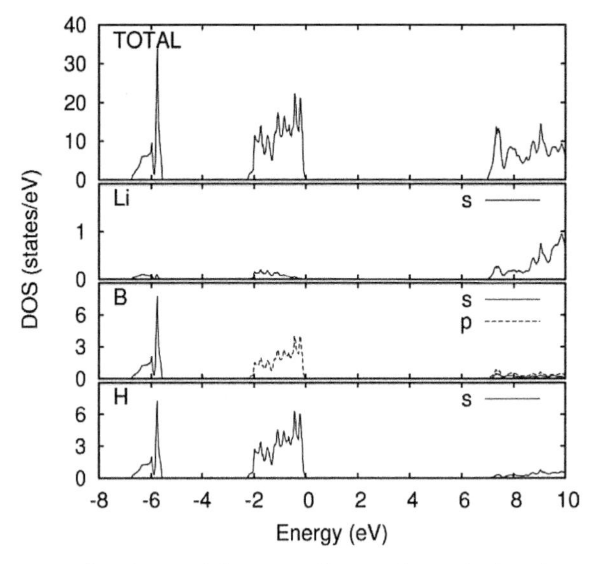

Figure 1.5 Total and partial densities of states for orthorhombic $LiBH_4$. The origin of the energy is set to be the top of valence states. Reprinted (figure) with permission from Ref. [34]. Copyright (2004) by the American Physical Society.

The electronic density of states (DOS) for $LiBH_4$ is shown in Fig. 1.5. The electronic structure is nonmetallic, with a calculated energy gap of 6.8 eV. Since the contribution of Li orbitals to the occupied states is rarely seen, Li atoms are thought to be ionized as Li^+ cations. The occupied states split into two peaks whose states are composed of B-2s, B-2p, and H-1s orbitals. A boron atom constructs sp^3 hybrids and most likely forms covalent bonds with the surrounding four H atoms, similar to a CH_4 molecule, where a deficient electron to form these bonds is compensated by a Li^+ cation.

To further analyze the bonding property of $LiBH_4$, the Born effective charge tensors Z^* are calculated using the linear response method [41], which provide a clear bonding picture for insulators since they are observable and well-defined quantities. The results are summarized in Table 1.3.

Table 1.3 Born effective charge tensors of orthorhombic $LiBH_4$, Z^*

Atom	Z^*_{xx}	Z^*_{yy}	Z^*_{zz}	$1/3\ \mathrm{Tr}[Z^*]$
Li	0.98	1.08	1.02	1.03
B	0.10	0.15	0.06	0.10
H1	−0.28	−0.18	−0.33	−0.26
H2	−0.32	−0.15	−0.41	−0.29
H3	−0.24	−0.45	−0.18	−0.29

Source: Reprinted (table) with permission from Ref. [34]. Copyright (2004) by the American Physical Society.
Note: The diagonal elements of tensors and their average values are given.

For Li, the diagonal elements of Z^* are almost isotopic and their average is close to the expected nominal value of +1. Because of the acoustic sum rule, BH_4 complexes are considered to be ionized as $[BH_4]^-$ anions. This supports an ionic interaction between Li and BH_4 expected from the electronic DOS analysis. If an ionic picture is also assumed for the internal bonding of $[BH_4]^-$, the effective charge should be +3 and −1 for B and H atoms, respectively. The absolute values of Z^* for B and H are, however, considerably smaller than these nominal values, and thus it is expected that the internal bonding of $[BH_4]^-$ is primarily covalent.

Figure 1.6 shows the Γ-phonon eigenmode frequencies. They are classified into three groups: group I, less than 500 cm^{-1}; group II,

1000–1300 cm^{-1}; and group III, 2250–2400 cm^{-1}. The eigenmodes in group I are composed of displacements of both Li^+ cations and $[BH_4]^-$ anions, whereas those in groups II and III originate from the internal H–B–H bending and B–H stretching vibrations of $[BH_4]^-$ complexes, respectively. For comparison, the normal mode frequencies of an isolated $[BH_4]^-$ anion are also shown in Fig. 1.6, which are in relatively good agreement with the phonon eigenmode frequencies in groups II and III. The molecular approximation holds well for $LiBH_4$, suggesting strong internal bonding of $[BH_4]^-$.

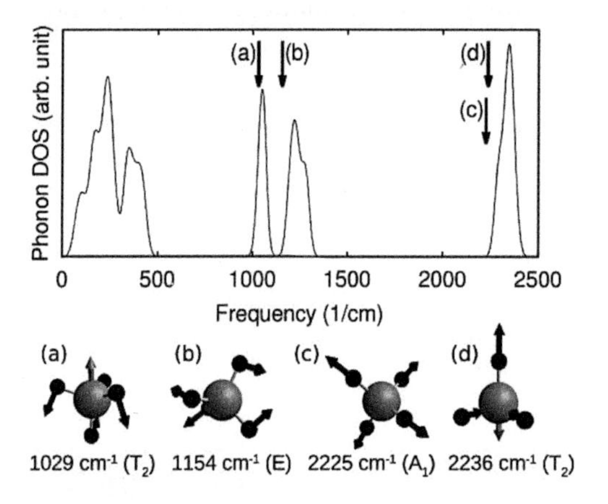

Figure 1.6 Phonon eigenmode frequencies of orthorhombic $LiBH_4$. The top panel shows phonon DOS calculated from the optical Γ-phonon modes, where Gaussian broadening with 30 cm^{-1} width is used. Arrows indicate the normal mode frequencies of a free $[BH_4]^-$ anion, whose vibration patterns and the symmetry labels for the point group T_d are given (a–d).

The heat of formation for the reaction

$$\mathrm{Li + B + 2H_2 \rightarrow LiBH_4} \tag{1.4}$$

is calculated as ΔH = –160 kJ/mol, including the ZPE contribution. The agreement with the experimental value of –194 kJ/mol [55] is reasonably good. The actual hydrogen desorption reaction proceeds through the following one, accompanied by the formation of LiH:

$$\mathrm{LiBH_4 \rightarrow LiH + B + \frac{3}{2}H_2}. \tag{1.5}$$

The enthalpy change for this reaction is predicted to be ΔH_{des} = 56 kJ/mol H_2, which agrees well with the experimental value, 69 kJ/mol H_2 [55]. Comparing this ΔH_{des} with the target range of ΔH mentioned in Section 1.3.1, it can be confirmed that $LiBH_4$ is thermodynamically too stable to desorb hydrogen at ambient conditions.

The theoretical analyses for the electronic, dielectric, and vibrational properties suggest that a $[BH_4]^-$ anion complex forms the stable covalent bonds similar to a CH_4 molecule and a deficient electron to form them is compensated by a Li^+ cation. Therefore, it is expected that the charge transfer from Li^+ to $[BH_4]^-$ is a key feature for the stability of $LiBH_4$.

1.4.2 Borohydrides with Multivalent Cations

Among borohydrides, alkali-metal borohydrides such as $LiBH_4$ and $NaBH_4$ are well known and their properties have been studied moderately. Although we can find borohydrides composed of not only alkali metals but also other types of metals in literature, little is known about the latter compounds. In this section, we briefly summarize the results of the theoretical analyses of borohydrides with the divalent cation $Ca(BH_4)_2$ [36] and the trivalent cation $Y(BH_4)_3$ [48].

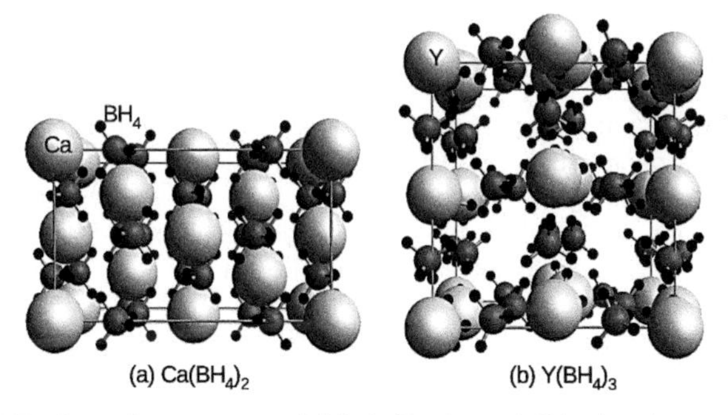

Figure 1.7 Crystal structures of (a) $Ca(BH_4)_2$ and (b) $Y(BH_4)_3$. Large white spheres indicate cation metal elements, (a) Ca and (b) Y. Middle gray and small black spheres represent B and H atoms, respectively.

Figure 1.7a shows the crystal structures of $Ca(BH_4)_2$, which is orthorhombic with the space group *Fddd* (No. 70). The calculation predicts a nearly ideal tetrahedral geometry for BH_4 complexes with $d_{B\text{-}H}$ = 1.23–1.24 Å and $\theta_{H\text{-}B\text{-}H}$ = 106°–113°. The structure of $Y(BH_4)_3$ is cubic, with the space group $Pa\overline{3}$ (No. 205), which is given in Fig. 1.7b. The arrangement of Y and BH_4 is basically a distorted ReO_3

type. The tetrahedral shape of BH_4 in $Y(BH_4)_3$ is also almost ideal, d_{B-H} = 1.24 Å and θ_{H-B-H} = 107°–113°.

Figure 1.8 depicts the electronic DOSs for $Ca(BH_4)_2$ and $Y(BH_4)_3$. For both compounds, the electronic structures are found to be nonmetallic and the valence states split into two peaks whose positions and widths are quite similar to those of $LiBH_4$ shown in Fig. 1.5. The metal cations make little contribution to the valence states.

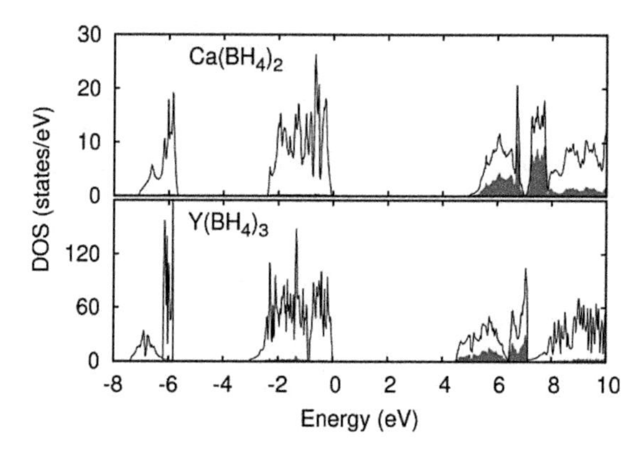

Figure 1.8 Electronic densities of states for $Ca(BH_4)_2$ and $Y(BH_4)_3$. The shaded gray parts represent the projected DOS for cation elements Ca and Y. The origin of the energy is set to be the top of valence states. Reprinted (figure) with permission from Ref. [36]. Copyright (2006) by the American Physical Society. Reprinted (figure) with permission from Ref. [48]. Copyright (2008) by the American Physical Society.

The Γ-phonon eigenmode frequencies of $Ca(BH_4)_2$ and $Y(BH_4)_3$ are given in Fig. 1.9. As found in $LiBH_4$, the eigenmodes of both compounds separate into three groups and those due to the internal H–B–H bending and B–H stretching vibrations of $[BH_4]^-$ are in the same range as $LiBH_4$, 1000–1300 cm^{-1} and 2250–2400 cm^{-1}.

These similarities of the electronic and vibrational properties imply that the bonding nature discussed for $LiBH_4$ in the previous section is a common feature for borohydrides regardless of cation elements: the internal bonding of $[BH_4]^-$ complexes has a covalent character and the deficient electron to form them is compensated by metal cations.

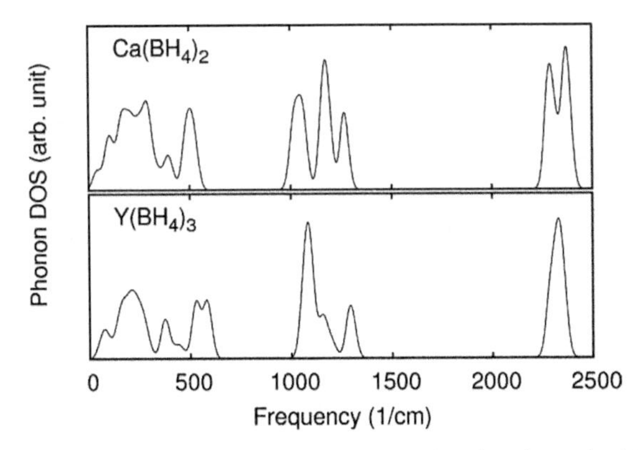

Figure 1.9 Phonon eigenmode frequencies of $Ca(BH_4)_2$ and $Y(BH_4)_3$. The phonon DOS is calculated from the optical Γ-phonon modes, where Gaussian broadening with 30 cm^{-1} width is used. Reprinted (figure) with permission from Ref. [36]. Copyright (2006) by the American Physical Society. Reprinted (figure) with permission from Ref. [48]. Copyright (2008) by the American Physical Society.

1.4.3 Thermodynamically Stability of Borohydrides

Our theoretical analyses suggest that the charge transfer from metal cations to $[BH_4]^-$ complexes is responsible for the stability of borohydrides. Because the charge transfer ability can be measured by the electronegativity, it will be a good indicator to measure their stabilities.

To confirm this, the stability of borohydrides composed of various cation elements has been predicted theoretically [35–37, 40, 48]. To compare the results of cations with different valences, the normalized heat of formation ΔH corresponding to the following reaction is considered:

$$\frac{1}{n}\mathrm{M} + \mathrm{B} + 2\mathrm{H}_2 \rightarrow \frac{1}{n}\mathrm{M(BH_4)}_n, \tag{1.6}$$

where n is valency of the cation element M. Though the definition of electronegativity is not unique, Pauling scaling [2, 44] is used in this study. Since this definition is based on the difference of binding energies between the elements, it will be suitable for the present purpose.

Figure 1.10 shows the predicted ΔH as a function of the Pauling electronegativity of cation elements χ_P. Most results are calculated by referring to the experimental crystallographic data for which the structural optimization is performed. For $Sc(BH_4)_3$, $Zn(BH_4)_2$, and $CuBH_4$, no experimental data are available. For the former two compounds, the structures of bromides and iodides are assumed to estimate the initial arrangement of the cation and $[BH_4]^-$ for structural optimization, since the ionic radii of Br^- (1.90 Å) and I^- (2.20 Å) are close to the effective radius of $[BH_4]^-$, 2.03 Å [46]. For $CuBH_4$, an orthorhombic structure (the same as that of $LiBH_4$) is adopted because of the similarity of ionic radii of Li^+ (0.59 Å) and Cu^+ (0.60 Å).

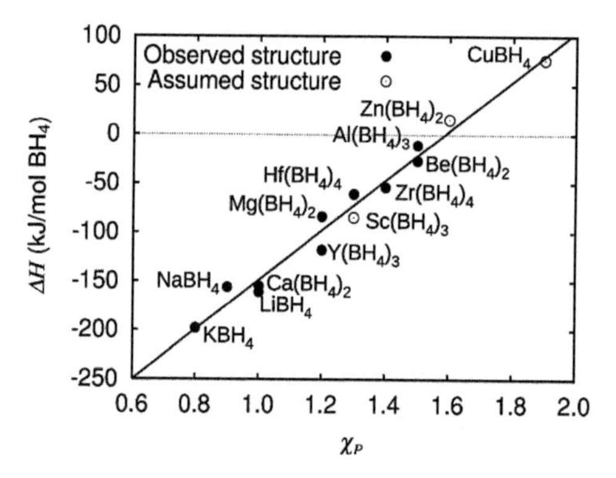

Figure 1.10 Normalized heat of formation for borohydrides ΔH as a function of the Pauling electronegativity of cation elements χ_P. The straight line indicates the result of the least-squares fitting.

In Fig. 1.10, a good correlation between ΔH and χ_P can be found. Assuming a linear relationship, the least-squares fitting yields

$$\Delta H = 253.7\chi_P - 402.3, \tag{1.7}$$

with an absolute mean error of 10.3 kJ/mol BH_4. The Pauling electronegativity of cation elements is a good indicator of the stability of borohydrides. From the relation 1.7, the borohydride is expected to become thermodynamically unstable for the cation element with $\chi_P \geq 1.6$.

A similar correlation between ΔH and χ_P has been found for other types of complex hydrides, such as alanates [16], and transition metal complex hydrides [39, 51].

1.4.4 Experimental Support

The hydrogen desorption property of borohydrides has also been investigated experimentally [40]. The samples of borohydrides with multivalent cations, $M(BH_4)_n$ (M = Mg, Zn, Sc, and Zr), are prepared using the mechanochemical synthesis technique according to the following reaction:

$$MCl_n + nLiBH_4 \rightarrow M(BH_4)_n + nLiCl. \quad (1.8)$$

The obtained samples are examined by powder X-ray diffraction measurement and Raman spectroscopy, which indicate that Reaction 1.8 successfully proceeds during the mechanical milling treatment.

The hydrogen desorption properties are studied by thermal desorption spectroscopy. For each sample, the desorption temperature T_d is defined as a temperature of the first desorption peak. In Fig. 1.11, T_d of the borohydrides, including alkali-metal ones, $LiBH_4$ and $NaBH_4$, are plotted as a function of the Pauling electronegativity χ_P. A good correlation can be found between both quantities: T_d decreases monotonically with increasing χ_P.

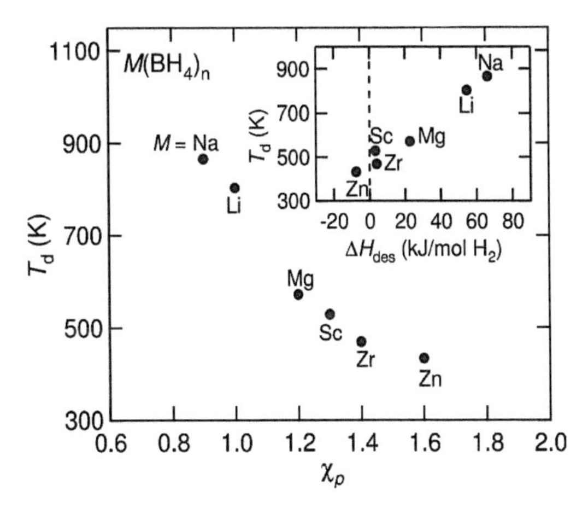

Figure 1.11 Hydrogen desorption temperature T_d as a function of the Pauling electronegativity of cation elements χ_P. The inset shows the correlation between T_d and estimated ΔH_{des}. Reprinted (figure) with permission from Ref. [40]. Copyright (2006) by the American Physical Society.

Since the actual desorption reactions of borohydrides are usually accompanied by the formation of hydrides and/or borides as seen in Reaction 1.5, the stability of these products has to be taken into account for the experimental results. To do so, the following reaction is assumed except for $Zn(BH_4)_2$ (due to the instabilities of Zn hydride and boride).

$$M(BH_4)_n \rightarrow MH_m + nB + \frac{4n-m}{2} H_2, \qquad (1.9)$$

where MH_m is a hydride phase observed experimentally. The enthalpy change for this reaction ΔH_{des} can be estimated using Eq. 1.7 for $M(BH_4)_n$ and the experimental value for MH_m. As shown in the inset of Fig. 1.11, there is also a good correlation between the observed T_d and the estimated ΔH_{des}.

These experimental results support the theoretical suggestion that the electronegativity of cation elements is a good indicator of the stability of borohydrides. From this consideration, it is also expected that the suppression of charge transfer by partial substitution of more electronegative elements for Li is effective in destabilizing $LiBH_4$. This effect has been examined for the double-cation system both theoretically [35] and experimentally [28].

1.5 Future Scope

In this chapter, two types of hydrogen storage materials, transition-metal hydrides and borohydrides, have been investigated theoretically. The first-principles calculation gives reliable results for them, which is beneficial in understanding their properties and designing new materials. The present analyses are, however, restricted to the thermodynamical properties assuming the equilibrium reaction. In actual applications, the reaction kinetics is also an important factor.

Several theoretical approaches have been proposed for the reaction pathway search [19, 20, 22, 29, 30]. These techniques are most likely useful, in particular, for conventional metal hydrides, to investigate the elemental processes of hydrogenation reactions such as hydrogen dissociation on the surface and hydrogen diffusion in the host materials. The path-integral molecular dynamics simulations help take into account the quantum mechanical effect for nuclei [6, 31].

For complex hydrides, the situation may be somewhat complicated since the dehydriding reactions are usually accompanied by decomposition into some products. The experimental effort to improve their kinetics is often accomplished by the catalytic reaction, as done for Ti-catalyzed $NaAlH_4$ by Bogdanović and Schwickardi [5]. Despite many successful applications, the mechanism of the catalytic effect has not been fully understood yet. The theoretical understanding of this mechanism will be an important research direction to enhance the reaction kinetics of complex hydrides.

Acknowledgments

Valuable collaboration and discussion with the groups of Drs. Fukumoto, Ohba, Aoki, Noritake, Matsumoto, and Towata at Toyota Central R&D Labs., Inc.; Drs. Takagi, Sato, Matsuo, Li, Nakamori, and S. Orimo at Tohoku University; and Dr. Züttel are highly appreciated. The studies for the borohydrides were partially supported by the New Energy and Industrial Technology Development Organization, Japan.

References

1. Akiba, E. and Iba, H. (1998). Hydrogen absorption by Laves phase related BCC solid solution, *Intermetallics*, **6**, pp. 461–470.
2. Allred, A. L. (1961). Electronegativity values from thermochemical data, *J. Inorg. Nucl. Chem.*, **17**, pp. 215.
3. Baroni, S., Giannozzi, P., and Testa, A. (1987). Green's-function approach to linear response in solids, *Phys. Rev. Lett.*, **58**, pp. 1861–1864.
4. Baroni, S., de Gironcoli, S., Dal Corso, A., and Giannozzi, P. (2001). Phonons and related crystal properties from density-functional perturbation theory, *Rev. Mod. Phys.*, **73**, pp. 515–562.
5. Bogdanović, B. and Schwickardi, M. (1997). Ti-doped alkali metal aluminium hydrides as potential novel reversible hydrogen storage materials, *J. Alloys Compd.*, **253–254**, pp. 1–9.
6. Cao, J. and Voth, G. A. (1994). The formulation of quantum statistical mechanics based on the Feynman path centroid density. I. Equilibrium properties, *J. Chem. Phys.*, **100**, pp. 5093–5105.

7. de Gironcoli, S., Baroni, S., and Resta, R. (1989). Piezoelectric properties of III-V semiconductors from first-principles linear-response theory, *Phys. Rev. Lett.*, **62**, pp. 2853–2856.
8. de Gironcoli, S. (1995). Lattice dynamics of metals from density-functional perturbation theory, *Phys. Rev. B*, **51**, pp. 6773–6776.
9. Filinchuk, Y., Chernyshov, D., and Cerny, R. (2008). Lightest borohydride probed by synchrotron X-ray diffraction: experiment calls for a new theoretical revision, *J. Phys. Chem. C*, **112**, pp. 10579–10584.
10. Friedel, J. (1969). *The Physics of Matal* (Cambridge University Press, London).
11. Fukai, Y. (1993). *The Metal-Hydrogen System*, Vol. 21 of Springer Series in Material Science (Spinger-Verlag, Berlin).
12. Gonze, X. and Vigneron, J. P. (1989). Density-functional approach to nonlinear-response coefficients of solids, *Phys. Rev. B*, **39**, pp. 13120–13128.
13. Gonze, X. (1995). Perturbation expansion of variational principles at arbitrary order, *Phys. Rev. A*, **52**, pp. 1086–1095.
14. Gonze, X. (1997). First-principles responses of solids to atomic displacements and homogeneous electric fields: implementation of a conjugate-gradient algorithm, *Phys. Rev. B*, **55**, pp. 10337–10354.
15. Gonze, X. and Lee, C. (1997). Dynamical matrices, Born effective charges, dielectric permittivity tensors, and interatomic force constants from density-functional perturbation theory, *Phys. Rev. B*, **55**, pp. 10355–10368.
16. Graetz, J. (2009). New approaches to hydrogen storage, *Chem. Soc. Rev.*, **38**, pp. 73–82.
17. Griessen, R. (1988). Heats of solution and lattice-expansion and trapping energies of hydrogen in transition metals, *Phys. Rev. B*, **38**, pp. 3690–3698.
18. Hamann, D. R., Wu, X., Rabe, K. M., and Vanderbilt, D. (2005). Metric tensor formulation of strain in density-functional perturbation theory, *Phys. Rev. B*, **71**, pp. 035117.
19. Henkelman, G. and Jónsson, H. (1999). A dimer method for finding saddle points on high dimensional potential surfaces using only first derivatives, *J. Chem. Phys.*, **111**, pp. 7010.
20. Henkelman, G. and Jónsson, H. (2000). Improved tangent estimate in the nudged elastic band method for finding minimum energy paths and saddle points, *J. Chem. Phys.*, **113**, pp. 9978.

21. Hohenberg, P. and Kohn, W. (1964). Inhomogeneous electron gas, *Phys. Rev.*, **136**, pp. 864.

22. Iannuzzi, M., Laio, A., and Parrinello, M. (2003). Efficient exploration of reactive potential energy surfaces using car-parrinello molecular dynamics, *Phys. Rev. Lett.*, **90**, pp. 238302.

23. Jones, R. O. (2015). Density functional theory: its origins, rise to prominence, and future, *Rev. Mod. Phys.*, **87**, pp. 897–923.

24. Knox, J. H. (1971). *Molecular Thermodynamics: An Introduction of Statistical Mechanics for Chemists* (John Wiley & Sons, USA).

25. Kohn, W. and Sham, L. J. (1965). Self-consistent equations including exchange and correlation effects, *Phys. Rev.*, **140**, pp. A1133–A1138.

26. Kubo, K., Kawaharazaki, Y., and Itoh, H. (2017). Development of large MH tank system for renewable energy storage, *Int. J. Hydorgen Energy*, **42**, pp. 22475–22479.

27. Laasonen, K., Pasquarello, A., Car, R., Lee, C., and Vanderbilt, D. (1993). Car-Parrinello molecular dynamics with Vanderbilt ultrasoft pseudopotentials, *Phys. Rev. B*, **47**, pp. 10142–10153.

28. Li, H. W., Orimo, S., Nakamori, Y., Miwa, K., Ohba, N., Towata, S., and Züttel, A. (2007). Materials designing of metal borohydrides: viewpoints from thermodynamical stabilities, *J. Alloys Compd.*, **446–447**, pp. 315.

29. Malek, R. and Mousseau, N. (2000). Dynamics of Lennard-Jones clusters: a characterization of the activation-relaxation technique, *Phys. Rev. E*, **62**, pp. 7723–7728.

30. Martoňák, R. R., Laio, A., and Parrinello, M. (2003). Predicting crystal structures: the Parrinello-Rahman method revisited, *Phys. Rev. Lett.*, **90**, pp. 075503.

31. Marx, D. and Parrinello, M. (1996). Ab initio path integral molecular dynamics: basic ideas, *J. Chem. Phys.*, **104**, pp. 4077–4082.

32. Miedema, A. R., Buschow, K. H. J., and Van Mal, H. H. (1976). Which intermetallic compounds of transition metals form stable hydrides?, *J. Less Common Met.*, **49**, pp. 463–472.

33. Miwa, K. and Fukumoto, A. (2002). First-principles study on 3D transition-metal dihydrides, *Phys. Rev. B*, **65**, pp. 155114.

34. Miwa, K., Ohba, N., Towata, S., Nakamori, Y., and Orimo, S. (2004). First-principles study on lithium borohydride $LiBH_4$, *Phys. Rev. B*, **69**, pp. 245120.

35. Miwa, K., Ohba, N., Towata, S., Nakamori, Y., and Orimo, S. (2005). First-principles study on copper substituted lithium borohydride, $(Li_{1-x}Cu_x)BH_4$, *J. Alloys Compd.*, **404–406**, pp. 140.

36. Miwa, K., Aoki, M., Noritake, T., Ohba, N., Nakamori, Y., Towata, S., Züttel, A., and Orimo, S. (2006). Thermodynamical stability of calcium borohydride $Ca(BH_4)_2$, *Phys. Rev. B*, **74**, pp. 155122.
37. Miwa, K., Ohba, N., Towata, S., Nakamori, Y., Züttel, A., and Orimo, S. (2007). First-principles study on thermodynamical stability of metal borohydrides: aluminum borohydride $Al(BH_4)_3$, *J. Alloys Compd.*, **446–447**, pp. 310.
38. Miwa, K. (2011). Prediction of Raman spectra with ultrasoft pseudopotentials, *Phys. Rev. B*, **84**, pp. 094304.
39. Miwa, K., Takagi, S., Matsuo, M., and Orimo, S. (2013). Thermodynamical stability of complex transition metal hydrides Mg_2FeH_6, *J. Phys. Chem. C*, **117**, pp. 8014–8019.
40. Nakamori, Y., Miwa, K., Ninomiya, A., Li, H.-W., Ohba, N., Towata, S., Züttel, A., and Orimo, S. (2006). Correlation between thermodynamical stabilities of metal borohydrides and cation electronegativites: first-principles calculations and experiments, *Phys. Rev. B*, **74**, pp. 045126.
41. Ohba, N., Miwa, K., Nagasako, N., and Fukumoto, A. (2001). First-principles study on structural, dielectric, and dynamical properties for three BN polytypes, *Phys. Rev. B*, **63**, pp. 115207.
42. Orimo, S., Nakamori, Y., Eliseo, J. R., Züttel, A., and Jensen, C. M. (2007). Complex hydrides for hydrogen storage, *Chem. Rev.*, **107**, pp. 4111–4132.
43. Paulatto, L., Mauri, F., and Lazzeri, M. (2013). Anharmonic properties from a generalized third-oder ab inito approach: theory and applications to graphite and graphene, *Phys. Rev. B*, **87**, pp. 214303.
44. Pauling, L. (1960). *The Nature of the Chemical Bonds* (Cornell University Press, New York).
45. Pickard, C. J. and Mauri, F. (2001). All-electron magnetic response with pseudopotentials: NMR chemical shifts, *Phys. Rev. B*, **63**, pp. 245101.
46. Pistorius, C. W. F. T. (1974). Melting and polymorphism of $LiBH_4$ to 45 kbar, *Z. Phys. Chem., Neue Folge*, **88**, pp. 253–263.
47. Sandrock, G. (1999). A panoramic overview of hydrogen storage alloys from a gas reaction point of view, *J. Alloys Compd.*, **293–295**, pp. 877–888.
48. Sato, T., Miwa, K., Nakamori, Y., Ohoyama, K., Li, H.-W., Noritake, T., Aoki, M., Towata, S., and Orimo, S. (2008). Experimental and computational studies on solvent-free rare-earth metal borohydrides $R(BH_4)_3$ (R=Y, Dy, and Gd), *Phys. Rev. B*, **77**, pp. 104114.

49. Snavely, C. A. and Vaughan, D. A. (1949). Hydrogen absorption by Laves phase related BCC solid solution, *J. Am. Ceram. Soc.*, **17**, pp. 313.

50. Soulié, J. P., Renaudin, G., Černý, R., and Yvon, K. (2002). Lithium borohydride $LiBH_4$: I. Crystal structure, *J. Alloys Compd.*, **346**, pp. 200–205.

51. Takagi, S., Humphries, T. D., Miwa, K., and Orimo, S. (2014). Enhanced tunability of thermodynamic stability of complex hydrides by the incorporation of H^- anions, *Appl. Phys. Lett.*, **104**, pp. 203901.

52. Vanderbilt, D. (1990). Soft self-consistent pseudopotentials in a generalized eigenvalue formalism, *Phys. Rev. B*, **41**, pp. 7892–7895.

53. Veithen, M., Gonze, X., and Ghosez, P. (2005). Nonlinear optical susceptibilities, Raman efficiencies, and electro-optic tensors from first-principles density functional perturbation theory, *Phys. Rev. B*, **71**, pp. 125107.

54. Züttel, A. (2003). Materials for hydrogen storage, *Mater. Today*, **6**, pp. 24–33.

55. Züttel, A., Wenger, P., Rentsch, S., Sudan, P., Mauron, P., and Emmenegger, C. (2003). $LiBH_4$ a new hydrogen storage material, *J. Power Sources*, **118**, pp. 1–7.

Chapter 2

Atomistic Analysis of Electrolytes: Redox Potentials and Electrochemical Reactions in a Lithium-Ion Battery

Kaito Miyamoto
Toyota Central R&D Laboratories, Inc., Nagakute, Aichi 480-1192, Japan
kaito@mosk.tytlabs.co.jp

2.1 Introduction

Lithium-ion batteries (LIBs) have now become one of the most important electrochemical devices due to their broad applicability in portable electronics, such as laptop computers and mobile phones, as well as electric and hybrid vehicles. Because of their widespread use, improvements in LIB performance, for example, in terms of power, longevity, and safety, have been pursued across the world. Currently, unfortunately, there are no batteries that satisfy all the requirements and research and development related to LIB are still ongoing.

To improve the battery performance, information regarding electrochemical reactions at the electrolyte/electrode interface is

Multiscale Simulations for Electrochemical Devices
Edited by Ryoji Asahi

ISBN 978-981-4800-71-6 (Hardcover), 978-0-429-29545-4 (eBook)
www.jennystanford.com

imperative, and theoretical approaches using electronic structure theory are quite effective in investigating such reactions since chemical reactions at the atomistic scale can be analyzed. Indeed, as shown in Refs. [6] and [37], many research studies have already utilized electronic structure calculations to clarify the electrochemistry in LIBs, and we feel that such calculations have now become one of the standard tools even for experimentalists. Therefore, it is useful to summarize such theoretical approaches using modern electronic structure methods, which is the main aim of this chapter.

Among theoretical approaches, we focus on two methods, the redox potential computations using a combination of density functional theory (DFT) [67] and polarizable continuum model (PCM) [18, 56, 83, 84] (the DFT/PCM method) and the energy diagram approach using a combination of the global reaction route mapping (GRRM) method [52] and the DFT/PCM method. Here, the energy diagram approach is the method to investigate elementary reaction pathways. We selected the two approaches from the points of view of ease and effectiveness, meaning even people unfamiliar with the electronic structure calculations should find it easy to use the methods and the methods should endure practical applications in LIB development. We should note that the aim of this chapter is not a comprehensive review of electronic structure approaches to LIB. For the people who are interested in it, you can refer to some excellent reviews, such as Refs. [6, 37].

In the next section, we summarize the redox potential calculation method using the DFT/PCM method and show its effectiveness by applying it to typical electrolyte solvents. Then, in Section 2.3, we describe the energy diagram method to investigate electrolyte decomposition, which is useful to understand the interface reactions between electrode and electrolyte in LIBs. Finally, we summarize this chapter in Section 2.4.

2.2 Redox Potential Computations Using the DFT/PCM Method

Information regarding oxidation and reduction potentials of solvents and salt ions is quite important not only for understanding electrochemical reactions occurring in batteries but also for designing new LIB electrolytes. To obtain such information, electronic structure calculations have now become a powerful tool in the field of LIB development. Among such methods, a combination of DFT and PCM, the so-called DFT/PCM method, is widely used in this field since the method provides standard redox potentials with good accuracy and acceptable computational cost [23, 24, 31, 32, 54, 57, 75].

In the beginning, standard redox potential calculations using the DFT/PCM method were performed in order to clarify reductive decomposition mechanisms of commonly used LIB-electrolyte solvents and additives [28, 34, 47, 74, 80, 82, 94, 96–98]. Balbuena and coworkers are quite active in this field. To the best of our knowledge, Li and Balbuena applied the DFT/PCM method to LIB-electrolyte solvent molecules for the first time in order to explain the difference between the reduction mechanisms of ethylene carbonate (EC) and propylene carbonate (PC) in 2000 [47]. In the subsequent papers, Balbuena and coworkers also contributed to clarify reductive decomposition mechanisms of EC and efficacy of vinylene carbonate (VC) as an electrolyte additive by conducting the redox potential calculations as well as reaction pathways search [96, 98]. The other notable work using the DFT/PCM method was done by Vollmer and coworkers. They explained reductive decomposition mechanisms of vinylethylene carbonate as well as EC and PC by comparing computationally obtained reduction potentials with experimental ones [94]. Very recently, Johansson et al. investigated the performance of a variety of computational methods using the reduction potentials of 10 famous electrolyte additives for materials design. They concluded that the combination of M06-2X [108] and 6-311++G** basis set [17, 41, 55] is most suited for the prediction of the properties of the additives if C-PCM [7, 40] is used as the PCM model since this combination reproduced the experimental reduction potentials well [34]. Reports regarding oxidation potential calculations are also numerous [10, 26, 30, 35, 36, 63, 76, 77, 101, 107]. Successful applications to redox shuttle additives are notable [13, 29, 95]. A number of reports exemplify the importance and popularity of the DFT/PCM method in the context of LIBs.

In this section, we summarize the basics behind the standard redox potential calculations using the DFT/PCM method and, then, look at the accuracy of the method by applying it to oxidation and reduction potentials of a variety of chemically different organic solvents.

2.2.1 Standard Redox Potential

Let's suppose the electrochemical equilibrium between oxidized (Ox) and reduced (Red) species is given by

$$\mathrm{Ox} + ne^- \leftrightarrow \mathrm{Red}, \tag{2.1}$$

where n is the number of electrons. The standard redox potential (E°_{SHE}) of Eq. 2.1 relative to the standard hydrogen electrode is obtained by using the Nernst equation [54, 87]:

$$\begin{aligned} E^\circ_{\mathrm{SHE}} &= -\frac{\Delta G^\circ}{nF} - E^\circ_{\mathrm{abs}}(\mathrm{SHE}) \\ &= -\frac{G^\circ_{\mathrm{Red}} - G^\circ_{\mathrm{Ox}}}{nF} - E^\circ_{\mathrm{abs}}(\mathrm{SHE}), \end{aligned} \tag{2.2}$$

where G°_{Ox} and G°_{Red} are standard Gibbs free energies of the Ox and Red species, respectively, and F is Faraday's constant (=23.061 kcal $\mathrm{mol^{-1}\,V^{-1}}$ [87]). The first term on the right-hand side in Eq. 2.2 is the so-called absolute electrode potential, which is the value relative to the potential of an electron at rest in vacuum [5, 9]. To convert the reference system from the vacuum level to the standard hydrogen electrode, the second term ($E^\circ_{\mathrm{abs}}(\mathrm{SHE})$) is subtracted.

$E^\circ_{\mathrm{abs}}(\mathrm{SHE})$ corresponds to the absolute electrode potential of the reaction $H^+ + e^- \leftrightarrow 1/2\ H_2$ in aqueous solution in the standard state and the reported values range from 4.05 V [54] to 4.78 V [9]. Marenich et al. suggest the use of 4.28 [33, 38, 39], 4.42 [21], or 4.44 V [85, 86] for $E^\circ_{\mathrm{abs}}(\mathrm{SHE})$ [54]. Indeed, these values are commonly used in the field of LIBs [10, 12, 63]. The difference between 4.28 and 4.44 (or 4.42) V mainly derives from the inclusion of the surface potential in the solvation free energy of the proton. Detailed discussions on this subject can be found in Ref. [54]. In this chapter, we employ 4.28 V for $E^\circ_{\mathrm{abs}}(\mathrm{SHE})$ [54].

A Li^+/Li reference electrode is widely used as a reference in the context of LIBs. Conversion from SHE to the Li^+/Li reference electrode is done using

$$E^\circ_{\mathrm{Li^+/Li}} = E^\circ_{\mathrm{SHE}} - E^\circ_{\mathrm{SHE}}(\mathrm{Li^+/Li}), \tag{2.3}$$

where $E^{\circ}_{\text{SHE}}(\text{Li}^{+}/\text{Li})$ is the standard redox potential of $Li^+ + e^- \leftrightarrow Li$ relative to SHE and −3.05 V for aqueous solutions [85]. By substituting Eq. 2.2 into Eq. 2.3 and using 4.28 V as $E^{\circ}_{\text{abs}}(\text{SHE})$, we get

$$E^{\circ}_{\text{Li}^{+}/\text{Li}} = -\frac{G^{\circ}_{\text{Red}} - G^{\circ}_{\text{Ox}}}{nF} - 1.23\,. \tag{2.4}$$

Therefore, once G°_{Ox} and G°_{Red} are obtained, the standard redox potential can be evaluated.

2.2.2 Standard Gibbs Free Energy Calculations Using the DFT/PCM Method

As shown in Eq. 2.2 or 2.4, the standard redox potential is computable if the standard Gibbs free energies ($G°$) of Ox and Red species are obtained. The DFT/PCM method is one of the common methods to evaluate $G°$. The DFT/PCM method is the method to compute the electronic structure of a solute (in this chapter, Ox and Red) in a solution. Since the method only treats solute quantum mechanically (at the DFT level) and others as a continuum medium as depicted in Fig. 2.1, $G°$ can be computed at almost the same computational cost as the DFT calculation in the gas phase.

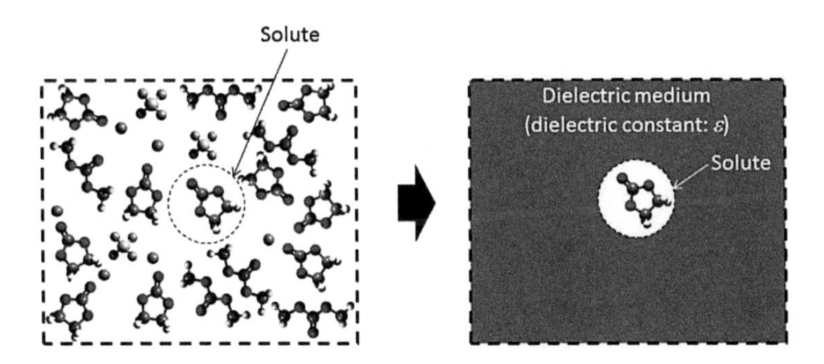

Figure 2.1 The PCM method approximates solvents around a solute by the continuum medium having a dielectric constant of ε.

Currently, a variety of solvation models have been reported for the PCM, including the integral equation formalism for the PCM [14, 73], conductor-like screening model [7, 40], and solvation model based on density (SMD) [53]. Among them, the SMD model is recommended for the evaluation of $G°$ since this model is known to reproduce the solvation free energy with good accuracy [32, 53], which directly affects the accuracy of the standard redox potential.

As for the computational method of DFT, the combination of B3LYP [8, 79] as the exchange-correlation functional and 6-311++G** [17, 41, 55] for the basis set is a reasonable choice, as it gives an average error of around 0.2 eV for the ionization potential and electron affinity of typical molecules [19, 20], which strongly affects the accuracy of the standard redox potential [32].

When we assume the one-electron redox potential of a molecule, the relationship between the standard redox potential (Eq. 2.2) and the gas-phase properties is given by

$$E^{\circ}_{\text{SHE}} = \frac{\Delta E^{\text{0K}}_{\text{gas}} + \Delta E^{\text{therm}}_{\text{gas}} + \Delta\Delta G^{\text{solv}}}{F} - E^{\circ}_{\text{abs}}(\text{SHE}), \qquad (2.5)$$

where $\Delta E^{\text{0K}}_{\text{gas}}$, $\Delta E^{\text{therm}}_{\text{gas}}$, and $\Delta\Delta G^{\text{solv}}$ are the energy differences at 0 K, thermal free energy difference, and solvation free energy difference between Ox and Red species, respectively. Normally, the contribution of $\Delta E^{\text{therm}}_{\text{gas}}$ to E°_{SHE} is quite small. Here, it is noteworthy that $\Delta E^{\text{0K}}_{\text{gas}}$ corresponds to the adiabatic ionization energy at 0 K for the computation of the oxidation potential and the adiabatic electron affinity at 0 K for the reduction potential computation. Since the error derived from the ionization potential and the electron affinity is around 0.2 eV, as noted, and that from $\Delta\Delta G^{\text{solv}}$ is also around 0.2 eV [32], we can expect an error of around or less than 0.4 V for the standard redox potential with the recommended computational condition (we denote this condition as B3LYP/6-311++G**/SMD). In the following sections, we use this condition for all the calculations. All the calculations are performed with Gaussian 09 software package [22].

2.2.3 Oxidation and Reduction Potentials of Typical Organic Solvents

In this section, we compare computationally obtained one-electron oxidation and reduction potentials (E°) using the DFT/PCM method with experimental oxidation and reduction potentials (E^{Ox} and E^{Red}) of typical electrolyte solvents [88–91]. Solvents considered in this section are summarized in Table 2.1. The main motivation of this section is to let readers know the applicability of the DFT/PCM method in redox potential evaluations. Here, it is noteworthy that E° is the value when redox species are in the standard state. Contrary to this, experimentally obtained values (E^{Ox} and E^{Red}) are not the values in the standard state and are defined by the potentials where the current density reached some criterion, for example, 1 mA cm^{-2}, using the voltammetry experiments. Since the criterion differs across reports and potential shape changes due to many factors, such as concentration and kinds of supporting electrolyte salts, kinds of solvents, and liquid junction potential, E^{Ox} and E^{Red} are not equal to E°. Nonetheless, it is valuable to understand the correspondence between these two potentials for a variety of chemically different solvents.

Table 2.1 Typical organic solvents and their dielectric constants (ε)

Molecule	ε^a
EC (ethylene carbonate)	90
PC (propylene carbonate)	65
BC (butylene carbonate)	53
DMC (dimethyl carbonate)	3.1
EMC (ethyl methyl carbonate)	3.0
DEC (diethyl carbonate)	2.8
GBL (γ-butyrolactone)	42
AN (acetonitrile)	36
MAN (methoxyacetonitrile)	21
MPN (3-methoxypropionitrle)	36
DMF (*N,N*-dimethylformamide)	37
NMO (*N*-methyloxazolidinone)	78
DMI (*N,N'*-dimethylimidazolidinone)	38
NM (nitromethane)	38
TMS (sulfolane)	43
DMSO (dimethylsulfoxide)	47

[a] The dielectric constants are taken from Refs. [88, 91].

Here it is noteworthy that, for computing the oxidation potential, Red and Ox in Eq. 2.1 correspond to the organic solvent molecule and its oxidized state, respectively. On the other hand, Ox and Red become the solvent molecule and its reduced state, respectively, for the reduction potential. To take into account the solvation effect using the SMD model, the dielectric constant (ε) of the solvent is necessary. The ε values used in this section are summarized in Table 2.1. The spin states of neutral molecules and their ionic states are singlet and doublet, respectively.

To evaluate $E°$ using the DFT/PCM method, equilibrium geometries of the redox species are necessary. They are usually obtained by geometry optimizations implemented in electronic structure calculation programs such as Gaussian 09 [22]. However, for such calculations, initial geometries that are reasonably similar to optimized ones must be provided as input data. Although experimental data may be available for the (neutral) organic molecules, it is difficult to obtain such information for the oxidized and reduced states. However, it is a reasonable choice to use optimized geometries of the neutral molecules as the initial geometries for oxidized and reduced forms since optimization starting from the neutral molecule mimics the dynamics of the redox reaction, that is, the molecular geometry relaxes after receiving an electron from (or releasing an electron to) the electrode. The optimized geometries of the solvents and their cations and anions are shown in Figs. 2.2–2.4.

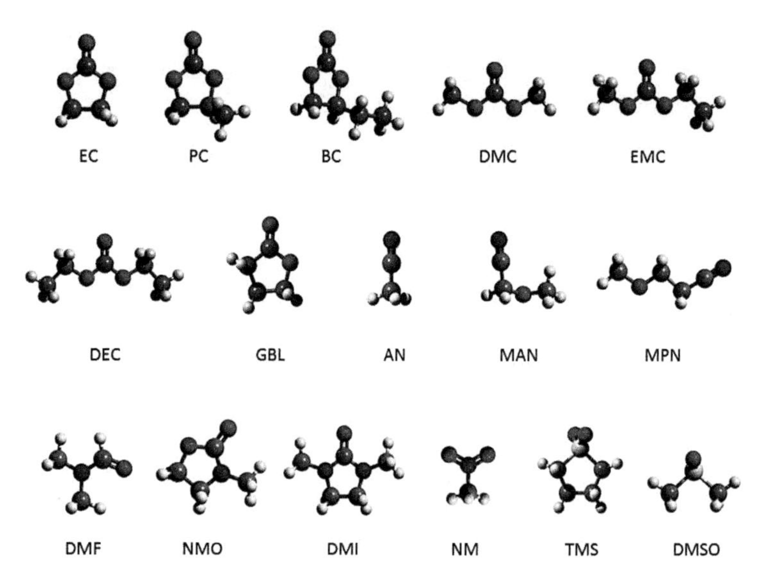

Figure 2.2 Geometries of organic solvents. Geometry optimizations were carried out with B3LYP/6-311++G**/SMD. We confirmed all the stationary were minima by carrying out frequency calculations. All calculations were performed with Gaussian 09.

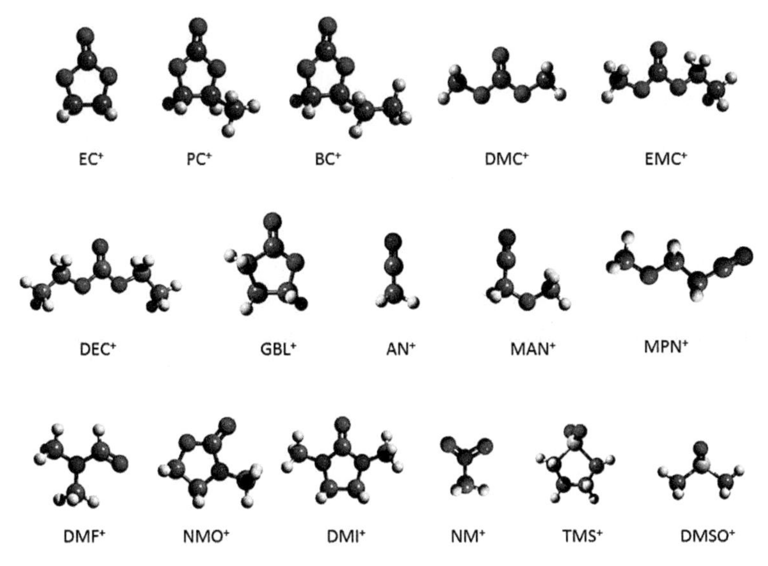

Figure 2.3 Geometries of oxidized forms of organic solvents. Geometry optimizations were carried out with B3LYP/6-311++G**/SMD. We confirmed all the stationary were minima by carrying out frequency calculations. All calculations were performed with Gaussian 09.

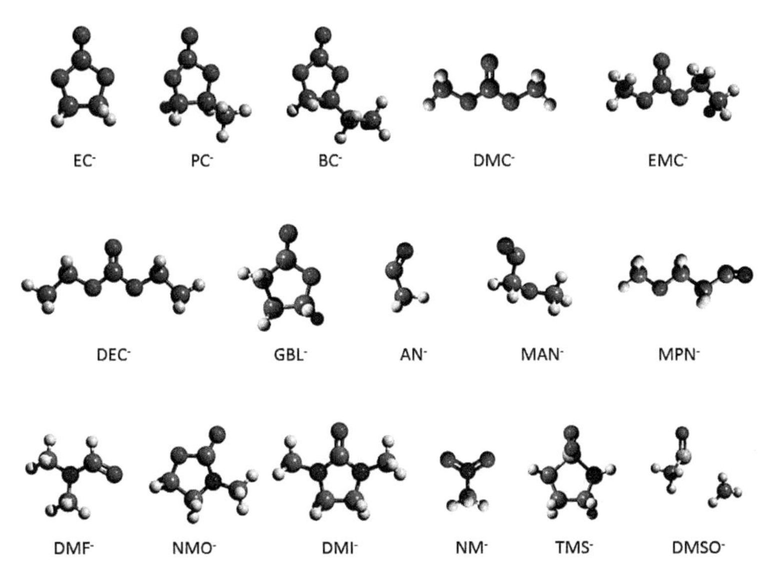

Figure 2.4 Geometries of reduced forms of organic solvents. Geometry optimizations were carried out with B3LYP/6-311++G**/SMD. We confirmed all the stationary were minima by carrying out frequency calculations. All calculations were performed with Gaussian 09.

2.2.3.1 One-electron oxidation potential

Table 2.2 compares computationally and experimentally obtained oxidation potentials of organic molecules. As an overall trend, we can see that $E°$s agree well with the corresponding E^{Ox}s except in the case of γ-butyrolactone (GBL) when taking into account the fact that the expected error of $E°$ is around 0.4 V.

Table 2.2 The oxidation potentials of organic solvents (V vs. Li^+/Li)

Molecule	$E°$	$E^{Ox,a}$	Error
EC (ethylene carbonate)	6.7	6.5[b]	0.2
PC (propylene carbonate)	6.6	6.9[b,c]	−0.3
BC (butylene carbonate)	6.4	7.5[b,c]	−1.1
DMC (dimethyl carbonate)	7.4	7.0[b]	0.4
EMC (ethyl methyl carbonate)	7.3	7.0[b]	0.3
DEC (diethyl carbonate)	7.3	7.0[b]	0.3
GBL (γ-butyrolactone)	6.3	8.5[b,c]	−2.2
AN (acetonitrile)	7.6	6.6[c]	1.0
MAN (methoxyacetonitrile)	6.2	6.3[c]	−0.1
MPN (3-methoxypropionitrle)	5.8	6.4[c]	−0.6
DMF (*N,N*-dimethylformamide)	5.3	4.9[c]	0.4
NMO (*N*-methyloxazolidinone)	5.1	5.0[c]	0.1
DMI (*N,N'*-dimethylimidazolidinone)	4.4	4.5[c]	−0.1
NM (nitromethane)	7.1	6.0[c]	1.1
TMS (sulfolane)	6.2	6.6[c]	−0.4
DMSO (dimethylsulfoxide)	5.2	4.8[c]	0.4

[a]Experimental values were reported as relative values to the saturated calomel electrode (SCE). To convert the SCE values to the relative values to the Li^+/Li reference electrode, we used $E^{Ox}_{Li+/Li} = E^{Ox}_{SCE} + 3.29$. For this conversion, 0.24 V vs. SHE [48] is used as the redox potential of the SCE.
[b]*Source*: [90].
[c]*Source*: [88].
Note: $E°$ and E^{Ox} denote the computed standard redox potential and the limiting oxidation potential (the experimental value), respectively.

To understand the reasons for the discrepancy between $E°$ and E^{Ox} of GBL, the polarization curve of an Et_4NBF_4/GBL electrolyte is shown in Fig. 2.5 [88]. In this experiment, E^{Ox} is defined as the potential at the current density of 1 mA cm^{-2} and, therefore, E^{Ox} is about 5.2 V versus saturated calomel electrode (SCE = 8.5 V vs. Li^+/Li). However, as can be seen in this figure, another oxidation peak is confirmed at about 3.2 V (=6.5 V vs. Li^+/Li). This implies two different oxidation reactions taking place in the range from 6.5 V to 8.5 V (vs. Li^+/Li). Since the peak at 6.5 V (vs. Li^+/Li) agrees well with $E°$, we can assume this peak corresponds to the oxidation of GBL. Regarding the reaction at E^{Ox} = 8.5 V, oxidation reactions of the supporting electrolyte salt, the complex of GBL, or the decomposed product of GBL may take place. Here, from this example, we should emphasize that there is a possibility that we can assign a redox peak by carefully investigating experimental data and comparing them with computed $E°$ even if $E°$ and E^{Ox} disagree. We obtain the mean-absolute error (MAE) of 0.4 V, provided that E^{Ox} of GBL is 6.5 V. This accuracy is what we expect for the DFT/PCM method.

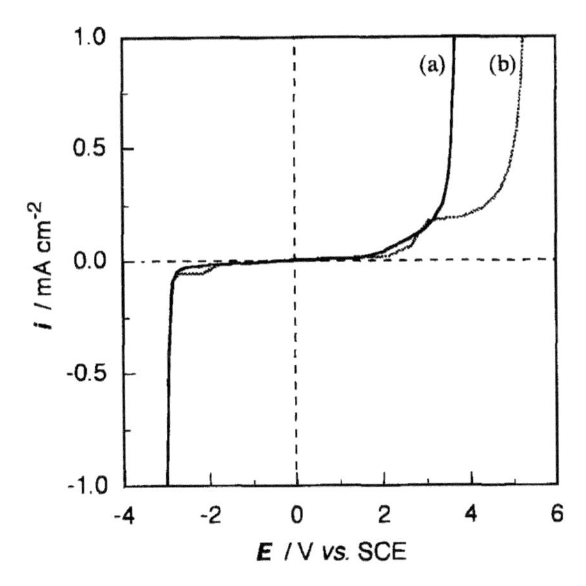

Figure 2.5 Polarization curves for (a) 0.65 mol dm^{-3} Et_4NBF_4/PC and (b) 0.65 mol dm^{-3} Et_4NBF_4/GBL electrolytes on a glassy carbon electrode. Republished with permission of the Electrochemical Society, from Ref. [88]; permission conveyed through Copyright Clearance Center, Inc.

2.2.3.2 One-electron reduction potential

Reduction potentials of electrolyte solvents are summarized in Table 2.3. Negative values of $E°$ can be confirmed for linear alkyl carbonates (dimethyl carbonate [DMC], ethyl methyl carbonate [EMC], and diethyl carbonate [DEC]), *N,N′*-dimethylimidazolidinone (DMI), and sulfolane (TMS). The negative value of $E°$ means that the assumed one-electron reduction of these solvents (the reaction where the solvents in Fig. 2.2 and their reduced forms in Fig. 2.4 become Ox and Red in Eq. 2.1) does not occur in the LIB. E^{Red} for these electrolytes probably corresponds to the reduction of the supporting electrolyte salts or the salts-induced solvent reduction. Indeed, the reductive decomposition of the tetraethylammonium cation, the supporting salt cation, is confirmed experimentally at around 0.3 V (vs. Li^+/Li) [88].

$E°$s of other solvents show excellent agreement with corresponding experimental values, that is, absolute errors of less than 0.3 V. This demonstrates that the DFT/PCM method is a powerful tool for investigating reduction potentials of organic solvents. We should note that, from the comparison between MAEs of the oxidation potentials and reduction potentials, the DFT/PCM method seems to predict the reduction potentials better than the oxidation potentials. However, as noted above, there is arbitrariness in the definition of the limiting oxidation and reduction potentials and, therefore, there is no such relationship.

In this section, the computationally obtained oxidation and reduction potentials of the typical electrolyte solvents were compared with the experimental values. Good agreements between the computed and experimental redox potentials were confirmed for most of the solvents, which demonstrates the usefulness of the DFT/PCM method. It is also important that, even if there are discrepancies between the computed and experimental values, we may draw valuable information by investigating experimental data carefully and comparing them with computed data as we exemplified using the oxidation potential of GBL and the reduction potentials of the chain carbonates (DMC, EMC, and DEC), DMI, and TMS.

Throughout this section, we assumed the one-electron redox reaction of one molecule. However, the method presented here does not have such a limitation, meaning it is also possible to handle more than two-electron transfer reactions and redox reactions of a molecular cluster. We may draw new insights into reactions inside LIBs by considering such reactions as shown in Refs. [10, 11, 46, 74, 96–98, 101, 102].

Table 2.3 The reduction potentials of organic solvents (V vs. Li^+/Li)

Molecule	E^o	$E^{Red,a}$	Error
EC (ethylene carbonate)	0.6	0.3[b]	0.3
PC (propylene carbonate)	0.5	0.3[b, c]	0.2
BC (butylene carbonate)	0.5	0.3[b, c]	0.2
DMC (dimethyl carbonate)	−1.1	0.3[b]	−1.4
EMC (ethyl methyl carbonate)	−1.1	0.3[b]	−1.4
DEC (diethyl carbonate)	−1.1	0.3[b]	−1.4
GBL (γ-butyrolactone)	0.4	0.3[c]	0.1
AN (acetonitrile)	0.2	0.5[c]	−0.3
MAN (methoxyacetonitrile)	0.4	0.6[c]	−0.2
MPN (3-methoxypropionitrle)	0.3	0.6[c]	−0.3
DMF (*N,N*-dimethylformamide)	0.1	0.3[c]	−0.2
NMO (*N*-methyloxazolidinone)	0.1	0.3[c]	−0.2
DMI (*N,N′*-dimethylimidazolidinone)	−0.9	0.3[c]	−1.2
NM (nitromethane)	2.2	2.1[c]	0.1
TMS (sulfolane)	−0.5	0.2[c]	−0.7
DMSO (dimethylsulfoxide)	0.7	0.4[c]	0.3

[a]Experimental values were reported as relative values to the SCE (saturated calomel electrode). To convert the SCE values to the relative values to the Li^+/Li reference electrode, we used $E^{Red}_{Li+/Li} = E^{Red}_{SCE} + 3.29$. For this conversion, 0.24 V vs. SHE [48] is used as the redox potential of the SCE.
[b]*Source*: [90].
[c]*Source*: [88].
Note: E^o and E^{Red} denote the computed standard redox potential and the limiting reduction potential (the experimental value), respectively.

2.3 Electrolyte Decomposition Analysis

Understanding of decomposition mechanisms and products of oxidized and reduced forms of electrolytes is crucial for LIB development. For example, reductive decomposition products form a passivating layer, the so-called solid/electrolyte interphase (SEI) [3, 16, 68], on negatively polarized graphite anode during the first charge, and it is widely known that the SEI plays a crucial role in the cycle life, power capability, and safety of the LIB [6, 103, 104]. Clarifications of oxidative decompositions are also vital and are closely related to the longevity and safety issues, including overcharge and thermal runaway [1, 2]. To obtain a better SEI and to stabilize the positive electrode, electrolyte additives are quite effective [1, 2]. For the development of better additives, understanding of electrochemical decomposition mechanisms of the additives is of importance.

So far, electronic structure calculations have contributed highly to the understanding of the reductive and oxidative decompositions of electrolytes and additives. The advantages of the electronic structure calculations are that reaction pathways at the atomistic level can be analyzed, which is quite challenging with only experimental analyses. One standard way to investigate the reaction pathways using the electronic structure calculations is to create energy diagrams by locating equilibrium and transition structures, for example, Figs. 2.7 and 2.8. In this chapter, we call this method the energy diagram method. In this method, finding transition structures is the most difficult process and is done by hand using the detailed knowledge about the target reactions or using transition structure search methods such as the synchronous transit-guided quasi-Newton method [70, 71] and the GRRM method [52].

As for reductive decomposition mechanisms, seminal works were done by Balbuena and coworkers, as we touched on in the previous section. They carried out reaction pathways search starting from reduced forms of EC [47, 96, 98], PC [97], and VC [98] using DFT in vacuum or the DFT/PCM method. Predicted decomposed products agree with experimental results, and reaction mechanisms of solvents (EC and PC) and efficacy of VC as an additive are discussed on the basis of the results. Other notable works in this field were done by Vollmer [94], Tasaki [81], and Han [25, 27, 28].

Detailed analyses of oxidative decompositions were done by Xing et al. for the first time [74, 99, 100, 102]. Borodin and coworkers investigated influences of anions, such as PF_6^-, ClO_4^-, and bis(fluorosulfonyl)imide, on the decomposition of a variety of solvents, including carbonate, sulfone, and alkyl phosphate solvents [11, 12, 101].

Since 2010, because of the great improvements in electronic structure calculation programs as well as computer performances, reports regarding investigations of reactions in electrolytes and at (ideal) electrolyte/electrode interfaces using ab initio molecular dynamics (AIMD) simulations have emerged [42–45, 58, 64, 78, 92, 93, 105]. Pioneering work in this field was done by Leung. He investigated reductive decompositions of EC with treating electrode and electrolyte molecules explicitly [42] and reported influences of modifications of carbon edge on the reductive decomposition of EC [42–44]. He also reported reactions at positive electrode/electrolyte interfaces [44, 45]. The other notable works were done by Tateyama and coworkers. They investigated efficacies of electrolyte additives [64, 92], SEI formation mechanisms [93], and reactions in highly concentrated electrolytes [78]. Ogata and coworkers developed the highly parallelized divide-and-conquer-type real-space grid DFT [59] and applied it to Li^+ transfer at electrolyte/SEI interfaces [58].

As for merits of AIMD simulations over the energy diagram method, one can directly observe reaction dynamics with explicitly considering environment effects, such as the boundary between two phases. On the other hand, AIMD simulations have a demerit in terms of huge computational costs. Related to this, simple but important reactions, for example, some decomposition mechanisms of one solvent molecule, may be overlooked due to the limited simulation time. Contrary to this, in the case of the energy diagram method, as will be shown, it is possible to investigate a variety of reaction routes if an appropriate reaction pathways search method is chosen. The energy diagram method and AIMD simulations are usually mutually complementary, and it is quite effective to make use of both merits as demonstrated in Refs. [64, 92, 93, 105].

In this section, we describe one simple way to investigate reaction pathways using the energy diagram method. The DFT/PCM method is used as the electronic structure calculation method, and the GRRM method [52] is used to find equilibrium and transition structures. The effectiveness of this combination will be demonstrated by applying it to the decomposition of EC^- and Li^+EC^-. About the AIMD simulations for LIB, see Chapter 5.

2.3.1 Global Reaction Route Mapping Method

Generally speaking, making the energy diagram is quite a challenging task since all the related reaction intermediates (equilibrium structures) and transition structures have to be predicted in advance. To complete this task, detailed knowledge about the reactions of target systems is required. Also optimizations of the transition structures from the reasonable initial geometries (the geometries at the quadratic region of the transition structures) using electronic structure calculation methods are not so easy, especially for the people unfamiliar with such calculations.

However, the advent of the GRRM method [52] has changed this situation. The GRRM program has two robust and efficient reaction route search methods, that is, the anharmonic downward distortion following (ADDF) [49, 60, 61] and the artificial force-induced reaction (AFIR) [50, 51] methods. One of the most important features of both methods is that information regarding reaction routes is unnecessary, that is, reaction intermediates and transition structures are found automatically. The ADDF method is particularly useful to explore decomposition mechanisms of a single molecule, and the AFIR method is suitable for investigating A + B $\rightarrow$ X (+Y) type of reactions. In this section, only the ADDF method is used to investigate decomposition mechanisms of EC^- and Li^+EC^-. For people interested in the AFIR method, see Refs. [50–52].

The most difficult part of the reaction pathways search is to find transition structures. The ADDF method locates transition structures by following the local maxima of anharmonic downward distortions (ADDs) from equilibrium structures. Here, the ADD is defined by the energy difference between the harmonic potential and the real potential as shown in Fig. 2.6. Since the harmonic potential is determined only by the information at the equilibrium geometry and the local maxima of ADD are determined by minimizing the real potential energy on the isoenergy hypersurface of the harmonic potential, the maximal ADD paths to the (approximate) transition structures can be followed without having any knowledge of target reactions.

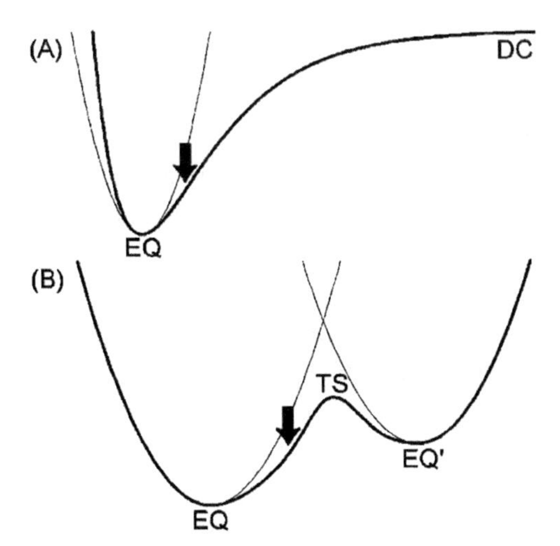

Figure 2.6 Typical potential energy curves (bold line) along (A) a dissociation path (DC) from an equilibrium structure (EQ) and (B) a reaction path from EQ to EQ′. TS denotes a transition structure. The thin line represents a harmonic potential around the EQ. Anharmonic downward distortion (ADD) is defined by the difference between the bold and thin lines indicated by arrows. Reprinted with permission from Ref. [61], Copyright (2006) American Chemical Society.

2.3.2 Reductive Decomposition of Ethylene Carbonate

Reaction pathways are explored by combining the DFT/PCM method and the ADDF method. As for the computational conditions of the DFT/PCM method, the same conditions as used in the previous section are employed except for the dielectric constant used in the SMD model.

In the field of LIBs, mixed solvent electrolytes—which are a mixture of EC and linear carbonates, such as DMC, EMC, and DEC—are used generally since a single EC solvent has high viscosity and a high melting point. The DFT/PCM method cannot handle mixed solvent effects precisely since the method approximates solvents as the dielectric medium having one dielectric constant, as shown in Fig. 2.1. However, such effects can be taken into account approximately by tuning the dielectric constant. In the field of LIBs, usually, the approximate dielectric constant of the mixed solvents is computed by the weighted average of the dielectric constants of pure solvents [28, 29, 80, 94]. In this section, we assume the mixture of EC and EMC at a volume ration of 1:2 as the (mixed) solvent and employ 40.9 as the dielectric constant. The weight is computed using molar ratio. But volume ratio is also often used. Here, however, it is noteworthy that influence of the dielectric constant on the solvation free energy is small if the value is larger than 20, as discussed in Ref. [15].

Our aim for this work is to identify the main decomposition mechanisms of EC^- and Li^+EC^-; therefore, information regarding the reaction routes having a considerably large activation barrier is unnecessary. The ADDF method can disregard such reaction routes. From Fig. 2.6, it is easy to expect the larger ADD to lead to the TS having a lower activation barrier. Therefore, the ADDF method can look for only the important reaction routes quite effectively by tracing only large ADDs. This method is the so-called large-ADDF method [52]. In this work, 10 large ADDs around each EQ are searched (LADD = 10). This parameter is determined by investigating the parameter dependence on the elementary reaction pathways from the equilibrium structure of EC^-. Reaction pathways are explored using the global reaction route mapping program version 1.22 [49, 60, 61] combined with Gaussian 09 [22].

Decomposition routes from EC^- obtained by the energy diagram method (the combination of the DFT/PCM method and the ADDF method) are shown in Fig. 2.7. From this analysis, open-chain anions ($CH_2CH_2OCO_2^-$:EQ1, 2, and 3) and CO_3^- as well as C_2H_4 gas are obtained as the decomposed products. This result is consistent with previous theoretical works [25, 42, 94, 96]. Again, we should emphasize that all EQs and TSs in Fig. 2.7 are automatically obtained only by providing the optimized geometry and spin state of EC^-. As for the validity of the calculations, $CH_2CH_2OCO_2^-$ is considered to be one of the reaction intermediates of organic components of the SEI, such as lithium butylene dicarbonate $(CH_2CH_2OCO_2Li)_2$ and lithium ethylene dicarbonate $(CH_2OCO_2Li)_2$ [4, 96]. CO_3^- immediately reduces to form a carbonate ion (CO_3^{2-}), which is also observed as one of the SEI components. C_2H_4 is observed as the main gas component of EC-based electrolytes [65, 66].

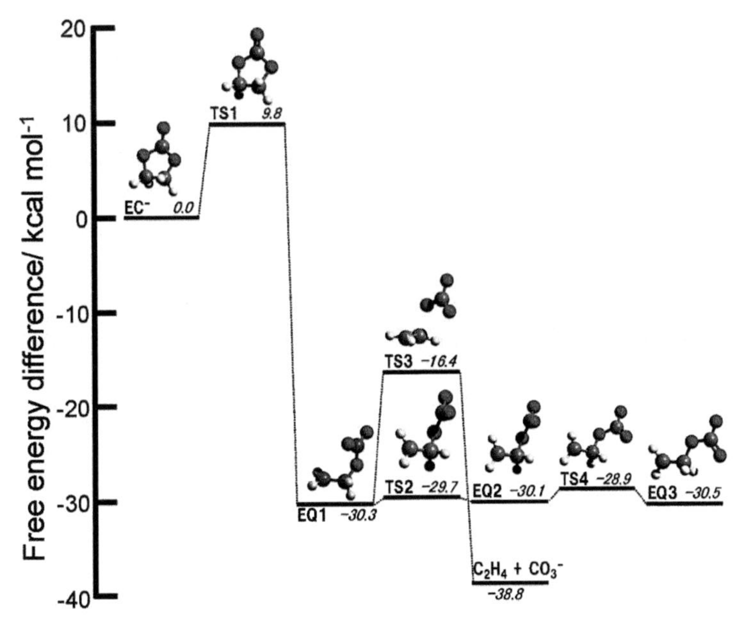

Figure 2.7 Decomposition pathways starting from an EC anion. All the calculations were performed with B3LYP/6-311++G**/SMD. For the SMD solvation model, a dielectric constant of 40.9 is used. Reaction routes were located using the global reaction route mapping program version 1.22 combined with Gaussian 09.

A much more complex diagram is obtained for the decomposition of Li^+EC^- as shown in Fig. 2.8. However, the decomposed products are the same as that from EC^-, that is, open-chain anions (EQ4, 5, and 6), $Li^+CO_3^-$, and C_2H_4. One of the important appealing points of this work is that Fig. 2.8 is the most detailed energy diagram starting from Li^+EC^- among the reported theoretical works [25, 27, 28, 94, 96], meaning the ADDF method is quite a powerful tool for the investigation of decomposition mechanisms. Very recently, Leung proposed two electron reduction mechanisms of EC as one of the main reduction routes [46]. Of course, the ADDF method can handle such reactions in the same manner as we did in this section.

Finally, we should touch on one problem regarding the experimental reduction potentials of EC and how the energy diagram can contribute to solve them. Although we showed that the experimental reduction potential of EC is 0.3 V versus Li^+/Li in

Table 2.3, this value changes with experimental conditions, as summarized in Ref. [74]. Indeed, the reported reduction potentials of EC range from 0.0 [91] to 1.36 V versus Li^+/Li [94, 106]. Among them, the reduction mechanism at 1.36 V has been proposed by Vollmer et al. [94]. They created the energy diagram of EC^- decomposition and found that if $CH_2CH_2OCO_2^-$ (EQ3 in Fig. 2.7) is chosen as the reduced form of EC, the reduction potential becomes 1.632 V (in our computational condition, we got a slightly higher value of 1.85 V). From the agreement between the computed and experimental values, they concluded that the reaction

$$EC + e^- \rightarrow EQ3 \tag{2.6}$$

would occur at this potential. This kind of phenomena may occur if, for example, the electrode works as a catalyst and the activation barriers along the ring-opening reaction disappear, as pointed out by Leung [44, 74]. This example suggests that we may assign unknown oxidation and reduction peaks by creating the energy diagram.

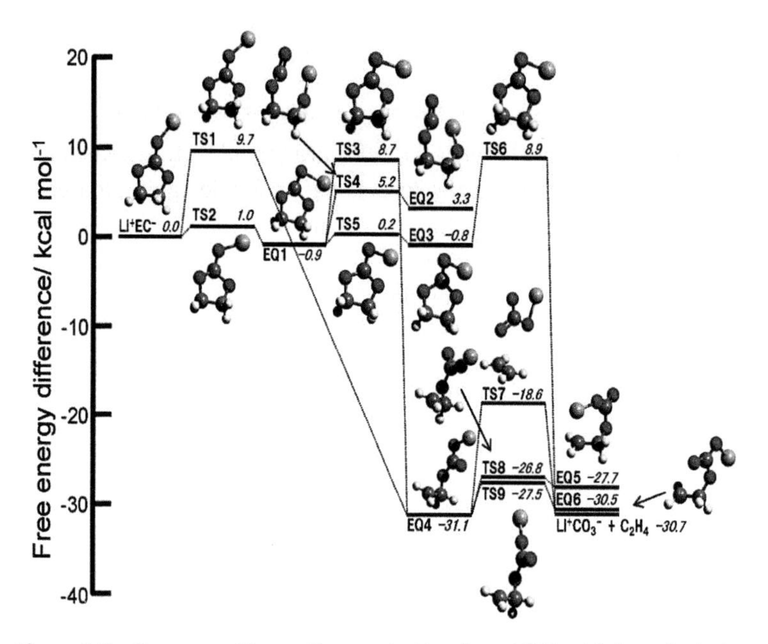

Figure 2.8 Decomposition pathways starting from Li^+EC^-. All the calculations were performed with B3LYP/6-311++G**/SMD. For the SMD solvation model, a dielectric constant of 40.9 is used. Reaction routes were located using global reaction route mapping program version 1.22 combined with Gaussian 09.

2.4 Summary and Future Scope

In this chapter, we discussed the methods to investigate redox reactions in LIBs. Firstly, we described the method to investigate the electrode potentials using the DFT/PCM method and showed the effectiveness and reliability of the method by applying it to redox potentials of typical organic solvents.

Then, we showed one of the effective and user-friendly methods to investigate decomposition mechanisms, that is, the combination of the DFT/PCM method and the ADDF method implemented in the GRRM program. We successfully created the energy diagrams of EC^- and Li^+EC^- and demonstrated the effectiveness of the method. The merit of the ADDF method is that once the initial (optimized) geometry and its spin state are provided, the method can find all the important reaction intermediates and transition structures automatically. Therefore, even people unfamiliar with such calculations can carry out the reaction pathways search quite easily. For the people who are interested in the GRRM program, see Ref. [62].

As shown in Section 2.2, the DFT/PCM method predicts the redox potentials of materials with good accuracy and within reasonable computational cost and, therefore, is useful for designing materials of electrochemical devices. Indeed, the method plays a crucial role in the Electrolyte Genome Project, where a high-throughput software infrastructure has been developed for screening battery electrolytes [15, 69, 72]. By considering the fact that electronic structure calculations have now become a common tool for materials design and the progress of high-throughput technology is quite fast, the high-throughput screening as used in the Electrolyte Genome Project would become common soon not only in the academic community but also in the industrial world.

Clarification of the reaction routes is vital for developing novel materials since it provides information about the chemistry that governs the performance of electrochemical devices. Therefore, we should try such analyses actively since now we have a practical method, that is, the combination of the DFT/PCM method and the ADDF method.

References

1. Abe, K. (2014). Nonaqueous electrolytes and advances in additives, in *Electrolytes for Lithium and Lithium-Ion Batteries* (Springer), pp. 167–207.
2. Amalraj, S. F., Sharabi, R., Sclar, H., and Aurbach, D. (2014). On the surface chemistry of cathode materials in Li-ion batteries, in *Electrolytes for Lithium and Lithium-Ion Batteries* (Springer), pp. 283–321.
3. Aurbach, D., Ein-Eli, Y., Chusid, O., Carmeli, Y., Babai, M., and Yamin, H. (1994). The correlation between the surface chemistry and the performance of Li-carbon intercalation anodes for rechargeable "rocking-chair" type batteries, *J. Electrochem. Soc.*, **141**, pp. 603–611.
4. Aurbach, D., Levi, M. D., Levi, E., and Schechter, A. (1997). Failure and stabilization mechanisms of graphite electrodes, *J. Phys. Chem. B*, **101**, pp. 2195–2206.
5. Böckris, J., Reddy, A., and Gamboa-Aldeco, M. (2000). *Modern Electrochemistry, 2A: Fundamentals of Electrodics* (Springer).
6. Balbuena, P. B. and Wang, Y. (2004). *Lithium-Ion Batteries: Solid-Electrolyte Interphase* (Imperial College Press).
7. Barone, V. and Cossi, M. (1998). Quantum calculation of molecular energies and energy gradients in solution by a conductor solvent model, *J. Phys. Chem. A*, **102**, pp. 1995–2001.
8. Becke, A. D. (1993). A new mixing of Hartree–Fock and local density-functional theories, *J. Chem. Phys.*, **98**, pp. 1372–1377.
9. Bockris, J. O. M. and Khan, S. U. M. (1993). *Surface Electrochemistry: A Molecular Level Approach* (Plenum Press, New York).
10. Borodin, O. and Jow, T. R. (2011). Quantum chemistry studies of the oxidative stability of carbonate, sulfone and sulfonate-based electrolytes doped with BF_4^-, PF_6^- anions, *ECS Trans.*, **33**, pp. 77–84.
11. Borodin, O., Behl, W., and Jow, T. R. (2013). Oxidative stability and initial decomposition reactions of carbonate, sulfone, and alkyl phosphate-based electrolytes, *J. Phys. Chem. C*, **117**, pp. 8661–8682.
12. Borodin, O. (2014). Molecular modeling of electrolytes, in *Electrolytes for Lithium and Lithium-Ion Batteries* (Springer), pp. 371–401.
13. Buhrmester, C., Moshurchak, L., Wang, R., and Dahn, J. (2006). The use of 2, 2, 6, 6-tetramethylpiperinyl-oxides and derivatives for redox shuttle additives in Li-ion cells, *J. Electrochem. Soc.*, **153**, pp. A1800–A1804.

14. Cances, E., Mennucci, B., and Tomasi, J. (1997). A new integral equation formalism for the polarizable continuum model: theoretical background and applications to isotropic and anisotropic dielectrics, *J. Chem. Phys.*, **107**, pp. 3032–3041.

15. Cheng, L., Assary, R. S., Qu, X., Jain, A., Ong, S. P., Rajput, N. N., Persson, K., and Curtiss, L. A. (2015). Accelerating electrolyte discovery for energy storage with high-throughput screening, *J. Phys. Chem. Lett.*, **6**, pp. 283–291.

16. Chusid, O. Y., Ely, E. E., Aurbach, D., Babai, M., and Carmeli, Y. (1993). Electrochemical and spectroscopic studies of carbon electrodes in lithium battery electrolyte systems, *J. Power Sources*, **43**, pp. 47–64.

17. Clark, T., Chandrasekhar, J., Spitznagel, G. W., and Schleyer, P. V. R. (1983). Efficient diffuse function-augmented basis sets for anion calculations. III. The 3-21+G basis set for first-row elements, Li–F, *J. Comput. Chem.*, **4**, pp. 294–301.

18. Cramer, C. J. and Truhlar, D. G. (1999). Implicit solvation models: equilibria, structure, spectra, and dynamics, *Chem. Rev.*, **99**, pp. 2161–2200.

19. Curtiss, L. A., Raghavachari, K., Redfern, P. C., and Pople, J. A. (1997). Assessment of Gaussian-2 and density functional theories for the computation of enthalpies of formation, *J. Chem. Phys.*, **106**, pp. 1063–1079.

20. Curtiss, L. A., Redfern, P. C., Raghavachari, K., and Pople, J. A. (1998). Assessment of Gaussian-2 and density functional theories for the computation of ionization potentials and electron affinities, *J. Chem. Phys.*, **109**, pp. 42–55.

21. Fawcett, W. R. (2008). The ionic work function and its role in estimating absolute electrode potentials, *Langmuir*, **24**, pp. 9868–9875.

22. Frisch, M. J., Trucks, G. W., Schlegel, H. B., Scuseria, G. E., Robb, M. A., Cheeseman, J. R., Scalmani, G., Barone, V., Petersson, G. A., Nakatsuji, H., Li, X., Caricato, M., Marenich, A., Bloino, J., Janesko, B. G., Gomperts, R., Mennucci, B., Hratchian, H. P., Ortiz, J. V., Izmaylov, A. F., Sonnenberg, J. L., Williams-Young, D., Ding, F., Lipparini, F., Egidi, F., Goings, J., Peng, B., Petrone, A., Henderson, T., Ranasinghe, D., Zakrzewski, V. G., Gao, J., Rega, N., Zheng, G., Liang, W., Hada, M., Ehara, M., Toyota, K., Fukuda, R., Hasegawa, J., Ishida, M., Nakajima, T., Honda, Y., Kitao, O., Nakai, H., Vreven, T., Throssell, K., J. A. Montgomery, J., Peralta, J. E., Ogliaro, F., Bearpark, M., Heyd, J. J., Brothers, E., Kudin, K. N., Staroverov, V. N., Keith, T., Kobayashi, R., Normand, J., Raghavachari, K., Rendell, A., Burant, J. C., Iyengar, S. S., Tomasi, J., Cossi, M., Millam, J. M., Klene, M.,

Adamo, C., Cammi, R., Ochterski, J. W., Martin, R. L., Morokuma, K., Farkas, O., Foresman, J. B., and Fox, D. J. (2009). *Gaussian 09 Rev. D.01.* (Gaussian, Inc., Wallingford CT).

23. Fu, Y., Liu, L., Yu, H.-Z., Wang, Y.-M., and Guo, Q.-X. (2005). Quantum-chemical predictions of absolute standard redox potentials of diverse organic molecules and free radicals in acetonitrile, *J. Am. Chem. Soc.*, **127**, pp. 7227–7234.
24. Fu, Y., Liu, L., Wang, Y.-M., Li, J.-N., Yu, T.-Q., and Guo, Q.-X. (2006). Quantum-chemical predictions of redox potentials of organic anions in dimethyl sulfoxide and reevaluation of bond dissociation enthalpies measured by the electrochemical methods, *J. Phys. Chem. A*, **110**, pp. 5874–5886.
25. Han, Y.-K., Lee, S. U., Ok, J.-H., Cho, J.-J., and Kim, H.-J. (2002). Theoretical studies of the solvent decomposition by lithium atoms in lithium-ion battery electrolyte, *Chem. Phys. Lett.*, **360**, pp. 359–366.
26. Han, Y.-K., Jung, J., Cho, J.-J., and Kim, H.-J. (2003). Determination of the oxidation potentials of organic benzene derivatives: theory and experiment, *Chem. Phys. Lett.*, **368**, pp. 601–608.
27. Han, Y.-K. and Lee, S. U. (2004). Performance of density functionals for calculation of reductive ring-opening reaction energies of Li+-EC and Li+-VC, *Theor. Chem. Acc.*, **112**, pp. 106–112.
28. Han, Y.-K. and Lee, S.-U. (2005). Density functional studies of ring-opening reactions of Li+-(ethylene carbonate) and Li+-(vinylene carbonate), *Bull. Korean Chem. Soc*, **26**, pp. 43–46.
29. Han, Y.-K., Jung, J., Yu, S., and Lee, H. (2009). Understanding the characteristics of high-voltage additives in Li-ion batteries: solvent effects, *J. Power Sources*, **187**, pp. 581–585.
30. Han, Y.-K., Yoo, J., and Yim, T. (2017). Computational screening of phosphite derivatives as high-performance additives in high-voltage Li-ion batteries, *RSC Adv.*, **7**, pp. 20049–20056.
31. Henry, J. B. and Mount, A. R. (2009). Calculation of the redox properties of aromatics and prediction of their coupling mechanism and oligomer redox properties, *J. Phys. Chem. A*, **113**, pp. 13023–13028.
32. Isegawa, M., Neese, F., and Pantazis, D. A. (2016). Ionization energies and aqueous redox potentials of organic molecules: comparison of DFT, correlated ab initio theory and pair natural orbital approaches, *J. Chem. Theory Comput.*, **12**, pp. 2272–2284.
33. Isse, A. A. and Gennaro, A. (2010). Absolute potential of the standard hydrogen electrode and the problem of interconversion of potentials in different solvents, *J. Phys. Chem. B*, **114**, pp. 7894–7899.

34. Jankowski, P., Wieczorek, W., and Johansson, P. (2017). SEI-forming electrolyte additives for lithium-ion batteries: development and benchmarking of computational approaches, *J. Mol. Model.*, **23**, pp. 6.

35. Johansson, P. (2006). Intrinsic anion oxidation potentials, *J. Phys. Chem. A*, **110**, pp. 12077–12080.

36. Johansson, P. (2006). Additions and corrections to "intrinsic anion oxidation potentials", *J. Phys. Chem. A*, **111**, pp. 1378–1379.

37. Jow, R. T., Xu, K., Borodin, O., and Ue, M. (2014). *Electrolytes for Lithium and Lithium-Ion Batteries* (Springer).

38. Kelly, C. P., Cramer, C. J., and Truhlar, D. G. (2006). Aqueous solvation free energies of ions and ion-water clusters based on an accurate value for the absolute aqueous solvation free energy of the proton, *J. Phys. Chem. B*, **110**, pp. 16066–16081.

39. Kelly, C. P., Cramer, C. J., and Truhlar, D. G. (2007). Single-ion solvation free energies and the normal hydrogen electrode potential in methanol, acetonitrile, and dimethyl sulfoxide, *J. Phys. Chem. B*, **111**, pp. 408–422.

40. Klamt, A. and Schüürmann, G. (1993). COSMO: a new approach to dielectric screening in solvents with explicit expressions for the screening energy and its gradient, *J. Chem. Soc., Perkin Trans. 2*, pp. 799–805.

41. Krishnan, R., Binkley, J. S., Seeger, R., and Pople, J. A. (1980). Self-consistent molecular orbital methods. XX. A basis set for correlated wave functions, *J. Chem. Phys.*, **72**, pp. 650–654.

42. Leung, K. and Budzien, J. L. (2010). Ab initio molecular dynamics simulations of the initial stages of solid-electrolyte interphase formation on lithium ion battery graphitic anodes, *Phys. Chem. Chem. Phys.*, **12**, pp. 6583–6586.

43. Leung, K., Qi, Y., Zavadil, K. R., Jung, Y. S., Dillon, A. C., Cavanagh, A. S., Lee, S.-H., and George, S. M. (2011). Using atomic layer deposition to hinder solvent decomposition in lithium ion batteries: first-principles modeling and experimental studies, *J. Am. Chem. Soc.*, **133**, pp. 14741–14754.

44. Leung, K. (2012). Electronic structure modeling of electrochemical reactions at electrode/electrolyte interfaces in lithium ion batteries, *J. Phys. Chem. C*, **117**, pp. 1539–1547.

45. Leung, K. (2012). First-principles modeling of the initial stages of organic solvent decomposition on $Li_xMn_2O_4(100)$ surfaces, *J. Phys. Chem. C*, **116**, pp. 9852–9861.

46. Leung, K. (2013). Two-electron reduction of ethylene carbonate: a quantum chemistry re-examination of mechanisms, *Chem. Phys. Lett.*, **568**, pp. 1–8.
47. Li, T. and Balbuena, P. B. (2000). Theoretical studies of the reduction of ethylene carbonate, *Chem. Phys. Lett.*, **317**, pp. 421–429.
48. Lide, D. R. (2007–2008). *CRC Handbook of Chemistry and Physics* (CRC Press, Boca Raton, FL).
49. Maeda, S. and Ohno, K. (2005). Global mapping of equilibrium and transition structures on potential energy surfaces by the scaled hypersphere search method: applications to ab initio surfaces of formaldehyde and propyne molecules, *J. Phys. Chem. A*, **109**, pp. 5742–5753.
50. Maeda, S. and Morokuma, K. (2010). Communications: a systematic method for locating transition structures of A+ B→ X type reactions, *J. Chem. Phys.*, **132**(24), pp. 241102.
51. Maeda, S. and Morokuma, K. (2011). Finding reaction pathways of type A+ B→ X: toward systematic prediction of reaction mechanisms, *J. Chem. Theory Comput.*, **7**, pp. 2335–2345.
52. Maeda, S., Ohno, K., and Morokuma, K. (2013). Systematic exploration of the mechanism of chemical reactions: the global reaction route mapping (GRRM) strategy using the ADDF and AFIR methods, *Phys. Chem. Chem. Phys.*, **15**, pp. 3683–3701.
53. Marenich, A. V., Cramer, C. J., and Truhlar, D. G. (2009). Universal solvation model based on solute electron density and on a continuum model of the solvent defined by the bulk dielectric constant and atomic surface tensions, *J. Phys. Chem. B*, **113**, pp. 6378–6396.
54. Marenich, A. V., Ho, J., Coote, M. L., Cramer, C. J., and Truhlar, D. G. (2014). Computational electrochemistry: prediction of liquid-phase reduction potentials, *Phys. Chem. Chem. Phys.*, **16**, pp. 15068–15106.
55. McLean, A. and Chandler, G. (1980). Contracted Gaussian basis sets for molecular calculations. I. Second row atoms, Z = 11–18, *J. Chem. Phys.*, **72**, pp. 5639–5648.
56. Mennucci, B. (2012). Polarizable continuum model, *Wiley Interdiscip. Rev.: Comput. Mol. Sci.*, **2**, pp. 386–404.
57. Namazian, M. and Coote, M. L. (2007). Accurate calculation of absolute one-electron redox potentials of some para-quinone derivatives in acetonitrile, *J. Phys. Chem. A*, **111**, pp. 7227–7232.
58. Ogata, S., Ohba, N., and Kouno, T. (2013). Multi-thousand-atom DFT simulation of Li-ion transfer through the boundary between the solid–

electrolyte interface and liquid electrolyte in a Li-ion battery, *J. Phys. Chem. C*, **117**, pp. 17960–17968.

59. Ohba, N., Ogata, S., Kouno, T., Tamura, T., and Kobayashi, R. (2012). Linear scaling algorithm of real-space density functional theory of electrons with correlated overlapping domains, *Comput. Phys. Commun.*, **183**, pp. 1664–1673.
60. Ohno, K. and Maeda, S. (2004). A scaled hypersphere search method for the topography of reaction pathways on the potential energy surface, *Chem. Phys. Lett.*, **384**, pp. 277–282.
61. Ohno, K. and Maeda, S. (2006). Global reaction route mapping on potential energy surfaces of formaldehyde, formic acid, and their metal-substituted analogues, *J. Phys. Chem. A*, **110**, pp. 8933–8941.
62. Ohno, K., Maeda, S., Osada, Y., and Morokuma, K. (2017). *GRRM: Global Reaction Route Mapping*. [cited 2018 2/28]; Available from: http://iqce.jp/GRRM/index_e.shtml.
63. Okoshi, M., Ishikawa, A., Kawamura, Y., and Nakai, H. (2015). Theoretical analysis of the oxidation potentials of organic electrolyte solvents, *ECS Electrochem. Lett.*, **4**, pp. A103–A105.
64. Okuno, Y., Ushirogata, K., Sodeyama, K., and Tateyama, Y. (2016). Decomposition of the fluoroethylene carbonate additive and the glue effect of lithium fluoride products for the solid electrolyte interphase: an ab initio study, *Phys. Chem. Chem. Phys.*, **18**, pp. 8643–8653.
65. Ota, H., Sakata, Y., Inoue, A., and Yamaguchi, S. (2004). Analysis of vinylene carbonate derived SEI layers on graphite anode, *J. Electrochem. Soc.*, **151**, pp. A1659–A1669.
66. Ota, H., Sakata, Y., Otake, Y., Shima, K., Ue, M., and Yamaki, J.-i. (2004). Structural and functional analysis of surface film on Li anode in vinylene carbonate-containing electrolyte, *J. Electrochem. Soc.*, **151**, pp. A1778–A1788.
67. Parr, R. G. and Yang, W. (1989). *Density-Functional Theory of Atoms and Molecules*, Vol. 16 of International Series of Monographs on Chemistry (Oxford University Press, New York).
68. Peled, E. (1979). The electrochemical behavior of alkali and alkaline earth metals in nonaqueous battery systems—the solid electrolyte interphase model, *J. Electrochem. Soc.*, **126**, pp. 2047–2051.
69. Pelzer, K. M., Cheng, L., and Curtiss, L. A. (2016). Effects of functional groups in redox-active organic molecules: a high-throughput screening approach, *J. Phys. Chem. C*, **121**, pp. 237–245.

70. Peng, C. and Bernhard Schlegel, H. (1993). Combining synchronous transit and quasi-Newton methods to find transition states, *Isr. J. Chem.*, **33**, pp. 449–454.

71. Peng, C., Ayala, P. Y., Schlegel, H. B., and Frisch, M. J. (1996). Using redundant internal coordinates to optimize equilibrium geometries and transition states, *J. Comput. Chem.*, **17**, pp. 49–56.

72. Qu, X., Jain, A., Rajput, N. N., Cheng, L., Zhang, Y., Ong, S. P., Brafman, M., Maginn, E., Curtiss, L. A., and Persson, K. A. (2015). The Electrolyte Genome project: a big data approach in battery materials discovery, *Comput. Mater. Sci.*, **103**, pp. 56–67.

73. Scalmani, G. and Frisch, M. J. (2010). Continuous surface charge polarizable continuum models of solvation. I. General formalism, *J. Chem. Phys.*, **132**, pp. 114110.

74. Scheers, J. and Johansson, P. (2014). Prediction of electrolyte and additive electrochemical stabilities, in *Electrolytes for Lithium and Lithium-Ion Batteries* (Springer), pp. 403–443.

75. Schmidt am Busch, M. and Knapp, E.-W. (2005). One-electron reduction potential for oxygen-and sulfur-centered organic radicals in protic and aprotic solvents, *J. Am. Chem. Soc.*, **127**, pp. 15730–15737.

76. Shao, N., Sun, X.-G., Dai, S., and Jiang, D.-e. (2011). Electrochemical windows of sulfone-based electrolytes for high-voltage Li-ion batteries, *J. Phys. Chem. B*, **115**, pp. 12120–12125.

77. Shao, N., Sun, X.-G., Dai, S., and Jiang, D.-e. (2012). Oxidation potentials of functionalized sulfone solvents for high-voltage Li-ion batteries: a computational study, *J. Phys. Chem. B*, **116**, pp. 3235–3238.

78. Sodeyama, K., Yamada, Y., Aikawa, K., Yamada, A., and Tateyama, Y. (2014). Sacrificial anion reduction mechanism for electrochemical stability improvement in highly concentrated li-salt electrolyte, *J. Phys. Chem. C*, **118**, pp. 14091–14097.

79. Stephens, P., Devlin, F., Chabalowski, C., and Frisch, M. J. (1994). Ab initio calculation of vibrational absorption and circular dichroism spectra using density functional force fields, *J. Phys. Chem.*, **98**, pp. 11623–11627.

80. Tasaki, K. (2005). Solvent decompositions and physical properties of decomposition compounds in Li-ion battery electrolytes studied by DFT calculations and molecular dynamics simulations, *J. Phys. Chem. B*, **109**, pp. 2920–2933.

81. Tasaki, K., Kanda, K., Kobayashi, T., Nakamura, S., and Ue, M. (2006). Theoretical studies on the reductive decompositions of solvents and

additives for lithium-ion batteries near lithium anodes, *J. Electrochem. Soc.*, **153**, pp. A2192–A2197.

82. Tasaki, K., Goldberg, A., and Winter, M. (2011). On the difference in cycling behaviors of lithium-ion battery cell between the ethylene carbonate-and propylene carbonate-based electrolytes, *Electrochim. Acta*, **56**, pp. 10424–10435.

83. Tomasi, J. and Persico, M. (1994). Molecular interactions in solution: an overview of methods based on continuous distributions of the solvent, *Chem. Rev.*, **94**, pp. 2027–2094.

84. Tomasi, J., Mennucci, B., and Cammi, R. (2005). Quantum mechanical continuum solvation models, *Chem. Rev.*, **105**, pp. 2999–3094.

85. Trasatti, S. (1986). The absolute electrode potential: an explanatory note (recommendations 1986), *Pure Appl. Chem.*, **58**, pp. 955–966.

86. Trasatti, S. (1987). Interfacial behaviour of non-aqueous solvents, *Electrochim. Acta*, **32**, pp. 843–850.

87. Truhlar, D. G., Cramer, C. J., Lewis, A., and Bumpus, J. A. (2004). Molecular modeling of environmentally important processes: reduction potentials, *J. Chem. Educ.*, **81**, pp. 596.

88. Ue, M., Ida, K., and Mori, S. (1994). Electrochemical properties of organic liquid electrolytes based on quaternary onium salts for electrical double-layer capacitors, *J. Electrochem. Soc.*, **141**, pp. 2989–2996.

89. Ue, M., Takeda, M., Takehara, M., and Mori, S. (1997). Electrochemical properties of quaternary ammonium salts for electrochemical capacitors, *J. Electrochem. Soc.*, **144**, pp. 2684–2688.

90. Ue, M. (2000). Solution chemistry of organic electrolytes, in *Lithium Ion Secondary Batteries*, 2nd Ed., Yoshio, M. and Kozawa, A., eds. (Nikkan Kogyo Shinbun, Tokyo), pp. 83–98.

91. Ue, M., Sasaki, Y., Tanaka, Y., and Morita, M. (2014). Nonaqueous electrolytes with advances in solvents, in *Electrolytes for Lithium and Lithium-Ion Batteries* (Springer), pp. 93–165.

92. Ushirogata, K., Sodeyama, K., Okuno, Y., and Tateyama, Y. (2013). Additive effect on reductive decomposition and binding of carbonate-based solvent toward solid electrolyte interphase formation in lithium-ion battery, *J. Am. Chem. Soc.*, **135**, pp. 11967–11974.

93. Ushirogata, K., Sodeyama, K., Futera, Z., Tateyama, Y., and Okuno, Y. (2015). Near-shore aggregation mechanism of electrolyte decomposition products to explain solid electrolyte interphase formation, *J. Electrochem. Soc.*, **162**, pp. A2670–A2678.

94. Vollmer, J. M., Curtiss, L. A., Vissers, D. R., and Amine, K. (2004). Reduction mechanisms of ethylene, propylene, and vinylethylene carbonates: a quantum chemical study, *J. Electrochem. Soc.*, **151**, pp. A178–A183.

95. Wang, R., Buhrmester, C., and Dahn, J. (2006). Calculations of oxidation potentials of redox shuttle additives for Li-ion cells, *J. Electrochem. Soc.*, **153**, pp. A445–A449.

96. Wang, Y., Nakamura, S., Ue, M., and Balbuena, P. B. (2001). Theoretical studies to understand surface chemistry on carbon anodes for lithium-ion batteries: reduction mechanisms of ethylene carbonate, *J. Am. Chem. Soc.*, **123**, pp. 11708–11718.

97. Wang, Y. and Balbuena, P. B. (2002). Theoretical insights into the reductive decompositions of propylene carbonate and vinylene carbonate: density functional theory studies, *J. Phys. Chem. B*, **106**, pp. 4486–4495.

98. Wang, Y., Nakamura, S., Tasaki, K., and Balbuena, P. B. (2002). Theoretical studies to understand surface chemistry on carbon anodes for lithium-ion batteries: how does vinylene carbonate play its role as an electrolyte additive?, *J. Am. Chem. Soc.*, **124**, pp. 4408–4421.

99. Xing, L., Li, W., Wang, C., Gu, F., Xu, M., Tan, C., and Yi, J. (2009). Theoretical investigations on oxidative stability of solvents and oxidative decomposition mechanism of ethylene carbonate for lithium ion battery use, *J. Phys. Chem. B*, **113**, pp. 16596–16602.

100. Xing, L., Wang, C., Li, W., Xu, M., Meng, X., and Zhao, S. (2009). Theoretical insight into oxidative decomposition of propylene carbonate in the lithium ion battery, *J. Phys. Chem. B*, **113**, pp. 5181–5187.

101. Xing, L., Borodin, O., Smith, G. D., and Li, W. (2011). Density functional theory study of the role of anions on the oxidative decomposition reaction of propylene carbonate, *J. Phys. Chem. A*, **115**, pp. 13896–13905.

102. Xing, L. and Borodin, O. (2012). Oxidation induced decomposition of ethylene carbonate from DFT calculations–importance of explicitly treating surrounding solvent, *Phys. Chem. Chem. Phys.*, **14**, pp. 12838–12843.

103. Xu, K. (2004). Nonaqueous liquid electrolytes for lithium-based rechargeable batteries, *Chem. Rev.*, **104**, pp. 4303–4418.

104. Xu, M., Xing, L., and Li, W. (2014). Interphases between electrolytes and anodes in lithium-ion battery, in *Electrolytes for Lithium and Lithium-Ion Batteries* (Springer), pp. 227–282.

105. Yu, J., Balbuena, P. B., Budzien, J., and Leung, K. (2011). Hybrid DFT functional-based static and molecular dynamics studies of excess electron in liquid ethylene carbonate, *J. Electrochem. Soc.*, **158**, pp. A400–A410.

106. Zhang, X., Kostecki, R., Richardson, T. J., Pugh, J. K., and Ross, P. N. (2001). Electrochemical and infrared studies of the reduction of organic carbonates, *J. Electrochem. Soc.*, **148**, pp. A1341–A1345.

107. Zhang, X., Pugh, J. K., and Ross, P. N. (2001). Computation of thermodynamic oxidation potentials of organic solvents using density functional theory, *J. Electrochem. Soc.*, **148**, pp. E183–E188.

108. Zhao, Y. and Truhlar, D. G. (2008). The M06 suite of density functionals for main group thermochemistry, thermochemical kinetics, noncovalent interactions, excited states, and transition elements: two new functionals and systematic testing of four M06-class functionals and 12 other functionals, *Theor. Chem. Acc.*, **120**, pp. 215–241.

Chapter 3

Electronic Structure Theory of Electrolyte/Electrode Interfaces

Ryosuke Jinnouchi, Kensaku Kodama, Eishiro Toyoda, and Yu Morimoto
Toyota Central R&D Laboratories, Inc., Nagakute, Aichi 480-1192, Japan
jryosuke@mosk.tytlabs.co.jp

3.1 Introduction

One of great challenges in theoretical interfacial electrochemistry is to establish a self-consistent physical model that can quantitatively predict responses of current density to electrode potential observed on electrolyte-electrode systems. Classical electric double-layer models [19, 44, 45, 50, 135] and kinetic equations [16, 29] have been proposed to explain observed potential-dependent capacitances and current-voltage curves on the basis of electrostatics and statistical mechanics. Those models, however, rely on empirical parameters that complexly depend on atomic scale natures of electrolyte and electrode materials. To better understand the microscopic mechanisms and make things more predictable, modern theoretical electrochemists have tackled this problem to develop more advanced

Multiscale Simulations for Electrochemical Devices
Edited by Ryoji Asahi

ISBN 978-981-4800-71-6 (Hardcover), 978-0-429-29545-4 (eBook)
www.jennystanford.com

molecular and electronic structure models on the basis of ab initio theory [7, 8, 32, 39, 51, 60, 65, 88, 96, 97, 101, 111, 114, 127, 128, 142, 153]. Many difficulties, however, arise, mainly because of the high computational demands to solve the equations governing the complex electrochemical interfaces.

In this chapter, after the essence of electrolyte-electrode systems has been explained, one of the modern theoretical methods is introduced [32, 65, 96, 101, 127]. This method combines the electronic structure theory described by the density functional theory (DFT) [55, 83] with a continuum electrolyte theory described by a modified Poisson–Boltzmann (MPB) theory [10, 11]. This approach paves the way for predicting essential electrochemical properties, such as capacitance, current-voltage curves, and potential-dependent spectroscopy. Although the method still relies on approximated descriptions of the electrolyte and empirical connection between the DFT and the MPB, this model includes the essence of physics at the electrochemical interfaces in a self-consistent manner. In Section 3.2, the essence of the electrochemical interface (i.e., the electric double layer) is briefly explained. In Section 3.3, the DFT-MPB model and its basic equations are explained. In Section 3.4, several application results are shown, and in Section 3.5, the future scope is described.

3.2 Electrolyte/Electrode Interface

3.2.1 Electrode Potential and Electric Double Layer

Here, Fig. 3.1A shows a three-electrode system, which consists of a working electrode, reference electrode, and counterelectrode. Because the counterelectrode is located just for supplying necessary current to the working electrode in order to keep the state of the reference electrode constant, the main focus is on the working electrode. The electrode potential E is defined as the difference in the Fermi energy between the working and reference electrodes as follows [147]:

$$E = -\frac{\mu_{\mathrm{e}} - \mu_{\mathrm{e}}^{\mathrm{ref}}}{e}, \tag{3.1}$$

where μ_{e} and $\mu_{\mathrm{e}}^{\mathrm{ref}}$ are the Fermi energies of the working and reference electrodes, respectively. Both μ_{e} and $\mu_{\mathrm{e}}^{\mathrm{ref}}$ are scaled with respect to the electrochemical potential in the bulk electrolyte far from the electrode surfaces. As in many experiments, it is assumed that the electrode potential E is controlled by the external circuit. Since μ_e^{ref} is kept constant, the change in E is equivalent to the change in μ_e. Because the time scale of the macroscopic change in the electrode potential controlled by the external circuit is much slower than atomic and electronic motions, atoms and electrons at the interface can be regarded to move under a constant Fermi energy condition.

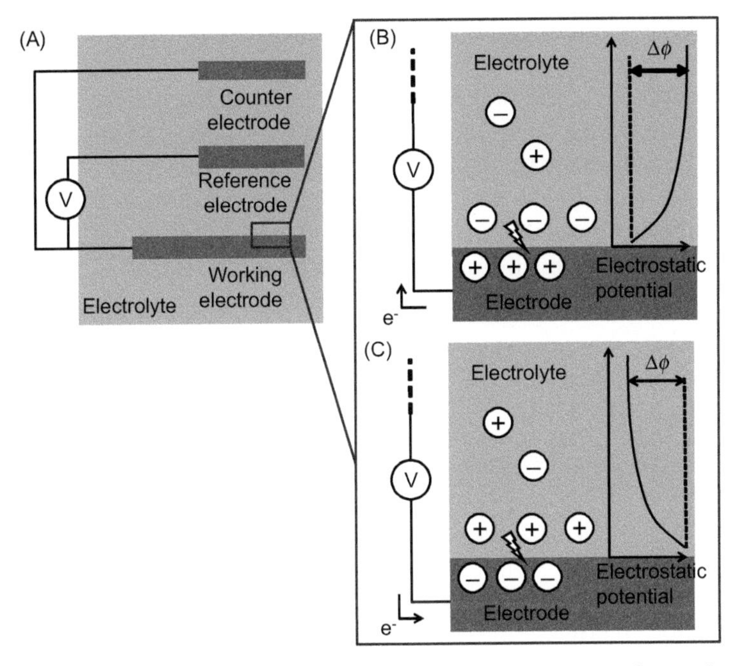

Figure 3.1 Schematic of a three-electrode system, consisting of a working electrode, reference electrode, and counterelectrode (A), and the interfacial structures near the working electrode (B and C). The electrostatic potential in (B) and (C) is defined to be positive when it destabilizes electrons.

To understand the microscopic interfacial structure, further attention is focused on a subsystem that includes the vicinity of the electrolyte/electrode interface. The electrons in this partial system are opened for the external circuit, and their number varies to realize the constant Fermi energy μ_e. Thus, the working electrode is not necessarily electrically neutral, and its surface charge changes with the change in μ_e. For example, as schematically shown in Fig. 3.1B, the working electrode is positively charged when μ_e is set to low. In this situation, anions in the electrolyte are distributed near the electrode and cations in the electrolyte are distributed far from the electrode. Thus, the interfacial electrolyte is negatively charged to cancel the positive charge induced on the electrode surface. The charge distribution in the electrolyte and electrode creates a polarization across the interface, and this polarization generates the electrostatic potential gap, lowering the Fermi energy. When μ_e is high, the working electrode is negatively charged and an opposite polarization raising μ_e is created, as shown in Fig. 3.1C. It should be also noted that due to polarization generation, the electrochemical potentials μ_+ and μ_- of the anion and the cation, respectively, become consistent with those in the bulk electrolyte. Hence, an equilibrium interfacial polarization is created to realize the externally controlled μ_e, μ_+, and μ_-. On the basis of the statistical mechanics, this equilibrium polarization structure can be described as a probability distribution within a grand canonical ensemble:

$$p(\xi, N_e, N_+, N_-) = \frac{e^{-\frac{E_{\text{tot}}(\xi, N_e, N_+, N_-) - \mu_e N_e - \mu_+ N_+ - \mu_- N_-}{k_B T}}}{\Xi} \tag{3.2}$$

and

$$\Xi = \sum_{N_e} \sum_{N_+} \sum_{N_-} \sum_{\xi} e^{-\frac{E_{\text{tot}}(\xi, N_e, N_+, N_-) - \mu_e N_e - \mu_+ N_+ - \mu_- N_-}{k_B T}}, \tag{3.3}$$

where N_e, N_+ and N_- are, respectively, the number of electrons, cations, and anions in this interfacial subsystem, E_{tot} is the total internal energy of the subsystem at a state specified by ξ, k_B is the Boltzmann constant, T is the temperature, and Ξ is the grand partition function. Here, ξ includes all variables affecting the total internal energy E_{tot}, such as wave functions and their occupation numbers, nuclear positions, and nuclear momentums.

The polarization is called the electric double layer, and its description is the central problem in the interfacial electrochemistry. As illustrated in Fig. 3.1, the electrostatic potential across the entire electric double layer determines its internal electrochemical potentials. This means that a self-consistent description of the entire electric double layer is necessary.

3.2.2 Grand Partition Function and Modeling of the Electric Double Layer

For quantitative evaluations of interfacial electrochemical properties, detailed formulations of the grand partition function are necessary. On the basis of the Born–Oppenheimer approximation [9], the grand partition function can be described as follows:

$$\Xi = \sum_{N_e}\sum_{N_+}\sum_{N_-}\int e^{-\frac{\sum_k \frac{|\mathbf{P}_k|^2}{2M_k} + F_e\left(N_e, \{\mathbf{R}_k\}\right) - \mu_e N_e - \mu_+ N_+ - \mu_- N_-}{k_B T}} \mathrm{d}\mathbf{P}_k \mathrm{d}\mathbf{R}_k, \quad (3.4)$$

where $\mathbf{R}_k$ and $\mathbf{P}_k$ are the nuclear positions and momentums, respectively, and F_e is the ground state potential energy at the electron number N_e and nuclear positions $\mathbf{R}_k$.

On the basis of this equation, molecular dynamics using force fields has been developed [129–131]. In this method, an atomistic model consisting of an interfacial subsystem and an electrolyte buffer on top of the subsystem is prepared and the atomic motions are simulated by solving the equations of motion numerically [2]. During the simulation, the number of atoms in the entire system does not change. However, if the size of the bulk buffer is large enough to suppress the fluctuations in the electrochemical potentials in the interfacial subsystem, the grand canonical distribution is correctly described during the simulation. To handle the large simulation system, the ground state potential energy F_e needs to be approximated by simple functions expressing the atomic interactions. The pioneering studies provided atomic-scale ionic distributions in the electric double layers formed at interfaces between aqueous electrolytes and metal electrodes [129–131]. The method is also applied to more complex electrolyte/electrode interfaces, including ionic liquids and polymer electrolytes [36, 74, 79]. This approach, however, does not explicitly handle electrons and, therefore, cannot evaluate the Fermi energy or the electrode potential. It should be also noted that the accuracy is always limited by the approximated force fields.

The most rigorous method to overcome these limitations is to evaluate the ground state potential energy by using first-principles electronic structure theory. DFT [55, 83] is one of the most promising electronic structure theories, and in this method the ground state potential energy is described by the Helmholtz free energy of an electronic subsystem as

$$F_e[\psi_{n\mathbf{k}\sigma}, f_{n\mathbf{k}\sigma}, \phi] = K + E_{xc} + E_{es} - TS_e. \quad (3.5)$$

Here

$$K = \sum_n \sum_{\mathbf{k}} \sum_{\sigma} f_{n\mathbf{k}\sigma} \int \psi^*_{n\mathbf{k}\sigma}(\mathbf{r}) \left(-\frac{1}{2} \nabla^2 \right) \psi_{n\mathbf{k}\sigma}(\mathbf{r}) d\mathbf{r}, \quad (3.6)$$

$$E_{xc} = \int f_{xc}\left(\rho_\uparrow, \rho_\downarrow, \nabla\rho_\uparrow, \nabla\rho_\downarrow\right) d\mathbf{r}, \quad (3.7)$$

and

$$E_{es} = \int \left[\rho_e(\mathbf{r}) + \rho_c(\mathbf{r})\right] \phi(\mathbf{r}) d\mathbf{r} - \frac{\varepsilon_0}{8\pi} \int \left|\nabla\phi(\mathbf{r})\right|^2 d\mathbf{r}, \quad (3.8)$$

where

$$\rho_e(\mathbf{r}) = \sum_\sigma \rho_\sigma(\mathbf{r}) \quad (3.9)$$

and

$$\rho_\sigma(\mathbf{r}) = \sum_{\mathbf{k}} \sum_n f_{n\mathbf{k}\sigma} \left|\psi_{n\mathbf{k}\sigma}(\mathbf{r})\right|^2. \quad (3.10)$$

Here K, E_{xc}, and E_{es} are the kinetic, exchange-correlation, and electrostatic energies, respectively; S_e is the electronic entropy; $\psi_{n\mathbf{k}\sigma}$ and $f_{n\mathbf{k}\sigma}$ are the wave function with band index n, point $\mathbf{k}$ in the Brillouin zone, and spin index σ, respectively; $f_{n\mathbf{k}\sigma}$ is its occupation number; ρ_e is the electron density; ρ_σ is the electron density with spin index σ; and ρ_c is the nuclear charge. Here, the exchange-correlation energy functional E_{xc} is described by the generalized gradient approximation [49, 116], but the extension to more advanced approximation is straightforward. The method can accurately describe the atomic interactions, and a molecular dynamics simulation employing the obtained atomic interactions can provide information on the interfacial double-layer structures, including electron distributions. Thus, the method can provide the Fermi energy. However, its application in the huge electric double layer involves overly expensive computational costs. The simple Gouy–Chapman model [19, 44], for example, indicates that a length of 3–35 nm is necessary to realize the screening of the electrostatic potential induced by the charged electrode in the electrolyte with 1–10^{-3} mol·L^{-1}. Furthermore, the Debye–Falkenhagen model [27] indicates that the relaxation time of the electric double layer is 0.05–50 ns. In addition to this large interfacial subsystem, a bulk buffer electrolyte region is also necessary to mimic the constant electrochemical potential condition.

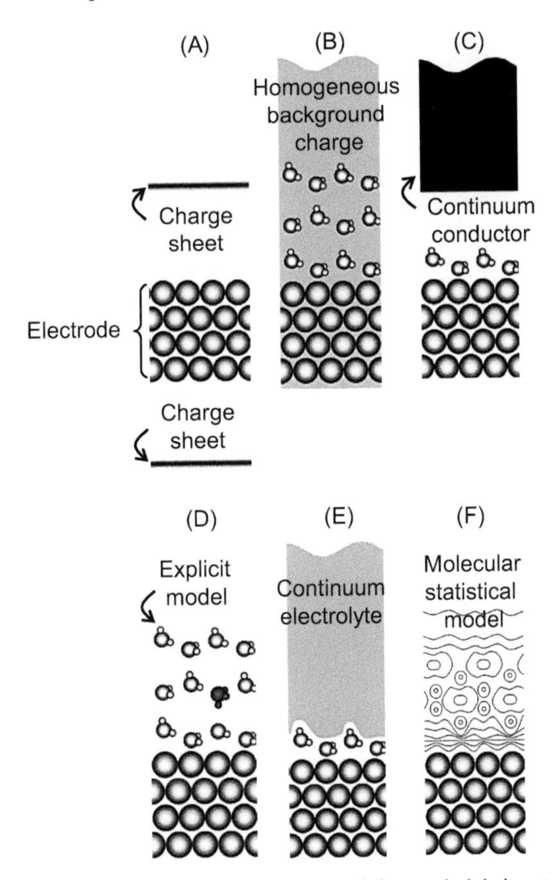

Figure 3.2 Models of electrified interfaces: (A) metal slab located between two countercharge sheets in vacuum, (B) metal slab located in a homogeneous background charge in vacuum, (C) metal slab and perfect conductor inducing the countercharge sheet on its surface, (D) metal slab and explicit molecular models including the counterions, (E) metal slab and continuum electrolyte model, and (F) metal slab and molecular distributions given by a statistical model of electrolyte. Reprinted from Ref. [76], Copyright (2018), with permission from Elsevier.

For this reason, several approximated approaches were developed. Sketches of the proposed models are shown in Fig. 3.2. In all approaches, the majority of the electrolyte region is replaced by simple models: (A) a charge sheet in vacuum [97], (B) a homogeneous background charge [142], (C) a charge sheet induced on the perfect conductor model [114], (D) a few explicit solvent molecules and ions [128], (E) ion distributions derived from the continuum electrolyte model [32, 39, 65, 70, 96, 101, 127], and (F) electrolyte distribution functions derived from the statistical model of molecular liquids [88, 111, 153]. Although the simplicity and accuracy of the double-layer models are different, their basic concept and the purpose of simplification are the same. The following sections introduce the model shown in Fig. 3.2E as an example.

Before this section ends, it needs to be stressed that many calculations have been carried out by simpler models that completely neglect the electric double layer and approximate the interfaces by electrically neutral surfaces [105, 119]. These simple models have provided valuable information for material designs [46, 47, 66, 73, 98, 133] but cannot provide electrochemical properties originating from electrified interfaces. Therefore, these details are not described in this chapter.

3.3 Density Functional Theory Combined with Modified Poisson–Boltzmann Theory

3.3.1 Model and Approximation in Grand Partition Function

Here, the model in Fig. 3.2E is called DFT-MPB [65, 70]. This method handles an interfacial subsystem opened for both electrons and ions, as schematically shown in Fig. 3.3. The number of electrons and ions in the subsystem can vary to match their electrochemical potentials with those in the external circuit and in the bulk electrolyte. In this method, the interfacial subsystem is further divided into two regions: (i) a region including the electrode and electrolyte near the electrode and (ii) the remaining electrolyte region. Region (i) is described by an electronic structure theory (DFT) [55, 83] that can accurately evaluate atomic interactions and also give information on the Fermi energy. Region (ii) is described by an MPB theory

[10, 11] that can efficiently predict equilibrium distributions of ions in the electrolyte. The atoms in the DFT region are located in vacuum similarly to conventional DFT calculations, while ions in the MPB region are located in the continuum solvent medium with a dielectric permittivity ε_b. To connect the two different mediums, the polarizable continuum model (PCM) [34, 35, 145] is adopted. As will be shown in the following section, the PCM smoothly connects two mediums and can describe the mean field of the solvent molecules quantitatively.

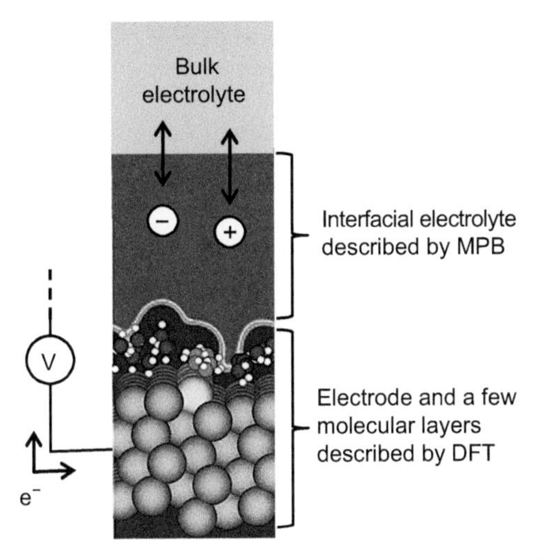

Figure 3.3 Interfacial subsystem described by the DFT-MPB method.

Here, the word "mean field" is a significant concept that solves the problems caused by the long length scales and time scales of the electrolyte. In this approach, the phase space of the ions and solvents inside region (ii) is decoupled from that in atoms and electrons in region (i) and the field averaged over the ionic and solvent motions in region (ii) is assumed to dominate the effects by region (ii) on region (i). This so-called static solvation approximation [18, 134, 140] has been successfully applied in the community of molecular science, and the same concept is adopted in the DFT-MPB method.

The procedure of approximation can be explained by using the grand partition function in Eq. 3.4 as follows. First, the summation over the electron number in Eq. 3.4 is decoupled by using the grand canonical DFT formalism [102, 115] as follows:

$$\Xi = \sum_{N_+}\sum_{N_-}\int e^{-\frac{\sum_k \frac{|\mathbf{P}_k|^2}{2M_k} + \Omega_e(\{\mathbf{R}_k\}) - \mu_+ N_+ - \mu_- N_-}{k_B T}} d\mathbf{P}_k d\mathbf{R}_k, \tag{3.11}$$

where

$$\Omega_e(\{\mathbf{R}_k\}) = -k_B T \ln \sum_{N_e} e^{-\frac{F_e(N_e, \{\mathbf{R}_k\}) - \mu_e N_e}{k_B T}}. \tag{3.12}$$

Second, the summations over the numbers of ions and integrals over the phase space of atoms in region (ii) are decoupled by applying the mean field approximation as follows:

$$\Xi = \int e^{-\frac{\sum_i \frac{|\mathbf{P}_i|^2}{2M_i} + \Omega(\{\mathbf{R}_i\})}{k_B T}} d\mathbf{P}_i d\mathbf{R}_i \tag{3.13}$$

and

$$\Omega(\{\mathbf{R}_i\}) = -k_B T \ln \sum_{N_+}\sum_{N_-}\int e^{-\frac{\sum_j \frac{|\mathbf{P}_j|^2}{2M_j} + \Omega_e(\{\mathbf{R}_i\}, \{\mathbf{R}_j\}) - \mu_+ N_+ - \mu_- N_-}{k_B T}} d\mathbf{P}_j d\mathbf{R}_j, \tag{3.14}$$

where i and j in Eqs. 3.13 and 3.14 denote the atoms in region (i) and region (ii), respectively. As clearly shown by the new grand partition function in Eq. 3.13, the subsystem can be regarded as a "closed" system, including only the explicit atoms in region (i) that move on the mean field Ω.

3.3.2 Equations on Ω and Details of the Calculation Scheme

3.3.2.1 Equations of the mean field Ω

The basic concept described by Eqs. 3.11–3.14 is employed in all the models shown in Fig. 3.2A–F. The difference is in the detailed formulation of the mean field Ω. In the DFT-MPB method, the mean field Ω is described as follows:

$$\Omega(\{\mathbf{R}_i\}) = F(\{\mathbf{R}_i\}) - \mu_e N_e - \mu_+ N_+ - \mu_- N_-. \tag{3.15}$$

Within the modified Poisson–Boltzmann model, assuming a lattice gas model proposed by Borukhov et al. [10, 11], $\mu_{\pm}$ are described as follows:

$$\mu_{\pm} = k_{\mathrm{B}}T\ln\frac{c_{\mathrm{b}}a^3}{1-2c_{\mathrm{b}}a^3}, \tag{3.16}$$

where c_{b} is the ion concentration of the bulk electrolyte and a is the size of ions.

F takes account of all the equilibrium interactions among species in the electronic and ionic subsystem and is described as a functional of wave functions $\psi_{n\mathbf{k}\sigma}$ for the electrons in region (i), the occupation numbers $f_{n\mathbf{k}\sigma}$ of the wave functions, the ion densities $\rho_{\pm}$ in region (ii), and the electrostatic potential ϕ in the whole system as follows:

$$F_{\mathrm{e}}\left[\psi_{n\mathbf{k}\sigma}, f_{n\mathbf{k}\sigma}, \rho_{+}, \rho_{-}, \phi\right] = K + E_{\mathrm{xc}} + E_{\mathrm{es}} + F_{\mathrm{ss,nes}} + F_{\mathrm{is,nes}} - TS_{\mathrm{e}} - TS_{\pm}, \tag{3.17}$$

where $F_{\mathrm{ss,nes}}$ refers to the non-electrostatic interaction energies between atoms in region (i) and solvent in region (ii), $F_{\mathrm{is,nes}}$ is the non-electrostatic interaction between atoms in region (i) and ions in region (ii), and $S_{\pm}$ is the entropy of the electrolyte in region (ii). The kinetic energy K and exchange-correlation energy E_{xc} are described as Eqs. 3.6 and 3.7, respectively. In contrast, the electrostatic energy is reformulated as follows:

$$E_{\mathrm{es}} = \int\left[\rho_{\mathrm{e}}(\mathbf{r}) + \rho_{\mathrm{c}}(\mathbf{r}) + \rho_{+}(\mathbf{r}) + \rho_{-}(\mathbf{r})\right]\phi(\mathbf{r})d\mathbf{r} - \int\frac{\varepsilon(\mathbf{r})}{8\pi}\left|\nabla\phi(\mathbf{r})\right|^2 d\mathbf{r}, \tag{3.18}$$

where $\varepsilon(\mathbf{r})$ is the position-dependent dielectric permittivity defined by the following equation:

$$\varepsilon(\mathbf{r}) = 1 + \frac{\varepsilon_{\infty}(\mathbf{r})-1}{2}\left[1 + \frac{1-\left(\rho_{\mathrm{e}}(\mathbf{r})/\rho_0\right)^{2\beta}}{1+\left(\rho_{\mathrm{e}}(\mathbf{r})/\rho_0\right)^{2\beta}}\right], \tag{3.19}$$

where β and ρ_0 are the constants determined to reproduce the experimental solvation free energies of molecules and ions isolated in a bulk solvent. The dielectric permittivity $\varepsilon(\mathbf{r})$ defined as Eq. 3.19 is close to unity when ρ_{e} is higher than ρ_0 and smoothly approaches $\varepsilon_{\infty}(\mathbf{r})$ when ρ_{e} decreases across ρ_0. Hence, $\varepsilon(\mathbf{r})$ is close to unity in the region near the explicit atoms in region (i) whereas it approaches

$\varepsilon_\infty(\mathbf{r})$ in the region far from the explicit atoms. In the original equation proposed by Gygi and Fattebert [34, 35], $\varepsilon_\infty(\mathbf{r})$ is set as the dielectric permittivity of the bulk electrolyte ε_b. This parameter setting is proper when both sides of the electrode come in contact with the continuum electrolyte. However, the original equation is inapplicable to the electrode whose bottom side is not in contact with the continuum electrolyte as shown in Fig. 3.3. To handle this asymmetric system, Jinnouchi and Anderson [65] introduced the position dependence in ε_∞ as follows:

$$\varepsilon_\infty(\mathbf{r}) = -\frac{\varepsilon_b - 1}{2}\left[1 - \mathrm{erf}\left(\frac{z - z_0}{\Delta_z}\right)\right] + \varepsilon_b, \qquad (3.20)$$

where z_0 is the position of the bottom of the electrode and Δ_z is a constant parameter. Equation 3.20 gives $\varepsilon_\infty = 1$ and ε_b at $z << z_0$ and $z >> z_0$, respectively. Hence, $\varepsilon(\mathbf{r})$ approaches unity below the electrode. In addition to the introduction of the position-dependent ε_∞, Jinnouchi and Anderson replaced ρ_e in Eq. 3.19 by a sum of spherical atomic electron densities ρ_{na} as follows:

$$\varepsilon(\mathbf{r}) = 1 + \frac{\varepsilon_\infty(\mathbf{r}) - 1}{2}\left[1 + \frac{1 - (\rho_{na}(\mathbf{r})/\rho_0)^{2\beta}}{1 + (\rho_{na}(\mathbf{r})/\rho_0)^{2\beta}}\right]. \qquad (3.21)$$

This modification enables the use of coarse numerical meshes necessary to solve the generalized Poisson equation, which will be described later, by the finite difference method [64, 65].

Before the non-electrostatic interaction models are described, the physical meaning of the electrostatic model described by Eqs. 3.18–3.21 is briefly described. As discussed by Jinnouchi and Anderson [65], in the electrostatic model described by Eqs. 3.18–3.21, the interactions between the explicit atoms in region (i) and solvent molecules in region (ii) are replaced by a continuum polarization induced at the isosurface of $\rho_e = \rho_0$. This continuum polarization mimics the actual polarization induced by explicit solvent molecules. For example, when liquid water comes in contact with a positively charged electrode, water molecules are oriented so their oxygen atoms are closer to the electrode. This molecular polarization is replaced by the continuum polarization in the DFT-MPB model.

As in the case of the PCM [125, 145], the non-electrostatic interactions $F_{ss,nes}$ between atoms in region (i) and solvent in region

(ii) are described as a summation of cavitation, dispersion, and repulsion free energies as follows:

$$F_{\mathrm{ss,nes}} = F_{\mathrm{ss,cav}} + F_{\mathrm{ss,dr}} + F_{\mathrm{ss,rep}}. \tag{3.22}$$

The cavitation free energy corresponds to the free energy necessary for creating a vacuum cavity inside the continuum medium and is described as a product of the molecular surface area S and the surface tension γ_{b} of the solvent. The molecular surface area S is described by a functional of the electron density as follows [125]:

$$S = \int s\left(\rho_{\mathrm{e}}, |\nabla\rho_{\mathrm{e}}|\right)d\mathbf{r}, \tag{3.23}$$

$$s\left(\rho_{\mathrm{e}}, |\nabla\rho_{\mathrm{e}}|\right) = \eta(\mathbf{r})\left[\vartheta_{\rho_0-\Delta/2}(\mathbf{r}) - \vartheta_{\rho_0+\Delta/2}(\mathbf{r})\right] \times \frac{|\rho_{\mathrm{e}}(\mathbf{r})|}{\Delta}, \tag{3.24}$$

and

$$\vartheta_{\rho_0}(\mathbf{r}) = \frac{1}{2}\left[\frac{\left(\rho_{\mathrm{e}}(\mathbf{r})/\rho_0\right)^{2\beta} - 1}{\left(\rho_{\mathrm{e}}(\mathbf{r})/\rho_0\right)^{2\beta} + 1} + 1\right]. \tag{3.25}$$

If $\eta(\mathbf{r})$ equals unity, Eq. 3.23 provides the area of isosurface $\rho_{\mathrm{e}} = \rho_0$. However, as in the problem in Eq. 3.19, this parameter setting is improper for the system shown in Fig. 3.3. To apply the same method to this asymmetric situation, Jinnouchi and Anderson introduced the following position-dependent $\eta(\mathbf{r})$:

$$\eta(\mathbf{r}) = -\frac{1}{2}\left[1 - \mathrm{erf}\left(\frac{z - z_0}{\Delta_z}\right)\right] + 1. \tag{3.26}$$

As in the case of Eq. 3.21, ρ_{e} in Eq. 3.25 can also be replaced by a sum of atomic electron densities ρ_{na} in order to decrease the necessary numerical meshes as follows:

$$\vartheta_{\rho_0}(\mathbf{r}) = \frac{1}{2}\left[\frac{\left(\rho_{\mathrm{na}}(\mathbf{r})/\rho_0\right)^{2\beta} - 1}{\left(\rho_{\mathrm{na}}(\mathbf{r})/\rho_0\right)^{2\beta} + 1} + 1\right]. \tag{3.27}$$

The dispersion and repulsion free energies are described, respectively, by the following equations:

$$F_{\mathrm{ss,dr}} = \sum_i \left(a_i^{\mathrm{dr}} S_i + b_i^{\mathrm{dr}}\right) \tag{3.28}$$

and

$$F_{\text{ss,rep}} = \sum_i \left(a_i^{\text{rep}} S_i + b_i^{\text{rep}} \right), \tag{3.29}$$

where a_i^{k} and b_i^{k} (k = dr or rep) are the constant parameters determined to reproduce the experimental solvation free energies. S_i is the atomic surface area obtained by partitioning the molecular surface as follows:

$$S_i = \int p_i(\mathbf{r}) s\left(\rho_{\text{e}}, \left|\nabla \rho_{\text{e}}\right|\right) d\mathbf{r}, \tag{3.30}$$

where p_i is a partitioning function and is, for example, described by an equation proposed by Delley [28] as follows:

$$p_i(\mathbf{r}) = \frac{\sum_{\mathbf{R}} f_i\left(\left|\mathbf{r} - \mathbf{R}_i - \mathbf{R}\right|\right)}{\sum_{\mathbf{R}} \sum_k f_k\left(\left|\mathbf{r} - \mathbf{R}_k - \mathbf{R}\right|\right)} \tag{3.31}$$

and

$$f_i(r) = \frac{\rho_{\text{na},i}(r)}{r^2}, \tag{3.32}$$

where $\rho_{\text{na},i}$ is the electron density of an isolated atom i. It should be noted that more sophisticated forms of the dispersion and repulsion interactions have been recently proposed to decrease the number of empirical parameters. See details of the equations in Refs. [3, 138].

The non-electrostatic interactions between explicit atoms in region (i) and ions in region (ii) are introduced in order to prevent unphysical incursions of the continuum ions into region (i). This term can be, for example, described by the following function:

$$F_{\text{is,nes}} = \int \left[\left|\rho_+(\mathbf{r})\right| + \left|\rho_-(\mathbf{r})\right| \right] \phi_{\text{rep}}(\mathbf{r}) d\mathbf{r}, \tag{3.33}$$

where ϕ_{rep} denotes the repulsive potential. Details of ϕ_{rep} are shown in the publications in Refs. [32, 65, 127].

The electronic entropy is described in a way similar to the conventional finite temperature DFT scheme [89, 90]. The entropy of the electrolyte is described on the basis of the lattice gas model as follows:

$$\begin{aligned} S_{\pm} = -\frac{k_{\text{B}}}{a^3} \int [& \left|\rho_+(\mathbf{r})\right| a^3 \ln\left(\left|\rho_+(\mathbf{r})\right| a^3\right) + \left|\rho_-(\mathbf{r})\right| a^3 \ln\left(\left|\rho_-(\mathbf{r})\right| a^3\right) \\ & + \left(1 - \left|\rho_+(\mathbf{r})\right| a^3 - \left|\rho_-(\mathbf{r})\right| a^3\right) \ln\left(1 - \left|\rho_+(\mathbf{r})\right| a^3 - \left|\rho_-(\mathbf{r})\right| a^3\right)] d\mathbf{r}. \end{aligned} \tag{3.34}$$

3.3.2.2 Minimization of Ω

Ω is formulated as a functional of distribution functions of all species composing the interfacial subsystem. The distribution functions are determined to minimize Ω. The minimization is realized by solving equations derived from the variational principle. A variation with the wave function $\psi_{n\mathbf{k}\sigma}$ provides the following Kohn–Sham equation:

$$\widehat{H}_\sigma \psi_{n\mathbf{k}\sigma}(\mathbf{r}) = \varepsilon_{n\mathbf{k}\sigma}\psi_{n\mathbf{k}\sigma}(\mathbf{r}), \tag{3.35}$$

where

$$\widehat{H}_\sigma = -\frac{1}{2}\nabla^2 + \phi(\mathbf{r}) + \frac{\delta E_{\mathrm{xc}}}{\delta \rho_\sigma}. \tag{3.36}$$

Here, δ denotes the variation. When the dielectric permittivity ε and surface areas S and S_i are defined as functionals of the electron density ρ_e, as in Eqs. 3.19 and 3.25, respectively, additional terms appear in the electronic Hamiltonian as follows:

$$\widehat{H}_\sigma = -\frac{1}{2}\nabla^2 + \phi(\mathbf{r}) + \frac{\delta E_{\mathrm{xc}}}{\delta \rho_\sigma} - \frac{1}{8\pi}\frac{\delta \varepsilon}{\delta \rho_\sigma}\left|\nabla \phi(\mathbf{r})\right|^2 + \frac{\delta F_{\mathrm{ss,nes}}}{\delta \rho_\sigma}. \tag{3.37}$$

A variation with the electrostatic potential ϕ provides the generalized Poisson equation

$$\nabla \cdot \left(\varepsilon(\mathbf{r})\nabla \phi(\mathbf{r})\right) = -4\pi\left[\rho_{\mathrm{e}}(\mathbf{r}) + \rho_{\mathrm{c}}(\mathbf{r}) + \rho_{+}(\mathbf{r}) + \rho_{-}(\mathbf{r})\right]. \tag{3.38}$$

Variations with the ion distribution functions ρ_+ and ρ_- provide the following modified Poisson–Boltzmann distributions:

$$\rho_{\pm}(\mathbf{r}) = \mu \frac{c_{\mathrm{b}} e^{\left(\pm\phi(\mathbf{r}) - \phi_{\mathrm{rep}}(\mathbf{r})\right)/k_{\mathrm{B}}T}}{1 - 2c_{\mathrm{b}}a^3 + 2c_{\mathrm{b}}a^3 \cosh\left(\phi(\mathbf{r})/k_{\mathrm{B}}T\right) e^{-\phi_{\mathrm{rep}}(\mathbf{r})/k_{\mathrm{B}}T}}. \tag{3.39}$$

A variation in the occupation number $f_{n\mathbf{k}\sigma}$ provides the following equation determining the occupation numbers:

$$T\frac{\partial S_{\mathrm{e}}}{\partial f_{n\mathbf{k}\sigma}} = \varepsilon_{n\mathbf{k}\sigma} - \mu_{\mathrm{e}}. \tag{3.40}$$

The variables $\psi_{n\mathbf{k}\sigma}$, $f_{n\mathbf{k}\sigma}$, $\rho_\pm$, and ϕ minimizing Ω are determined by simultaneously solving Eqs. 3.35–3.40. The solution can be realized by using a conventional iterative self-consistent field (SCF) scheme used in many DFT calculations [28, 89, 90]. By using the obtained variables, Ω is calculated by Eqs. 3.15–3.17. Furthermore, on the basis of Eq. 3.13, the mean forces acting on the nuclei in region

(i) are obtained as first derivatives of Ω with respect to the nuclear positions $\mathbf{R}_i$ as follows:

$$\nabla_{\mathbf{R}_i}\Omega = -\mathbf{F}_{\mathrm{HF}} - \mathbf{F}_{\mathrm{Pulay}} - \mathbf{F}_{\mathrm{solv,nes}}, \tag{3.41}$$

where

$$\mathbf{F}_{\mathrm{HF}} = -\int \nabla_{\mathbf{R}_i}\rho_{\mathrm{c}}(\mathbf{r})\phi(\mathbf{r})\mathrm{d}\mathbf{r}, \tag{3.42}$$

$$\mathbf{F}_{\mathrm{Pulay}} = -\sum_{n}\sum_{\mathbf{k}}\sum_{\sigma} f_{n\mathbf{k}\sigma}\int [\nabla_{\mathbf{R}_i}\psi^{*}_{n\mathbf{k}\sigma}(\mathbf{r})\left(\widehat{H}_{\sigma} - \varepsilon_{n\mathbf{k}\sigma}\right)\psi_{n\mathbf{k}\sigma}(\mathbf{r})$$
$$+\psi^{*}_{n\mathbf{k}\sigma}(\mathbf{r})\left(\widehat{H}_{\sigma} - \varepsilon_{n\mathbf{k}\sigma}\right)\nabla_{\mathbf{R}_i}\psi_{n\mathbf{k}\sigma}(\mathbf{r})]\mathrm{d}\mathbf{r}, \tag{3.43}$$

and

$$\mathbf{F}_{\mathrm{solv,nes}} = \frac{1}{8\pi}\int \nabla_{\mathbf{R}_i}\rho_{\mathrm{na}}(\mathbf{r})\frac{\mathrm{d}\varepsilon}{\mathrm{d}\rho_{\mathrm{na}}}\left|\nabla\phi(\mathbf{r})\right|^2\mathrm{d}\mathbf{r} - \int \nabla_{\mathbf{R}_i}\rho_{\mathrm{na}}(\mathbf{r})\frac{\delta F_{\mathrm{ss,nes}}}{\delta\rho_{\mathrm{na}}}\mathrm{d}\mathbf{r}$$
$$-\int\left[\left|\rho_{+}(\mathbf{r})\right| + \left|\rho_{-}(\mathbf{r})\right|\right]\nabla_{\mathbf{R}_i}\phi_{\mathrm{rep}}(\mathbf{r})\mathrm{d}\mathbf{r}. \tag{3.44}$$

The first and second terms in Eq. 3.41 correspond to Hellmann–Feynman [37] and Pulay forces [117], respectively. The third term is derived from the non-electrostatic solvation interaction free energy term. When the dielectric permittivity ε and surface areas S and S_i are defined as functionals of the electron density ρ_{e}, the first and second terms in Eq. 3.44 disappear, as follows:

$$\mathbf{F}_{\mathrm{solv,nes}} = -\int\left[\left|\rho_{+}(\mathbf{r})\right| + \left|\rho_{-}(\mathbf{r})\right|\right]\nabla_{\mathbf{R}_i}\phi_{\mathrm{rep}}(\mathbf{r})\mathrm{d}\mathbf{r}. \tag{3.45}$$

Ω and its first derivatives enable us to use conventional analytical and numerical schemes employed in many DFT calculations for predicting thermodynamic, kinetic, and spectroscopic properties. Equilibrium positions of the explicit atoms in region (i) can be, for example, obtained by minimizing Ω with respect to the nuclear positions $\mathbf{R}_i$ by using a quasi-Newton method [33], and vibration frequencies at the equilibrium positions can be also obtained from Hessian matrices calculated from the first derivatives numerically. The Ω at the equilibrium positions and vibration frequencies can be used to calculate the total free energy of the interfacial subsystem on the basis of the harmonic oscillator model and/or the ideal gas model as follows [134]:

$$G = \Omega\left(\left\{\mathbf{R}_i^0\right\}\right) + H_{\mathrm{n}} - TS_{\mathrm{n}}, \tag{3.46}$$

where $\mathbf{R}_i^0$ denotes the equilibrium positions and H_{n} and S_{n} are the enthalpy and entropy contributions by the nuclear motions, respectively. The vibration frequencies and dynamic dipole moments can be used to predict the vibrational spectroscopy. In a similar manner, saddle points on the mean field Ω can be determined by conventional methods, such as the nudged elastic band method [52, 53], and the determined transition states can provide kinetic properties of electrochemical reactions. A significant difference from the standard DFT calculation is that all the properties are calculated within the constant Fermi energy condition. An activation barrier, for example, is calculated as the difference between the free energies of the transition state and the reactant state at an identical Fermi energy. It should be, however, mentioned that a constant Fermi energy calculation can be realized by sequentially executing the standard constant electron number calculations. The method is explained in the next section.

3.3.2.3 Constant Fermi energy calculation

In the first calculation on an electrified surface described by the model shown in Fig. 3.2A done by Lozovoi and Alavi [97], the constant Fermi energy calculation was realized by an SCF calculation under a variable electron number condition. This approach, however, causes a severe charge sloshing problem that destabilizes the SCF procedure. Later, Taylor et al. [142] proposed an alternative scheme that can be executed by sequential standard SCF calculations under constant electron number conditions.

To explain this scheme, calculations of the reaction energy and activation barrier are considered as examples. In the proposed scheme, calculations with several different electron numbers, N_{e}, N_{e}', and N_{e}'', are executed to locate stationary points for the reactant and product states and a saddle point for the transition state. By the calculations, one can collect mean fields (Ω_s, Ω'_s, Ω''_s) and Fermi energies ($\mu_{\mathrm{e},s}$, $\mu'_{\mathrm{e},s}$, $\mu''_{\mathrm{e},s}$) (s = R, P, or TS) at the electron numbers (N_{e}, N_{e}', and N_{e}''), where R, P, and TS denote reactant, product and transition states, respectively. The Fermi energies can be converted to the electrode potentials ($E_{\mathrm{e},s}$, $E'_{\mathrm{e},s}$, $E''_{\mathrm{e},s}$) through Eq. 3.1. The collected data are interpolated by quadratic functions as follows:

$$\Omega_s(E) = a_s E^2 + b_s E + c_s \tag{3.47}$$

and

$$N_{e,s}(E) = e_s E^2 + f_s E + g_s\,. \tag{3.48}$$

Then, the reaction energy and the activation barrier are calculated as $\Omega_P(E) - \Omega_R(E)$ and $\Omega_{TS}(E) - \Omega_R(E)$, respectively. Furthermore, the number of electrons transferred by the reaction can be calculated as $N_{e,P}(E) - N_{e,R}(E)$.

A typical example of the calculated results is shown in Fig. 3.4 for the case of the following proton and electron transfer reaction on the Pt(111) electrode [71]:

$$OH(ad) + H^+(aq) + e^- \rightarrow H_2O(aq), \tag{R1}$$

where (ad) and (aq) mean adsorbed and aqueous phases, respectively. The results indicate that the mean fields and electron numbers are accurately interpolated by the quadratic functions.

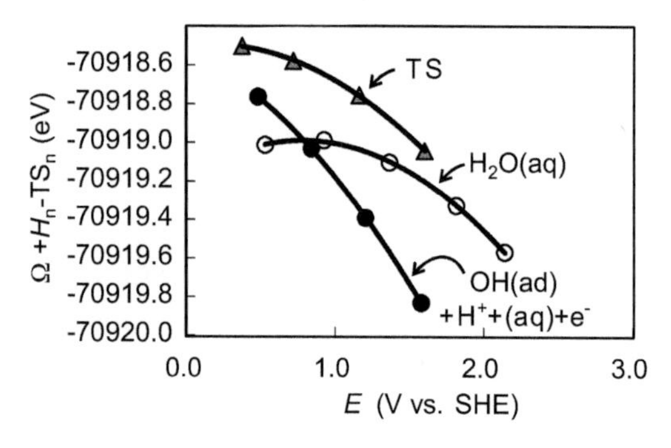

Figure 3.4 An example of the reaction energies and activation barriers evaluated by the constant Fermi energy scheme. The *y* axis in this figure includes contributions of the nuclear motions H_n–TS_n calculated by the ideal gas model and the harmonic oscillator model. Reprinted from Ref. [76], Copyright (2018), with permission from Elsevier.

3.3.2.4 Electrosorption valency value and symmetry factor

This section describes an essential relationship between Ω and N_e [68, 71]. The first derivative of Ω with respect to the Fermi energy μ_e provides the following equation:

$$\frac{\partial \Omega}{\partial \mu_e} = \frac{\partial F}{\partial \mu_e} - N_e - \mu_e \frac{\partial N_e}{\partial \mu_e} = \frac{\partial N_e}{\partial \mu_e}\frac{\partial F}{\partial N_e} - N_e - \mu_e \frac{\partial N_e}{\partial \mu_e}\,. \tag{3.49}$$

This equation can be further simplified by using the Janak theorem [61],

$$\frac{\partial F}{\partial N_{\mathrm{e}}} = \mu_{\mathrm{e}}, \tag{3.50}$$

as follows:

$$\frac{\partial \Omega}{\partial \mu_{\mathrm{e}}} = -N_{\mathrm{e}}. \tag{3.51}$$

Using Eq. 3.51, the first derivative of the reaction energy and activation barrier with respect to the electrode potential E can be shown to be identical to the numbers of transferred electrons to form the product and transition states, respectively, from the reactant as follows:

$$\frac{1}{e}\frac{\partial(\Omega_s - \Omega_{\mathrm{R}})}{\partial E} = N_{\mathrm{e},s} - N_{\mathrm{e,R}}, (s = \mathrm{P\ or\ TS}). \tag{3.52}$$

This equation provides essential macroscopic thermodynamic and kinetic equations on responses of the reaction free energy and activation free energy to the electrode potential. To derive the equations, the free energy difference is formulated by a thermodynamic integration scheme as follows:

$$G_s - G_{\mathrm{R}} = \int_{\zeta_{\mathrm{R}}}^{\zeta_s} \frac{\mathrm{d}}{\mathrm{d}\zeta}\left[-k_{\mathrm{B}}T\ln\int e^{-\frac{\sum_i \frac{|\mathbf{P}_i|^2}{2M_i} + \Omega(\zeta,\{\mathbf{R}_i\}) + pV}{k_{\mathrm{B}}T}} \mathrm{d}\mathbf{P}_i \Pi_{\mathbf{R}_i \neq \zeta} \mathrm{d}\mathbf{R}_i \mathrm{d}V\right]\mathrm{d}\zeta, \tag{3.53}$$

where ζ indicates the reaction coordinate and ζ_{S} and ζ_{R} are the coordinates at the states s (s = P or TS) and R respectively. The first derivative of Eq. 3.53 with respect to the electrode potential E provides the following equation:

$$\begin{aligned}\frac{1}{e}\frac{\partial(G_s - G_{\mathrm{R}})}{\partial E} &= \int_{\zeta_{\mathrm{R}}}^{\zeta_s} \frac{\mathrm{d}}{\mathrm{d}\zeta}\left[\frac{1}{Q(\zeta)}\int\left(-\frac{1}{e}\frac{\partial\Omega}{\partial E}\right) e^{-\frac{\sum_i \frac{|\mathbf{P}_i|^2}{2M_i} + \Omega(\zeta,\{\mathbf{R}_i\}) + pV}{k_{\mathrm{B}}T}} \mathrm{d}\mathbf{P}_i \Pi_{\mathbf{R}_i \neq \zeta} \mathrm{d}\mathbf{R}_i \mathrm{d}V\right]\mathrm{d}\zeta \\ &= \int_{\zeta_{\mathrm{R}}}^{\zeta_s} \frac{\mathrm{d}}{\mathrm{d}\zeta}\left[\frac{1}{Q(\zeta)}\int N_{\mathrm{e}}(\zeta,\{\mathbf{R}_i\}) e^{-\frac{\sum_i \frac{|\mathbf{P}_i|^2}{2M_i} + \Omega(\zeta,\{\mathbf{R}_i\}) + pV}{k_{\mathrm{B}}T}} \mathrm{d}\mathbf{P}_i \Pi_{\mathbf{R}_i \neq \zeta} \mathrm{d}\mathbf{R}_i \mathrm{d}V\right]\mathrm{d}\zeta \\ &= N_{\mathrm{e}}(\zeta_s) - N_{\mathrm{e}}(\zeta_{\mathrm{R}}),\end{aligned} \tag{3.54}$$

where $Q(\zeta)$ is the partition function defined as follows:

$$Q(\zeta) = \int e^{-\frac{\sum_i \frac{|\mathbf{P}_i|^2}{2M_i} + \Omega(\zeta,\{\mathbf{R}_i\}) + pV}{k_B T}} \mathrm{d}\mathbf{P}_i \Pi_{\mathbf{R}_i \neq \zeta} \mathrm{d}\mathbf{R}_i \mathrm{d}V. \tag{3.55}$$

The equation for the product state (s = P) corresponds to the following well-known thermodynamic relationship between the electrosorption valency value $\gamma$and the first derivative of the reaction free energy,

$$\gamma = N_e(\zeta_P) - N_e(\zeta_R) = \frac{1}{e}\frac{\partial(G_P - G_R)}{\partial E}, \tag{3.56}$$

and the equation for the transition state (s = TS) also provides a well-known relationship [58] between the symmetry factor β and the number of electrons transferred to form the transition and product states as follows:

$$\beta = \frac{1}{e}\frac{1}{N_e(\zeta_P) - N_e(\zeta_R)}\frac{\partial(G_{TS} - G_R)}{\partial E} = \frac{N_e(\zeta_{TS}) - N_e(\zeta_R)}{N_e(\zeta_P) - N_e(\zeta_R)}. \tag{3.57}$$

In summary, the theory of microscopic interfacial subsystem based on the grand canonical formalism consistently provides the macroscopic responses of free energies to the electrode potential.

3.4 Applications

The DFT-MPB method has been applied to a wide variety of electrochemical reactions at interfaces between aqueous electrolytes and transition metal electrodes. The applications have provided valuable information on the reaction mechanisms and new materials. In this section, several application results are described to exemplify the usefulness of the proposed method.

3.4.1 Equilibrium Surface Phase Diagram

The stable surface structure strongly depends on the electrode potential, and surface reactivity is significantly affected by the structural change [100]. Hence, identification of stable surface structures is essential for designing active and stable electrocatalysts. The grand canonical DFT methods illustrated in Fig. 3.2 have been applied to predict surface phase diagrams, and those applications have indicated that these new methods can provide results consistent with the experiments particularly on close-packed single-crystal surfaces [64, 68, 70, 71, 107, 121, 141]. Here, the Pt(111) electrode, which is one of the most studied electrode surfaces through both experiments and theories [14, 21, 22, 30, 31, 40–42, 68, 78, 80, 84, 93, 99, 100, 105, 108, 110, 124, 126, 128], is introduced as an example.

Figure 3.5A shows the experimentally measured cyclic voltammogram of Pt(111) in 0.1 mol·L^{-1} $HClO_4$ electrolyte. Four distinct oxidation and reduction current waves are observed: (i) highly symmetric oxidation and reduction currents in $0.05 \leq E < 0.40$ V, (ii) small and constant oxidation and reduction currents in $0.40 \leq E < 0.60$ V, (iii) highly symmetric oxidation and reduction currents in $0.60 \leq E < 0.80$ V, and (iv) asymmetric oxidation and reduction currents in $0.80 \leq E < 1.10$ V. These current densities are assigned to the following electrochemical reactions:

$$\mathrm{H(ad)} \leftrightarrow \mathrm{H^+(aq)} + \mathrm{e^-} \tag{R2}$$

Double-layer charging and discharging (no chemical reaction)

$$\mathrm{H_2O(aq)} \leftrightarrow \mathrm{OH(ad)} + \mathrm{H^+(aq)} + \mathrm{e^-} \tag{R3}$$

$$\mathrm{OH(ad)} \leftrightarrow \mathrm{O(ad)} + \mathrm{H^+(aq)} + \mathrm{e^-} \tag{R4}$$

The corresponding surface phase diagram is shown in Fig. 3.5A [22, 42, 100, 108, 120]. The sharp spikes superposed on the broad current waves in $0.60 \leq E < 0.80$ V are believed to stem from order-disorder phase transitions as discussed in many studies [86, 123], but the topic is out of the scope of this chapter.

Several research groups have theoretically evaluated the reaction free energies of these electrochemical reactions and compared the theoretical surface phase diagrams with the experimental one. An example is shown in Fig. 3.5B. In this figure, the free energy of each surface state relative to that of the surface covered only by H_2O is plotted as a function of the electrode potential. In this figure, the stable surface structure is identified as the structure with the lowest free energy and the information provides a surface phase diagram comparable to the experiment. As indicated by the diagram summarized on top of the free energy results in Fig. 3.5B, the theory gives results close to the experiment, indicating that the DFT–MPB theory can generate reasonable potential-dependent free energies.

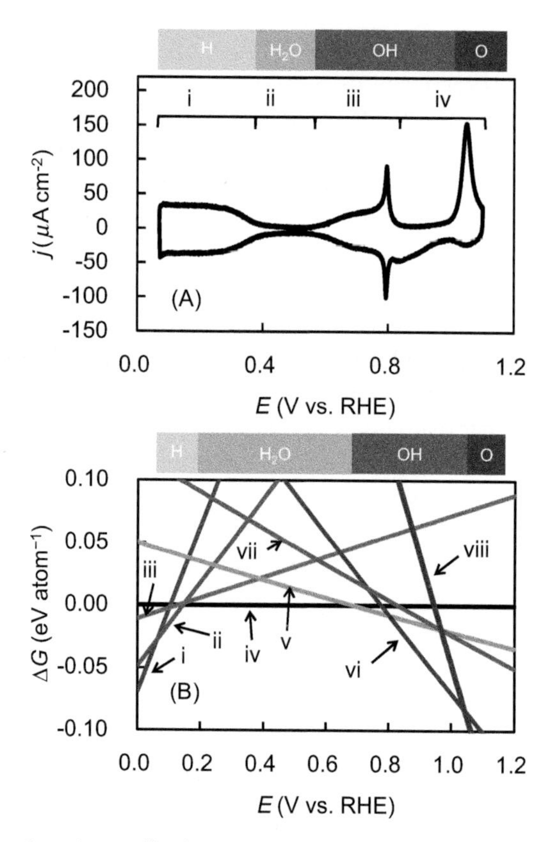

Figure 3.5 Experimentally obtained voltammogram of Pt(111) in 0.1 mol·L^{-1} $HClO_4$ solution and surface phase diagram (A) and theoretically obtained interfacial free energies (B). (B) (i) H(ad) at 0.50 ML, (ii) H(ad) at 0.25 ML, (iii) H(ad) at 0.08 ML, (iv) H_2O(ad), (v) OH(ad) at 0.08 ML, (vi) OH(ad) at 0.33 ML, (vii) O(ad) at 0.08 ML, and (viii) O(ad) at 0.50 ML. Here, ML denotes monolayer. 1 ML indicates the number of adsorbed atoms or molecules are the same as the number of surface atoms.

Theoretical evaluations of the surface phase diagram can provide new information that cannot be obtained by experiments. A typical example is shown of the specific adsorption of sulfuric acid anion on Pt(111) surface. Since the invention of the first fuel cells by Sir Grove in 1839, the H_2SO_4/Pt interfacial system has widely attracted the attention of electrochemists and surface scientists. Particularly, the H_2SO_4/Pt(111) interface has been studied for over 25 years as a model system of fuel cell catalysts. Intensive experimental studies have been done on this system by electrochemical measurements [21, 54, 84], infrared spectroscopy [14, 30, 31, 93, 109, 110, 124], in situ scanning tunneling microscopy [13, 40], and radio tracer measurement. There was, however, a longstanding issue on the assignment of the observed infrared absorption bands and the identification of stable interfacial anion species on the surface [14, 30, 31, 68, 69, 93, 109, 110, 124, 126].

To answer the question, Jinnouchi et al. [68] applied the DFT-MPB method to this interfacial system. The calculated interfacial free energies and the phase diagram are summarized in Fig. 3.6. The free energy diagram indicates that the sulfuric acid anion is initially adsorbed as bisulfate (HSO_4) when the electrode potential is raised from 0.1 V. Bisulfate is, however, not strongly bound to the Pt surface and is easily displaced by sulfate (SO_4) when the potential is slightly raised. The sulfate is bound so strongly with the Pt surface that this species remains on the surface in a wide potential range and hinders the formation of OH(ad) and O(ad). Hence, OH(ad) and O(ad) do not appear in the phase diagram in this potential range. The theoretical phase diagram reasonably explains the experimentally observed cyclic voltammograms [21, 54, 84].

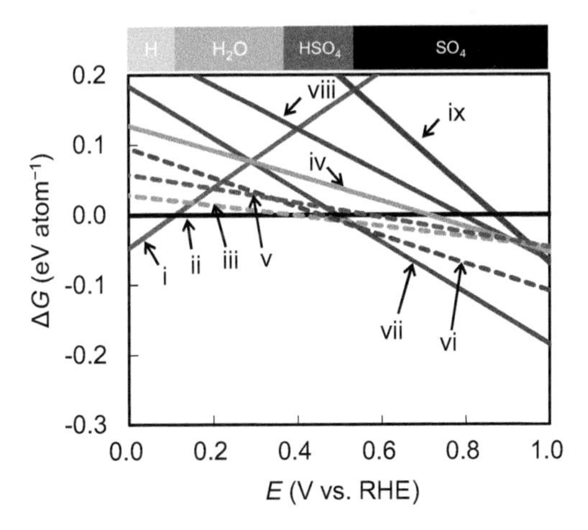

Figure 3.6 Theoretically obtained interfacial free energies of Pt(111) in 0.1 mol·L^{-1} H_2SO_4 solution. (i) H(ad) at 0.25 ML, (ii) H_2O(ad), (iii) HSO_4(ad) at 0.10 ML, (iv) HSO_4(ad) at 0.20 ML, (v) H···SO_4(ad) at 0.10 ML, (vi) SO_4(ad) at 0.10 ML, (vii) SO_4(ad) at 0.20 ML, (viii) OH(ad) at 0.33 ML,and (ix) O(ad) at 0.25 ML. Data are taken from Ref. [68].

3.4.2 Electrosorption Valency Value

Similar information on the equilibrium surface phase diagram can be also derived by DFT calculations on electrically neutral surfaces [47, 73, 78, 98, 105, 106, 119, 121, 133]. The theoretical phase diagrams, however, often contain non-negligible errors generated mainly from inaccuracies in bond strengths given by approximated exchange-correlation functionals [49, 116], and the conclusions derived only from the phase diagrams are sometimes unreliable. In these situations, other properties, such as the amount of charge transfer (i.e., electrosorption valency value γ), can provide conclusive information, and the information can be obtained only by using DFT calculations on electrified interfaces.

The electrosorption valency values γ of the specific adsorption of sulfuric acid anions calculated by the DFT-MPB method are compared with the experimentally measured one [84] in Fig. 3.7. The experimental result indicates that γ changes from −0.5 to −1.7 with a rise in the electrode potential, and the theory indicates that γ for bisulfate (HSO_4) is around −1 to −0.5 whereas that for sulfate (SO_4) is −1.8 to −1.7. As discussed in the previous section, the theoretical interfacial free energies indicate that the sulfuric acid anion is adsorbed as bisulfate in the low-potential region and sulfate in the high-potential region. Taking account of this free energy diagram, the theory is judged to quantitatively reproduce the experimentally observed transient in γ and the conclusion on the stable species of the adsorbed sulfuric anion is judged to be correct.

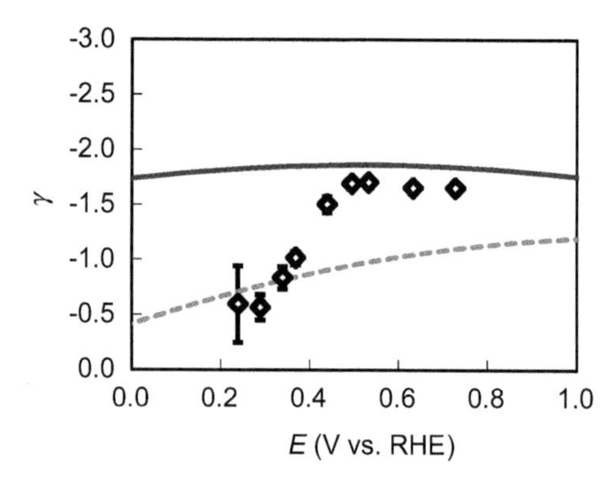

Figure 3.7 Theoretically obtained electrosorption valency values. The diagonal dots are the experimental data taken from Ref. [84]. The solid and dashed lines are for the bisulfate and sulfate, respectively. Data are taken from Ref. [68].

3.4.3 Potential-Dependent Spectroscopy

Potential-dependent spectroscopies have been utilized in experiments to clarify interfacial structures and reaction mechanisms [6, 14, 30, 31, 85, 91–95, 104, 109, 110, 122, 124, 139, 143]. Electronic structure calculations have provided essential information on these issues, too [5, 8, 25, 48, 57, 68, 69, 87, 112, 118, 143]. Before the development of the DFT methods for electrified interfaces, calculations were mainly carried out on electrically neutral surfaces under a constant electric field. Although these calculations provided new insights into observed potential-dependent spectroscopies, quantitative comparisons were difficult because an electric field cannot be easily converted into an electrode potential. As discussed in Section 3.2, the DFT calculations on electrified interfaces can directly provide the information on the electrode potential, and therefore, potential-dependent spectra comparable with the experiments can be obtained by these methods. Two examples are shown in this section.

The first example is of the infrared spectroscopy of sulfuric acid anion adsorbed on the Pt(111) surface. As described in Section 3.4.1, the assignment of observed absorption bands was a longstanding issue in the community of interfacial electrochemistry. In experiments, two absorption bands are observed, one at 1250 cm^{-1} and one at 950 cm^{-1} [14, 30, 31, 93, 109, 110, 124]. The former band has a larger intensity and is blue-shifted, with a slope of 58–130 $cm^{-1} \cdot V^{-1}$, when the electrode potential rises. In contrast, the lower band has a smaller intensity and its frequency does not strongly depend on the electrode potential. These experimental results are reproduced well by the theoretically obtained potential-dependent vibration frequencies of adsorbed sulfate as summarized in Figs. 3.8A and 3.8B [68, 69]. The result again supports the sulfate as the dominant adsorbate.

A theoretical analysis further provides the mechanisms underlying the experimentally observed absorption intensities and potential dependences. The adsorbed sulfate has two additional vibrational modes, one at 1020 cm^{-1} and the other at 997 cm^{-1}, as shown in Fig. 3.8C. These two vibrational modes, however, do not have large dynamic dipole moments normal to the surface because of their vibrational motions parallel to the surface. Therefore, these two modes are infrared-inactive by the surface selection rule. The large dynamic dipole moment of the highest frequency mode also reasonably explains its larger absorption intensity and stronger potential dependence.

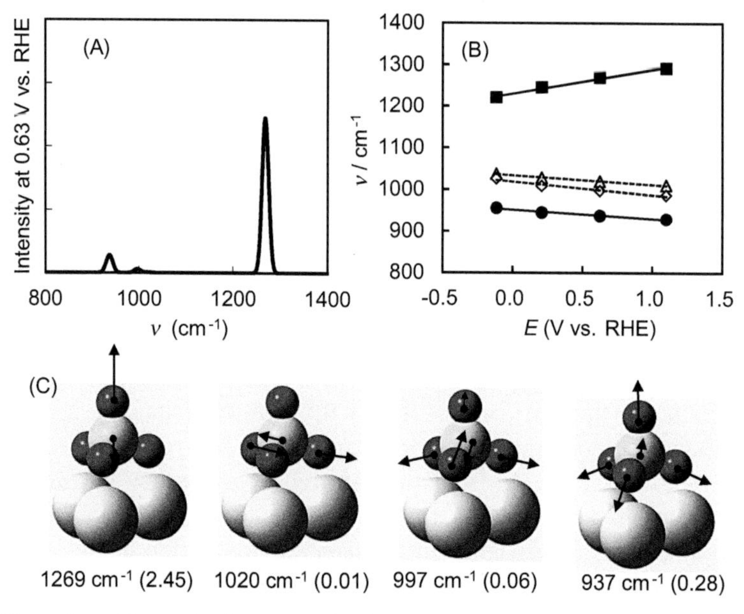

Figure 3.8 Theoretically obtained infrared spectra (A), potential-dependent frequencies (B), and vibrational modes of adsorbed sulfate (SO_4) on a Pt(111) electrode. The values in the parentheses in (C) show the calculated dynamic dipole moments (unit is e^2) normal to the surface. Data are taken from Ref. [68].

The second example is on the X-ray absorption near-edge structure (XANES) of the Pt(111) surface. In situ XANES has been employed to examine the surface (hydr)oxide structures of Pt electrocatalysts used for low-temperature fuel cells [85, 104, 122, 139, 144] because of their significant influence on the catalytic activity and durability [26, 100]. The DFT calculations on electrified interfaces can again provide information comparable to the experiments. Figure 3.9 summarizes theoretically obtained L_3-edge spectra of the Pt(111) surface covered with several (hydr)oxides. As observed in the experiments [144], the first peak located at 11547–11553 eV evolves with the surface oxidations and the absorption at 11553–11563 eV decreases with the formation of subsurface oxides.

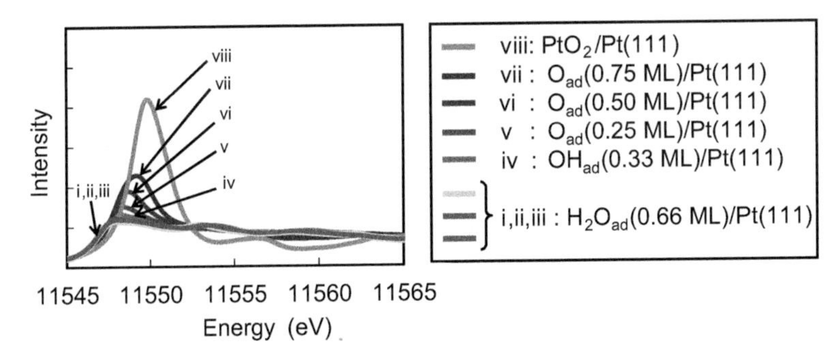

Figure 3.9 Theoretically obtained X-ray absorption near edge structure of the Pt(111) surface.

3.4.4 Kinetics and Symmetry Factor

The activation barrier of electron and ion transfer reactions is one of the most essential properties provided by the DFT calculations on electrified interfaces [17, 62, 71, 128, 142]. The first evaluation of the potential-dependent activation barriers was conducted on elementary reactions of methanol oxidation reactions on Pt surfaces by the model shown in Fig. 3.2B [17]. Later, by using the models shown in Fig. 3.2C–E, calculations have been executed on several electrochemical reactions [59, 71, 128]. In this section, the results of activation barriers are shown on the following oxidation and reduction reactions on Pt(111) surface obtained by using the DFT-MPB method (the model in Fig. 3.2E) as an example.

$$H_2O(aq) \leftrightarrow OH(ad) + H^+(aq) + e^- \tag{R5}$$

$$OH(ad) \leftrightarrow O(ad) + H^+(aq) + e^- \tag{R6}$$

Before the calculated results are described, a controversy on the mechanism of surface oxidations of Pt lasting over a half century [23, 24, 43, 85, 108, 151] is briefly introduced. As indicated by the cyclic voltammogram of Pt(111) shown in Fig. 3.5A, the reaction R5 generates symmetric oxidation and reduction currents, whereas R6 generates asymmetric ones. The origin of the latter asymmetric currents is a longstanding issue. Usually, asymmetric currents observed in voltammograms are attributed to slow kinetics; because reaction rates are slow, oxidation and reduction reactions are delayed from the scanned electrode potential. Thus, oxidation and reduction current peaks are shifted toward higher- and lower-potential regions, respectively, and asymmetric shapes appear in the voltammogram. Suppose the rate equation is described by the following the Butler–Volmer equation [16, 29] with a symmetry factor $\beta = 0.5$:

$$j = j_0 \left[\exp\left(\frac{e\beta(E - E_0)}{k_B T} \right) - \exp\left(-\frac{e(1-\beta)(E - E_0)}{k_B T} \right) \right] \tag{3.58}$$

Then the shifts in the peaks are estimated to be 120 mV per 1 order of magnitude change in the scan rate. In Eq. 3.58, j_0 and E_0 are the exchange current density and reversible potential, respectively. If β is increased from 0.5, the shift of the oxidation current peak decreases and that of the reduction current peak increases. Since R6 is a reaction involving one electron transfer, the oxidation and reduction currents are expected to shift in this manner. However, the experiments exhibit totally different results [24, 41, 149]. The shifts in the current peaks are quite small both for the oxidation and reduction reactions and the asymmetric behavior does not disappear even when the scan rate is considerably slowed. Very interestingly, regardless of the surface morphologies, similar asymmetric trends are observed on a wide variety of Pt surfaces, as reported in many studies [24, 71, 80]. The asymmetric behavior can never be described by the rate equation 3.58; therefore, many researchers have speculated that some irreversible chemical processes should be involved in these oxidation and reduction reactions [24, 43]. A typical model is a so-called place-exchange model [23, 24, 85, 108] that assumes the formation of subsurface (hydr)oxides. However, recent experiments and theoretical calculations indicate that the subsurface oxides are not formed at this potential range [56, 151].

To solve the problem, the DFT-MPB method was applied to examine the activation barriers of R5 and R6. In addition to these two electron transfer reactions, the following chemical reaction reported in theoretical and experimental studies on Pt surfaces in ultra-high-vacuum conditions [38, 77, 103, 148] was also examined by the DFT-MPB method.

$$2\text{OH(ad)} \leftrightarrow \text{O(ad)} + \text{H}_2\text{O(aq)} \tag{R7}$$

The calculated activation barriers are summarized in Fig. 3.10A and are compared with the results determined to reproduce the experimental voltammogram summarized in Fig. 3.10B. In these figures, the activation free energies ΔG_f* and ΔG_b* are plotted as functions of the reaction free energy ΔG. For the electron transfer reactions R5 and R6, ΔG equals $-e(E - E_0)$. Although the calculated results do not quantitatively agree with the experimental ones, several consistent trends are observed as listed below.

- The activation barriers for R5 and R6 linearly depend on the reaction free energy, indicating that their reaction rates can be described by the Butler–Volmer equation (Eq. 3.58).
- The forward and backward activation barriers of R5 behave symmetrically, whereas those for R6 and R7 behave asymmetrically; the activation barrier of the backward reaction of R6 strongly depends on ΔG and that of the forward reaction of R7 strongly depends on ΔG.

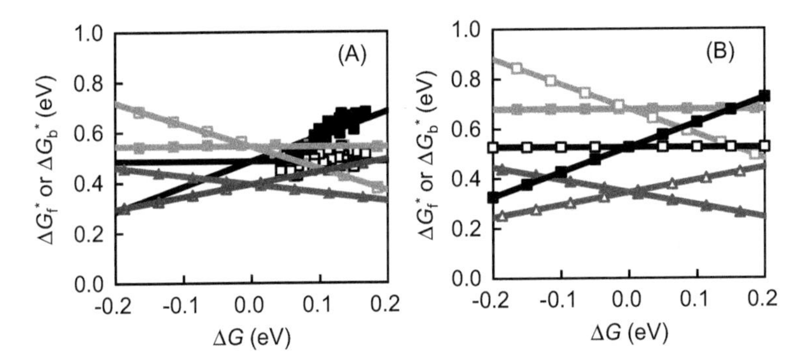

Figure 3.10 The theoretically calculated activation barriers for the forward (ΔG_f^*) and backward (ΔG_b^*) reactions R5–R7 (A) and those determined to reproduce the experimental voltammogram (B) as functions of the reaction free energy ΔG. Darker-gray triangles and solid lines, lighter-gray squares and solid lines, and black squares and solid lines indicate the results for R5, R6, and R7, respectively. The closed and open dots are the results for forward and backward reactions, respectively. Reprinted from Ref. [71], with the permission of AIP Publishing.

The trends reasonably explain the experimentally observed asymmetric behaviors as summarized in Fig. 3.11. Both oxidation and reduction peak potentials (E_{ox} and E_{red}) depend only weakly on the scan rate v as reported in the experiments, indicating that both oxidation and reduction rates strongly depend on the electrode potential. The strong potential dependence of the reaction rates stems from the asymmetric activation barriers. When the electrode potential is positively scanned from the low-potential region, the activation barrier for the forward reaction of R7 considerably drops while that for R6 does not because of the asymmetricity. Hence, O(ad) is formed through R7. In contrast, when the electrode potential is negatively scanned from the high-potential region, the activation barrier for the backward reaction of R6 significantly drops while that for R7 does not. Therefore, O(ad) is removed through R6. Because both formation and removal reactions occur through the pathways that have strong potential dependence, both rates strongly depend on the electrode potential.

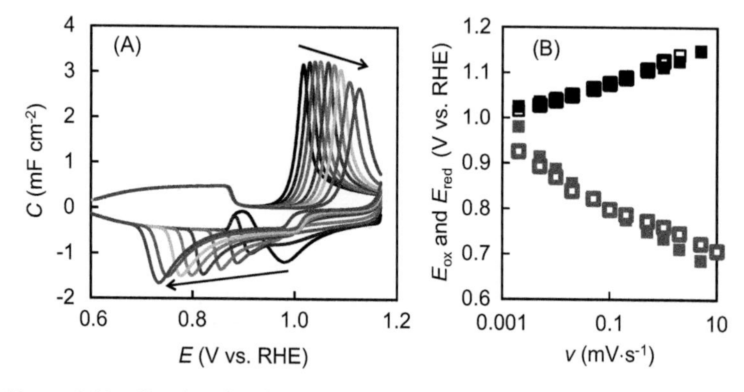

Figure 3.11 Simulated cyclic voltammograms on Pt(111) in 0.1 mol·L^{-1} $HClO_4$ solution and oxidation and reduction peak potentials as functions of scan rate v. The scan rates are 0.002, 0.005, 0.01, 0.02, 0.05, 0.1, 0.2, 0.5, or 1 (mV·s^{-1}). Arrows in (A) indicate the increasing scan rate. Black and gray squares in (B) show the oxidation and reduction peak potentials, respectively. The closed squares show the simulated results, and the open squares show the experimental data taken from Ref. [41]. Simulation data are taken from Ref. [71]. Reprinted from Ref. [71], with the permission of AIP Publishing.

In short, the activation barriers based on the DFT-MPB calculations indicate that the asymmetric oxidation and reduction currents indeed involve irreversible processes. However, differently from the experimental speculations, the theory can explain the asymmetric behaviors without assuming the presence of subsurface oxides and indicate that the irreversibility derives from asymmetric responses of the activation barriers of two different O(ad) formation reactions R6 and R7. This newly proposed mechanism is not experimentally proven yet. However, this application result exemplifies that evaluations of potential-dependent activation barriers can provide new insights into the reaction mechanisms.

3.4.5 Applications to New Materials

This final section introduces some applications of the DFT-MPB method in new materials. The first example is on electrocatalysts for oxygen reduction reaction (ORR) in low-temperature fuel cells. A pioneering study on this topic was conducted by Nørskov's group [105]. The method originally adopts the DFT calculation on electrically neutral surfaces. Because of its simplicity, this method has been applied to many Pt alloys and carbon catalysts [46, 106, 113, 133, 136, 154]. Later, on the basis of a very similar concept, DFT calculations on electrified interfaces have been executed on the same issue [59, 67, 141]. In most of these methods, the major ORR pathway is assumed to be as follows:

$$O_2(g) + H^+(aq) + e^- \rightarrow HO_2(ad) \tag{R8}$$

$$HO_2(ad) \rightarrow OH(ad) + O(ad) \tag{R9}$$

$$O(ad) + H^+(aq) + e^- \rightarrow OH(ad) \tag{R10}$$

$$OH(ad) + H^+(aq) + e^- \rightarrow H_2O(aq) \tag{R11}$$

The activation barriers are explicitly or implicitly calculated by using the DFT calculations, and kinetic mean field simulations that employ the predicted activation barriers are executed to calculate current-voltage curves of ORR on various catalysts. The predicted current density is often summarized in a volcano plot versus a descriptor of the stability of OH(ad) or O(ad), as exemplified in Fig. 3.12 for the cases of Pt alloys and core-shell catalysts [67]. The volcano plot derives from the so-called scaling relationship [1], which states that a surface-stabilizing adsorbate also stabilizes other adsorbates. For example, a Pt shell on Au, denoted as Pt@Au in Fig. 3.12, stabilizes all the intermediates of HO_2(ad), OH(ad), and O(ad). Therefore, these intermediates are easily formed on this surface. However, because of the strong stabilization, the intermediates OH(ad) and O(ad) cannot be removed from the surface easily, and as a result, the net reaction rate is considerably limited by the removal reactions R10 and R11. By destabilizing the intermediates, the rate-limiting removal steps are accelerated; therefore, the net reaction rate is raised. However, excessive destabilization causes decelerations of their formation steps. Hence, the net reaction rate drops again. The volcano plot indicates that the catalytic surface can be optimized by balancing the formation and removal reaction rates. The optimizations of surface morphology and composition executed by using the DFT calculations provide useful information on designs of new electrocatalysts as reported in Refs. [46, 47, 67, 72, 73, 75, 105, 106, 113, 133, 136, 146, 154]. One example is shown in Fig. 3.12 [146]. The DFT calculation indicates that the $Pt_{85}Ta_{15}$ alloy is highly active for ORR, and the experiment indicates that the $Pt_{85}Ta_{15}$ alloy thin film deposited on the $TaB_2(0001)$ single crystal possesses high catalytic activity.

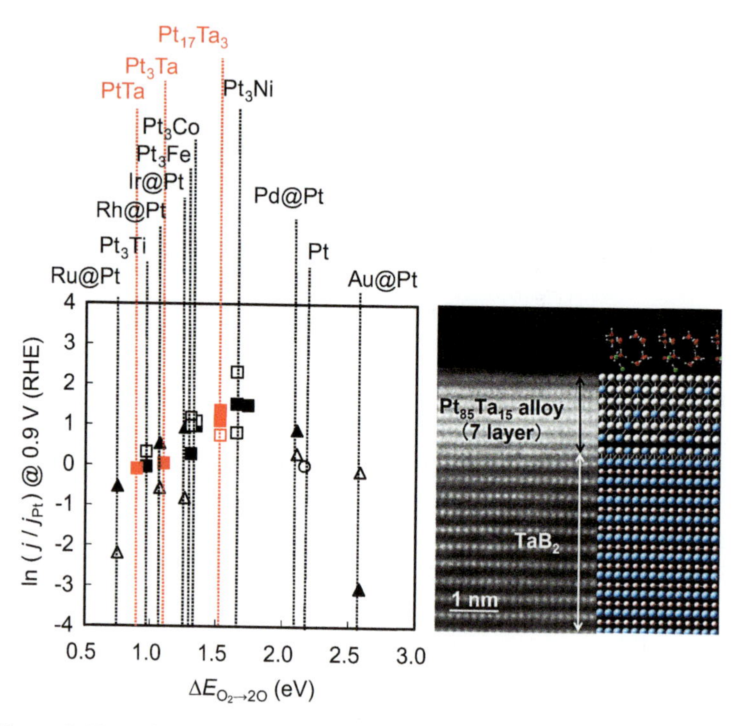

Figure 3.12 Volcano plot of the theoretical and experimental ORR activity on Pt alloys and core-shell catalysts versus theoretically obtained adsorption energy of 2O relative to O_2. The figure on the left shows a new $Pt_{85}Ta_{15}$ thin film catalyst on a $TaB_2(0001)$ single-crystal surface. Closed and open dots indicate theoretical and experimental results, respectively. Data are taken from Refs. [67, 146].

Another example of the application is shown on an ionomer suitable for cathode catalysts of low-temperature fuel cells. As indicated by reactions R8–R11, ORR needs supplies of protons from the electrolyte. Ionomer is a proton-conducting polymer electrolyte realizing the proton supplies to the catalyst surfaces. A typical ionomer is Nafion shown in Fig. 3.13. Nafion is a perfluorinated polymer that contains sulfonic acid groups as counterions of protons. Because of its high proton conductivity in a wide range of relative humidity and its relatively high chemical and mechanical durability [12, 132, 155], this polymer electrolyte is widely used in the low-temperature fuel cells. However, when this polymer is used as the ionomer for the cathode catalysts, a serious problem occurs; the sulfonic acid group is specifically adsorbed on the Pt surface and inhibits ORR [81, 137]. To put it the other way around, this means that ORR activity can be enhanced if this anion is replaced by a species less strongly adsorbed on the Pt surface.

Nafion

$O-CF_2-CF(CF_3)-O-CF_2-CF_2-SO_3H$

NBC4

$O-CF_2-CF(CF_3)-O-CF_2-CF_2-SO_2NHSO_2-(CF_2)_3-SO_2NHSO_2-(CF_2)_3-CF_3$

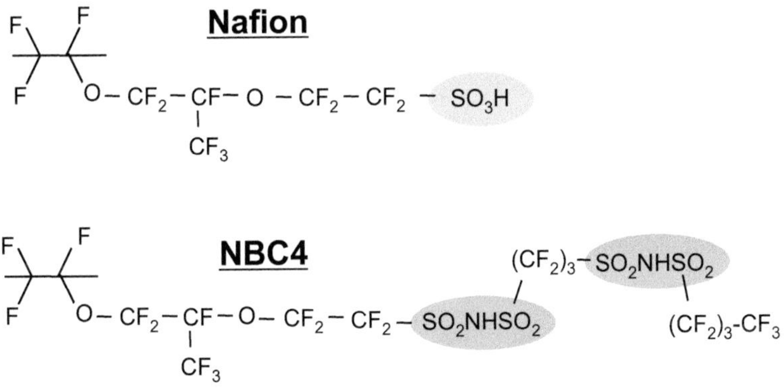

Figure 3.13 Structures of Nafion and NBC4. The latter polymer contains sulfonimide anions ($SO_2NSO_2^-$) instead of sulfonic anions (SO_3^-). Republished with permission of the Electrochemical Society, from Ref. [82]; permission conveyed through Copyright Clearance Center, Inc.

As exemplified by the fundamental study described in Section 3.4.1, the DFT-MPB method can predict the stability of anions on s Pt electrode. Thus, the method was applied to evaluate the adsorptivity of a new anion—sulfonimide—shown in Fig. 3.13 [82]. Figure 3.14 summarizes the theoretical result. The theory indicates that the sulfonimide is adsorbed less strongly on the Pt electrode than the sulfonic acid group. The result is supported by the experimentally obtained cyclic voltammograms shown in Fig. 3.15A, where the current peaks that are attributed to adsorption and desorption of anions are considerably suppressed if the anions are changed from the sulfonic acid group to the sulfonimide group. Furthermore, experiments indicate that the ionomer with a sulfonimide anion exhibits higher ORR activity, as shown in Fig. 3.15B.

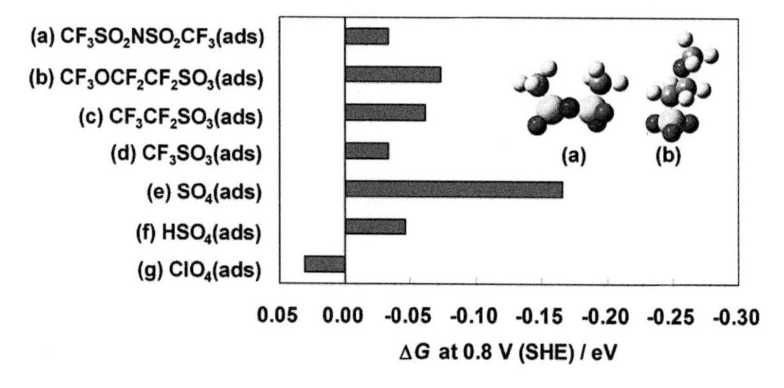

Figure 3.14 Theoretically obtained adsorption free energies of several anions on the Pt(111) electrode. The inset (a and b) shows molecular models of sulfonimide and sulfonic anions, respectively. Republished with permission of the Electrochemical Society, from Ref. [82]; permission conveyed through Copyright Clearance Center, Inc.

This discovery can have a great impact on the fuel cells community. As described in the previous example of the application, numerous challenges have been overcome to enhance ORR activity by optimizing surface composition of the catalyst. Optimization is, however, often limited by durability problems. As indicated by many experiments, secondary elements in Pt alloys and core shells are thoroughly dissolved in the electrolyte during operations of fuel cells [4, 15, 20, 63, 152]; therefore, catalysts with optimal compositions cannot be used in reality. Hence, alternative approaches are desired in this community, and designs of ionomers can be a new strategy.

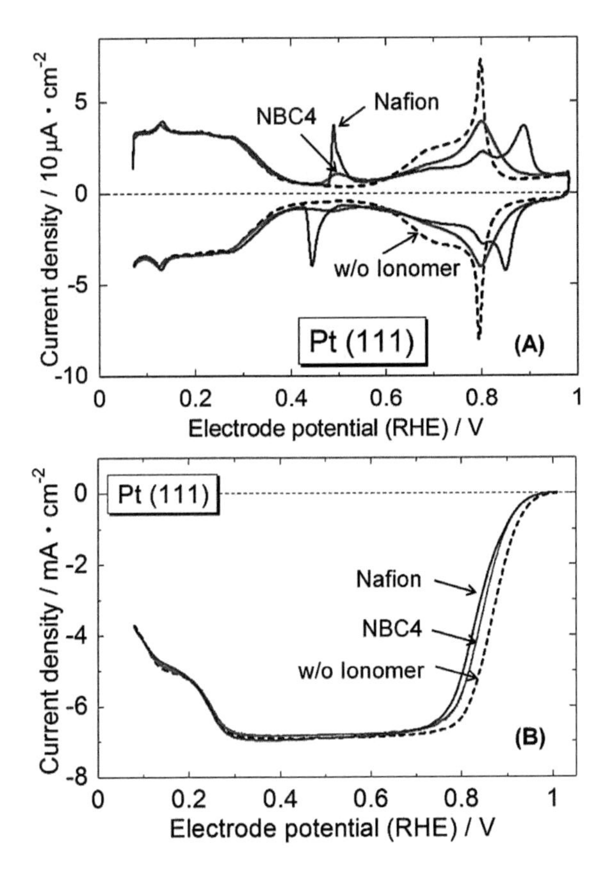

Figure 3.15 Experimental voltammograms of Pt(111) with and without the ionomers. Republished with permission of the Electrochemical Society, from Ref. [82]; permission conveyed through Copyright Clearance Center, Inc.

3.5 Future Scope

Basic concepts and a practical method of electronic structure calculations on electrified interfaces are described in this chapter. On adopting a grand canonical formalism, a variety of electrochemical properties become predictable, as illustrated by several calculation results. In this last section, however, a remaining problem in the DFT calculations on electrified interfaces is described.

Besides the problems caused by inaccuracies in bond strengths generated by the approximated exchange-correlation functional, one of the most serious problems is in their ad hoc treatment on entropy contributions from nuclear motions of reactants, products, and transition states [17, 62, 67, 75, 105, 128, 150]. In most of the calculations, for example, the nuclear contributions are approximated by the harmonic oscillator model and/or ideal gas model as described in Section 3.2. However, motions of the interfacial species might not be described by either of these simple statistical models. On one hand, interfacial species are only loosely bound to their equilibrium positions and can diffuse from electrolyte to electrode or vice versa. Therefore, if the nuclear motions are approximated by the harmonic oscillator model, their entropy will be considerably underestimated. On the other hand, interfacial species strongly interact with the surrounding solvent molecules and electrode surface and they cannot freely move around the phase space. Hence, if the motions are approximated by the ideal gas model, their entropy is considerably overestimated. The difference in the entropy between two extreme models is estimated to be more than 0.4 eV per water molecule, for example. This energetic difference corresponds to a difference of 6 orders of magnitude in the reaction rate at room temperature. Because of the lack of a method to determine the entropy of nuclear motions, all currently available models assume an empirically determined constant activation entropy (i.e., prefactor) for any reactions on any materials. The assumption not only generates inaccuracies in the predictions but also limits variables that can be optimized in material designs. In the designs of ORR electrocatalysts, for example, prefactors of all the elementary reactions are set as an empirical constant, and only the bond strengths between the reaction intermediates and electrode surfaces are optimized, as shown in Section 3.4.5. However, as indicated by the estimation described previously, great uncertainty still remains in entropy terms, and there might be considerable room for the optimization of these terms, for example, by controlling the surface morphology and electrolyte structure.

For the evaluation of the entropy term, rigorous simulations of nuclear motions will be necessary and for this purpose, full atomistic molecular dynamics simulations may be necessary. However, as described in Section 3.2, molecular dynamics using conventional force fields cannot handle reactive interfaces on one hand. First-principles molecular dynamics, on the other hand, cannot handle the long length scales and time scales of the electric double layer. How can this dilemma be overcome? We, the authors, believe that this is the most significant issue that should be resolved by the next generation of theory handling electrochemical interfacial systems.

References

1. Abild-Pedersen, F., Greeley, J., Studt, F., Rossmeisl, J., Munter, T. R., Moses, P. G., Skúlason, E., Bligaard, T., and Nørskov, J. K. (2007). Scaling properties of adsorption energies for hydrogen-containing molecules on transition-metal surfaces, *Phys. Rev. Lett.*, **99**, pp. 16105.
2. Allen, M. P. and Tildesley, D. J. (1989). *Computer Simulation of Liquids* (Clarendon Press).
3. Andreussi, O., Dabo, I., and Marzari, N. (2012). Revised self-consistent continuum solvation in electronic-structure calculations, *J. Chem. Phys.*, **136**, pp. 064102.
4. Antolini, E., Salgado, J. R. C., and Gonzalez, E. R. (2006). The stability of Pt–M (M=first row transition metal) alloy catalysts and its effect on the activity in low temperature fuel cells: a literature review and tests on a Pt–Co catalyst, *J. Power Sources*, **160**, pp. 957–968.
5. Bagus, P. S., Nelin, C. J., Müller, W., Philpott, M. R., and Seki, H. (1987). Field-induced vibrational frequency shifts of CO and CN chemisorbed on Cu(100), *Phys. Rev. Lett.*, **58**, pp. 559–562.
6. Beden, B., Bewick, A., and Lamy, C. (1983). A study by electrochemically modulated infrared reflectance spectroscopy of the electrosorption of formic acid at a platinum electrode, *J. Electroanal. Chem. Interfacial Electrochem.*, **148**, pp. 147–160.
7. Bonnet, N., Morishita, T., Sugino, O., and Otani, M. (2012). First-principles molecular dynamics at a constant electrode potential, *Phys. Rev. Lett.*, **109**, pp. 266101.
8. Bonnet, N., Dabo, I., and Marzari, N. (2014). Chemisorbed molecules under potential bias: detailed insights from first-principles vibrational spectroscopies, *Electrochim. Acta*, **121**, pp. 210–214.
9. Born, M. and Oppenheimer, R. (1927). Zur Quantentheorie der Molekeln, *Ann. Phys.*, **389**, pp. 457–484.

10. Borukhov, I., Andelman, D., and Orland, H. (1997). Steric effects in electrolytes: a modified Poisson-Boltzmann equation, *Phys. Rev. Lett.*, **79**, pp. 435–438.
11. Borukhov, I., Andelman, D., and Orland, H. (2000). Adsorption of large ions from an electrolyte solution: a modified Poisson–Boltzmann equation, *Electrochim. Acta*, **46**, pp. 221–229.
12. Borup, R., Meyers, J., Pivovar, B., Kim, Y. S., Mukundan, R., Garland, N., Myers, D., Wilson, M., Garzon, F., Wood, D., Zelenay, P., More, K., Stroh, K., Zawodzinski, T., Boncella, J., McGrath, J. E., Inaba, M., Miyatake, K., Hori, M., Ota, K., Ogumi, Z., Miyata, S., Nishikata, A., Siroma, Z., Uchimoto, Y., Yasuda, K., Kimijima, K.-i., and Iwashita, N. (2007). Scientific aspects of polymer electrolyte fuel cell durability and degradation, *Chem. Rev.*, **107**, pp. 3904–3951.
13. Braunschweig, B. and Daum, W. (2009). Superstructures and order–disorder transition of sulfate adlayers on Pt(111) in sulfuric acid solution, *Langmuir*, **25**, pp. 11112–11120.
14. Braunschweig, B., Mukherjee, P., Dlott, D. D., and Wieckowski, A. (2010). Real-time investigations of Pt(111) surface transformations in sulfuric acid solutions, *J. Am. Chem. Soc.*, **132**, pp. 14036–14038.
15. Brushett, F. R., Duong, H. T., Ng, J. W. D., Behrens, R. L., Wieckowski, A., and Kenis, P. J. A. (2010). Investigation of Pt, Pt3Co, and Pt3Co/Mo cathodes for the ORR in a microfluidic H-2/O-2 fuel cell, *J. Electrochem. Soc.*, **157**, pp. B837–B845.
16. Butler, J. A. V. (1924). Studies in heterogeneous equilibria. Part II. The kinetic interpretation of the nernst theory of electromotive force, *Trans. Faraday Soc.*, **19**, pp. 729–733.
17. Cao, D., Lu, G.-Q., Wieckowski, A., Wasileski, S. A., and Neurock, M. (2005). Mechanisms of methanol decomposition on platinum: a combined experimental and ab initio approach, *J. Phys. Chem. B*, **109**, pp. 11622–11633.
18. Chandrasekhar, J., Smith, S. F., and Jorgensen, W. L. (1985). Theoretical examination of the SN2 reaction involving chloride ion and methyl chloride in the gas phase and aqueous solution, *J. Am. Chem. Soc.*, **107**, pp. 154–163.
19. Chapman, D. L. (1913). Contribution to the theory of electrocapillarity, *Philos. Mag.*, **25**, pp. 475–481.
20. Chen, S., Gasteiger, H. A., Hayakawa, K., Tada, T., and Shao-Horn, Y. (2010). Platinum-alloy cathode catalyst degradation in proton exchange membrane fuel cells: nanometer-scale compositional and morphological changes, *J. Electrochem. Soc.*, **157**, pp. A82–A97.

21. Clavilier, J., Faure, R., Guinet, G., and Durand, R. (1980). Preparation of monocrystalline Pt microelectrodes and electrochemical study of the plane surfaces cut in the direction of the {111} and {110} planes, *J. Electroanal. Chem.*, **107**, pp. 205–209.

22. Climent, V., Gómez, R., Orts, J. M., and Feliu, J. M. (2006). Thermodynamic analysis of the temperature dependence of OH adsorption on Pt(111) and Pt(100) electrodes in acidic media in the absence of specific anion adsorption, *J. Phys. Chem. B*, pp. 11344–11351.

23. Conway, B. E., Barnett, B., Angerstein-Kozlowska, H., and Tilak, B. V. (1990). A surface-electrochemical basis for the direct logarithmic growth law for initial stages of extension of anodic oxide films formed at noble metals, *J. Chem. Phys.*, **93**, pp. 8361–8372.

24. Conway, B. E. (1995). Electrochemical oxide film formation at noble metals as a surface-chemical process, *Prog. Surf. Sci.*, **49**, pp. 331–452.

25. Dabo, I., Wieckowski, A., and Marzari, N. (2007). Vibrational recognition of adsorption sites for CO on platinum and platinum–ruthenium surfaces, *J. Am. Chem. Soc.*, **129**, pp. 11045–11052.

26. Darling, R. M. and Meyers, J. P. (2003). Kinetic model of platinum dissolution in PEMFCs, *J. Electrochem. Soc.*, **150**, pp. A1523–A1527.

27. Debye, P. and Falkenhagen, H. (1928). Frequency dependence of the conductivity and dielectric constant of a strong electrolyte, *Phys. Z.*, **29**, pp. 121–132.

28. Delley, B. (1990). An all-electron numerical method for solving the local density functional for polyatomic molecules, *J. Chem. Phys.*, **92**, pp. 508–517.

29. Erdey-Gruz, T. and Volmer, M. (1930). Zur theorie der Wasserstoffüberspannung, *Z. Phys. Chem.*, **150**, pp. 203–213.

30. Faguy, P. W., Markovic, N., Adzic, R. R., Fierro, C. A., and Yeager, E. B. (1990). A study of bisulfate adsorption on Pt(111) single crystal electrodes using in situ Fourier transform infrared spectroscopy, *J. Electroanal. Chem. Interfacial Electrochem.*, **289**, pp. 245–262.

31. Faguy, P. W., Marinković, N. S., and Adžić, R. R. (1996). Infrared spectroscopic analysis of anions adsorbed from bisulfate-containing solutions on Pt(111) electrodes, *J. Electroanal. Chem.*, **407**, pp. 209–218.

32. Fang, Y.-H., Wei, G.-F., and Liu, Z.-P. (2013). Theoretical modeling of electrode/electrolyte interface from first-principles periodic continuum solvation method, *Catal. Today*, **202**, pp. 98–104.

33. Farkas, Ö. and Schlegel, H. B. (1999). Methods for optimizing large molecules. II. Quadratic search, *J. Chem. Phys.*, **111**, pp. 10806–10814.

34. Fattebert, J.-L. and Gygi, F. (2002). Density functional theory for efficient ab initio molecular dynamics simulations in solution, *J. Comput. Chem.*, **23**, pp. 662–666.

35. Fattebert, J.-L. and Gygi, F. (2003). First-principles molecular dynamics simulations in a continuum solvent, *Int. J. Quantum Chem.*, **93**, pp. 139–147.

36. Fedorov, M. V. and Kornyshev, A. A. (2008). Ionic liquid near a charged wall: structure and capacitance of electrical double layer, *J. Phys. Chem. B*, **112**, pp. 11868–11872.

37. Feynman, R. P. (1939). Forces in molecules, *Phys. Rev.*, **56**, pp. 340–343.

38. Fisher, G. B. and Sexton, B., A. (1980). Identification of an adsorbed hydroxyl species on the Pt(111) Surface, *Phys. Rev. Lett.*, **44**, pp. 683–686.

39. Fisicaro, G., Genovese, L., Andreussi, O., N. Marzari, N., and Goedecker, S. (2016). A generalized Poisson and Poisson-Boltzmann solver for electrostatic environments, *J. Chem. Phys.*, **144**, pp. 014103.

40. Funtikov, A. M., Linke, U., Stimminga, U., and Vogel, R. (1995). An in-situ STM study of anion adsorption on Pt(111) from sulfuric acid solutions, *Surf. Sci.*, **324**, pp. L343–L348.

41. Gómez-Marín, A. M., Clavilier, J., and Feliu, J. M. (2013). Sequential Pt(111) oxide formation in perchloric acid: an electrochemical study of surface species inter-conversion, *J. Electroanal. Chem.*, **688**, pp. 360–370.

42. Gómez, R., Orts, M., and Feliu, J. M. (2004). Effect of temperature on hydrogen adsorption on Pt(111), Pt(110), and Pt(100) electrodes, *J. Phys. Chem. B*, pp. 228–238.

43. Gilroy, D. and Conway, B. E. (1968). Surface oxidation and reduction of platinum electrodes: coverage, kinetic and hysteresis studies, *Can. J. Chem.*, **46**, pp. 875.

44. Gouy, M. G. (1910). Sur la constitution de la charge électrique à la surface d'un électrolyte, *J. Phys. Theor. Appl.*, **9**, pp. 457–468.

45. Grahame, D. C. (1947). The electrical double layer and the theory of electrocapillarity, *Chem. Rev.*, **41**, pp. 441–501.

46. Greeley, J., Rossmeisl, J., Hellman, A., and Nørskov, J. K. (2007). Theoretical trends in particle size effects, *Z. Phys. Chem.*, **221**, pp. 1209–1220.

47. Greeley, J., Stephens, I. E. L., Bondarenko, A. S., Johansson, T. P., Hansen, H. A., Jaramillo, T. F., Rossmeisl, J., Chorkendorff, I., and Nørskov, J. K. (2009). Alloys of platinum and early transition metals as oxygen reduction electrocatalysts, *Nat. Chem.*, **1**, pp. 552–556.

48. Hamada, I. and Morikawa, Y. (2008). Density-functional analysis of hydrogen on Pt(111): electric field, solvent, and coverage effects, *J. Phys. Chem. C*, **112**, pp. 10889–10898.

49. Hammer, B., Hansen, L. B., and Nørskov, J. K. (1999). Improved adsorption energetics within density-functional theory using revised perdew-burke-ernzerhof functionals, *Phys. Rev. B*, **59**, pp. 7413–7421.

50. Helmholtz, H. (1853). Ueber einige Gesetze der Vertheilung elektrischer Ströme in körperlichen Leitern mit Anwendung auf die thierisch-elektrischen Versuche, *Ann. Phys. Chem.*, **165**, pp. 211–233.

51. Henderson, M. A. (2002). The interaction of water with solid surfaces: fundamental aspects revisited, *Surf. Sci. Rep.*, **46**, pp. 1–308.

52. Henkelman, G. and Jónsson, H. (2000). Improved tangent estimate in the nudged elastic band method for finding minimum energy paths and saddle points, *J. Chem. Phys.*, **113**, pp. 9978–9985.

53. Henkelman, G., Uberuaga, B. P., and Jónsson, H. (2000). A climbing image nudged elastic band method for finding saddle points and minimum energy paths, *J. Chem. Phys.*, **113**, pp. 9901–9904.

54. Herrero, H., Mostany, J., Feliu, J. M., and Lipkowski, J. (2002). Thermodynamic studies of anion adsorption at the Pt(111) electrode surface in sulfuric acid solutions, *J. Electroanal. Chem.*, **534**, pp. 79–89.

55. Hoenberg, P. and Kohn, W. (1964). Inhomogeneous electron gas, *Phys. Rev.*, **136**, pp. B864–B871.

56. Holby, E. F., Greeley, J., and Morgan, D. (2012). Thermodynamics and hysteresis of oxide formation and removal on platinum (111) surfaces, *J. Phys. Chem. C*, pp. 9942–9946.

57. Holloway, S. and Nørskov, J. K. (1984). Changes in the vibrational frequencies of adsorbed molecules due to an applied electric field, *J. Electroanal. Chem. Interfacial Electrochem.*, **161**, pp. 193–198.

58. Hush, N. S. (1958). Adiabatic rate processes at electrodes. I. Energy-charge relationships, *J. Chem. Phys.*, **28**, pp. 962–972.

59. Ikeshoji, T. and Otani, M. (2017). Toward full simulation of the electrochemical oxygen reduction reaction on Pt using first-principles and kinetic calculations, *Phys. Chem. Chem. Phys.*, **19**, pp. 4447–4453.

60. Izvekov, S., Mazzolo, A., VanOpdorp, K., and Voth, G. A. (2001). Ab initio molecular dynamics simulation of the Cu(110)–water interface, *J. Chem. Phys.*, **114**, pp. 3248–3257.

61. Janak, J. F. (1978). Proof that dE/dni=ei in density-functional theory, *Phys. Rev. B*, **18**, pp. 7165–7168.

62. Janik, M. J., Taylor, C. D., and Neurock, M. (2009). First-principles analysis of the initial electroreduction steps of oxygen over Pt(111), *J. Electrochem. Soc.*, **156**, pp. B126–B135.

63. Jia, Q., Li, J., Caldwell, K., Ramaker, D. E., Ziegelbauer, J. M., Kukreja, R. S., Kongkanand, A., and Mukerjee, S. (2016). Circumventing metal dissolution induced degradation of Pt-alloy catalysts in proton exchange membrane fuel cells: revealing the asymmetric volcano nature of redox catalysis, *ACS Catal.*, **6**, pp. 928–938.

64. Jinnouchi, R. and Anderson, A. B. (2008). Aqueous and surface redox potentials from self-consistently determined Gibbs energies, *J. Phys. Chem. C*, **112**, pp. 8747–8750.

65. Jinnouchi, R. and Anderson, A. B. (2008). Electronic structure calculations of liquid-solid interfaces: combination of density functional theory and modified Poisson-Boltzmann theory, *Phys. Rev. B*, **77**, pp. 245417.

66. Jinnouchi, R., Toyoda, E., Hatanaka, T., and Morimoto, Y. (2010). First principles calculations on site-dependent dissolution potentials of supported and unsupported Pt particles, *J. Phys. Chem. C*, **114**, pp. 17557–17568.

67. Jinnouchi, R., Kodama, K., Hatanaka, T., and Morimoto, Y. (2011). First principles based mean field model for oxygen reduction reaction, *Phys. Chem. Chem. Phys.*, **13**, pp. 21070–21083.

68. Jinnouchi, R., Hatanaka, T., Morimoto, Y., and Osawa, M. (2012). First principles study of sulfuric acid anion adsorption on a Pt(111) electrode, *Phys. Chem. Chem. Phys.*, pp. 3208–3218.

69. Jinnouchi, R., Hatanaka, T., Morimoto, Y., and Osawa, M. (2013). Stark effect on vibration frequencies of sulfate on Pt(111) electrode, *Electrochim. Acta*, **101**, pp. 254–261.

70. Jinnouchi, R., Kodama, K., and Morimoto, Y. (2014). DFT calculations on H, OH and O adsorbate formations on Pt(111) and Pt(332) electrodes, *J. Electroanal. Chem.*, **716**, pp. 31–44.

71. Jinnouchi, R., Kodama, K., Suzuki, T., and Morimoto, Y. (2015). Kinetically induced irreversibility in electro-oxidation and reduction of Pt surface, *J. Chem. Phys.*, **142**, pp. 184709.

72. Jinnouchi, R., Nagoya, A., Kodama, K., and Morimoto, Y. (2015). Solvation effects on OH adsorbates on stepped Pt surfaces, *J. Phys. Chem. C*, **119**, pp. 16743–16753.

73. Jinnouchi, R., Kodama, K., Suzuki, T., and Morimoto, Y. (2016). DFT calculations on electro-oxidations and dissolutions of Pt and Pt-Au nanoparticles, *Catal. Today*, **262**, pp. 100–109.

74. Jinnouchi, R., Kudo, K., Kitano, N., and Morimoto, Y. (2016). Molecular dynamics simulations on O2 permeation through Nafion ionomer on platinum surface, *Electrochim. Acta*, **188**, pp. 767–776.

75. Jinnouchi, R., Kodama, K., Nagoya, A., and Morimoto, Y. (2017). Simulated volcano plot of oxygen reduction reaction on stepped Pt surfaces, *Electrochim. Acta*, **230**, pp. 470–478.

76. Jinnouchi, R., Kodama, K., and Morimoto, Y. (2018). Electronic structure calculations on electrolyte–electrode interfaces: successes and limitations, *Curr. Opin. Electrochem.*, **8**, pp. 103–109.

77. Karlberg, G. and Wahnström, G. (2004). Density-functional based modeling of the intermediate in the water production reaction on Pt(111), *Phys. Rev. Lett.*, **92**, pp. 136103.

78. Karlberg, G. S., Jaramillo, T. F., Skúlason, E., Rossmeisl, J., Bligaard, T., and Nørskov, J. K. (2007). Cyclic voltammograms for H on Pt(111) and Pt(100) from first principles, *Phys. Rev. Lett.*, **99**, pp. 126101.

79. Kislenko, S. A., Samoylov, I. S., and Amirov, R. H. (2009). Molecular dynamics simulation of the electrochemical interface between a graphite surface and the ionic liquid [BMIM][PF6], *Phys. Chem. Chem. Phys.*, **11**, pp. 5584–5590.

80. Kodama, K., Jinnouchi, R., Suzuki, T., Hatanaka, T., and Morimoto, Y. (2012). Extraordinarily small Tafel slope for oxide formation reaction on Pt (111) surface, *Electrochim. Acta*, **78**, pp. 592–596.

81. Kodama, K., Jinnouchi, R., Suzuki, T., Murata, H., Hatanaka, T., and Morimoto, Y. (2013). Increase in adsorptivity of sulfonate anions on Pt (111) surface with drying of ionomer, *Electrochem. Commun.*, **36**, pp. 26–28.

82. Kodama, K., Shinohara, a., Hasegawa, N., Shinozaki, K., Jinnouchi, R., Suzuki, T., Hatanaka, T., and Morimoto, Y. (2014). Catalyst poisoning property of sulfonimide acid ionomer on Pt (111) surface, *J. Electrochem. Soc.*, **161**, pp. F649–F652.

83. Kohn, W. and Sham, L. J. (1965). Self-consistent equations including exchange and correlation effects, *Phys. Rev.*, **140**, pp. A1133–A1138.

84. Kolics, A. and Wieckowski, A. (2001). Adsorption of bisulfate and sulfate anions on a Pt(111) electrode, *J. Phys. Chem. B*, **105**, pp. 2588–2595.

85. Kongkanand, A. and Ziegelbauer, J. M. (2012). Surface platinum electrooxidation in the presence of oxygen, *J. Phys. Chem. C*, **116**, pp. 3684–3693.

86. Koper, M. T. M., Jansen, A. P. J., van Santen, R. A., Lukkien, J. J., and Hilbers, P. A. J. (1998). Monte Carlo simulations of a simple model for the electrocatalytic CO oxidation on platinum, *J. Chem. Phys.*, **109**, pp. 6051.

87. Koper, M. T. M., van Santen, R. A., Wasileski, S. A., and Weaver, M. J. (2000). Field-dependent chemisorption of carbon monoxide and nitric oxide on platinum-group (111) surfaces: quantum chemical calculations compared with infrared spectroscopy at electrochemical and vacuum-based interfaces, *J. Chem. Phys.*, **113**, pp. 4392–4407.

88. Kovalenko, A. and Hirata, F. (1999). Self-consistent description of a metal–water interface by the Kohn–Sham density functional theory and the three-dimensional reference interaction site model, *J. Chem. Phys.*, **110**, pp. 10095–10112.

89. Kresse, G. and Furthmüller, J. (1996). Efficiency of ab-initio total energy calculations for metals and semiconductors using a plane-wave basis set, *Comput. Mater. Sci.*, **6**, pp. 15–50.

90. Kresse, G. and Furthmüller, J. (1996). Efficient iterative schemes for ab initio total-energy calculations using a plane-wave basis set, *Phys. Rev. B*, **54**, pp. 11169–11185.

91. Kunimatsu, K., Golden, W. G., Seki, H., and Philpott, M. R. (1985). Carbon monoxide adsorption on a platinum electrode studied by polarization modulated FT-IRRAS. 1. Carbon monoxide adsorbed in the double-layer potential region and its oxidation in acids, *Langmuir*, **1**, pp. 245–250.

92. Kunimatsu, K., Seki, H., Golden, W. G., Gordon, J. G., and Philpott, M. R. (1985). Electrode/electrolyte interphase study using polarization modulated ftir reflection-absorption spectroscopy, *Surf. Sci.*, **158**, pp. 596–608.

93. Lachenwitzer, A., Li, N., and Lipkowski, J. (2002). Determination of the acid dissociation constant for bisulfate adsorbed at the Pt(111) electrode by subtractively normalized interfacial Fourier transform infrared spectroscopy, *J. Electroanal. Chem.*, **532**, pp. 85–98.

94. Lambert, D. K. (1983). Observation of the first-order Stark effect of Co on Ni(110), *Phys. Rev. Lett.*, **50**, pp. 2106–2109.

95. Lambert, D. K. (1984). Stark effect of adsorbate vibrations, *Solid State Commun.*, **51**, pp. 297–300.

96. Letchworth-Weaver, K. and Arias, T. A. (2012). Joint density functional theory of the electrode-electrolyte interface: application to fixed electrode potentials, interfacial capacitances, and potentials of zero charge, *Phys. Rev. B*, **86**, pp. 75140.

97. Lozovoi, A. Y., Alavi, A., Kohanoff, J., and Lynden-Bell, R. M. (2001). Ab initio simulation of charged slabs at constant chemical potential, *J. Chem. Phys.*, **115**, pp. 1661–1669.

98. Man, I. C., Su, H.-Y., Calle-Vallejo, F., Hansen, H. A., Martínez, J. I., Inoglu, N. G., Kitchin, J., Jaramillo, T. F., Nørskov, J. K., and Rossmeisl, J. (2011). Universality in oxygen evolution electrocatalysis on oxide surfaces, *ChemCatChem*, **3**, pp. 1159–1165.

99. Marković, N. M., Lucas, C. A., Grgur, B. N., and Ross, P. N. (1999). Surface electrochemistry of CO and H2/CO mixtures at Pt(100) interface: electrode kinetics and interfacial structures, *J. Phys. Chem. B*, **103**, pp. 9616–9623.

100. Marković, N. M. and Ross, P. N. J. (2002). Surface science studies of model fuel cell electrocatalysts, *Surf. Sci. Rep.*, **45**, pp. 117–229.

101. Mathew, K., Sundararaman, R., Letchworth-Weaver, K., Arias, T. A., and Hennig, R. G. (2014). Implicit solvation model for density-functional study of nanocrystal surfaces and reaction pathways, *J. Chem. Phys.*, **140**, pp. 084106.

102. Mermin, N. D. (1965). Thermal properties of the inhomogeneous electron gas, *Phys. Rev.*, **137**, pp. A1441–A1443.

103. Michaelides, A. and Hu, P. (2001). Catalytic water formation on platinum: a first-principles study, *J. Am. Chem. Soc.*, **123**, pp. 4235–4242.

104. Mukerjee, S., Srinivasan, S., Soriaga, M. P., and McBreen, J. (1995). Role of structural and electronic properties of Pt and Pt alloys on electrocatalysis of oxygen reduction an in situ XANES and EXAFS investigation, *J. Electrochem. Soc.*, **142**, pp. 1409–1422.

105. Nørskov, J. K., Rossmeisl, J., Logadottir, A., Lindqvist, L., Kitchin, J. R., Bligaard, T., and Jónsson, H. (2004). Origin of the overpotential for oxygen reduction at a fuel-cell cathode, *J. Phys. Chem. B*, **108**, pp. 17886–17892.

106. Nørskov, J. K., Bligaard, T., Rossmeisl, J., and Christensen, C. H. (2009). Towards the computational design of solid catalysts, *Nat. Chem.*, **1**, pp. 37–46.

107. Nagoya, A., Jinnouchi, R., Kodama, K., and Morimoto, Y. (2015). DFT calculations on H, OH and O adsorbate formations on Pt(322) electrode, *J. Electroanal. Chem.*, **757**, pp. 116–127.

108. Nagy, Z. and You, H. (2002). Applications of surface X-ray scattering to electrochemistry problems, *Electrochim. Acta*, **47**, pp. 3037–3055.

109. Nart, F. C., Iwasita, T., and Weber, M. (1994). Vibrational spectroscopy of adsorbed sulfate on Pt(111), *Electrochim. Acta*, **39**, pp. 961–968.

110. Nichols, R. J. (1992). *Adsorption of Molecules at Metal Electrodes* (VCH, Weinheim, New York), pp. 347–352.

111. Nishihara, S. and Otani, M. (2017). Hybrid solvation models for bulk, interface, and membrane: reference interaction site methods coupled with density functional theory, *Phys. Rev. B*, **96**, pp. 115429.

112. Okamoto, Y. (2006). Comparison of hydrogen atom adsorption on Pt clusters with that on Pt surfaces: a study from density-functional calculations, *Chem. Phys. Lett.*, **429**, pp. 209–213.

113. Okamoto, Y. and Sugino, O. (2010). Hyper-volcano surface for oxygen reduction reactions over noble metals, *J. Phys. Chem. C*, **114**, pp. 4473–4478.

114. Otani, M. and Sugino, O. (2006). First-principles calculations of charged surfaces and interfaces: a plane-wave nonrepeated slab approach, *Phys. Rev. B*, **73**, pp. 115407.

115. Parr, R. G. and Yang, W. (1989). *Density-Functional Theory of Atoms and Molecules*, Vol. 16 of International Sof Monographs on Chemistry (Oxford University Press, New York).

116. Perdew, J. P., Burke, K., and Ernzerhof, M. (1996). Generalized gradient approximation made simple, *Phys. Rev. Lett.*, **77**, pp. 3865–3868.

117. Pulay, P. (1969). Ab initio calculation of force constants and equilibrium geometries in polyatomic molecules, *Mol. Phys.*, **17**, pp. 197–204.

118. Ray, N. K. and Anderson, A. B. (1982). Variations in carbon-oxygen and platinum-carbon frequencies for carbon monoxide on a platinum electrode, *J. Phys. Chem.*, **86**, pp. 4851–4852.

119. Roques, J. R. M. and Anderson, A. B. (2004). Theory for the potential shift for OHads formation on the Pt skin on Pt3Cr(111) in acid, *J. Electrochem. Soc.*, **151**, pp. E85–E91.

120. Ross, P. N., Lucas, C. A., and Markovic, N. M. (1999). The adsorption and oxidation of carbon monoxide at the Pt(111) / electrolyte interface: atomic structure and surface relaxation, *Surf. Sci.*, **425**, pp. L381–L386.

121. Rossmeisl, J., Nørskov, J. K., Taylor, C. D., Janik, M. J., and Neurock, M. (2006). Calculated phase diagrams for the electrochemical oxidation

and reduction of water over Pt(111), *J. Phys. Chem. B*, **110**, pp. 21833–21839.

122. Russell, A. E. and Rose, A. (2004). X-ray absorption spectroscopy of low temperature fuel cell catalysts, *Chem. Rev.*, **104**, pp. 4613–4636.

123. Saravanan, C., Koper, M. T. M., Markovic, N. M., Head-Gordon, M., and Ross, P. N. (2002). Modeling base voltammetry and CO electrooxidation at the Pt(111)-electrolyte interface: Monte Carlo simulations including anion adsorption, *Phys. Chem. Chem. Phys.*, **4**, pp. 2660–2666.

124. Sawatari, Y., Inukai, J., and Ito, M. (1993). The structure of bisulfate and perchlorate on a Pt(111) electrode surface studied by infrared spectroscopy and ab-initio molecular orbital calculation, *J. Electron. Spectrosc. Relat. Phenom.*, **64–65**, pp. 515–522.

125. Scherlis, D., A., Fattebert, J.-L., Gygi, F., Cococcioni, M., and Marzari, N. (2006). A unified electrostatic and cavitation model for first-principles molecular dynamics in solution, *J. Chem. Phys.*, **124**, pp. 74103.

126. Shingaya, Y. and Ito, M. (1996). Interconversion of a bisulfate anion into a sulfuric acid molecule on a Pt(111) electrode in a 0.5 M H_2SO_4 solution, *Chem. Phys. Lett.*, **256**, pp. 438–444.

127. Sinstein, M., Scheurer, C., Matera, S., Blum, V., Reuter, K., and Oberhofer, H. (2017). An efficient implicit solvation method for full potential DFT, *J. Chem. Theory Comput.*, **13**, pp. 5582–5603.

128. Skúlason, E., Karlberg, G. S., Rossmeisl, J., Bligaard, T., Greeley, J., Jónsson, H., and Nørskov, J. K. (2007). Density functional theory calculations for the hydrogen evolution reaction in an electrochemical double layer on the Pt(111) electrode, *Phys. Chem. Chem. Phys.*, **9**, pp. 3241–3250.

129. Spohr, E. (1989). Computer simulation of the water/platinum interface, *J. Phys. Chem.*, **93**, pp. 6171–6180.

130. Spohr, E. (1999). Molecular simulation of the electrochemical double layer, *Electrochim. Acta*, **44**, pp. 1697–1705.

131. Spohr, E. (2002). Molecular dynamics simulations of water and ion dynamics in the electrochemical double layer, *Solid State Ionics*, **150**, pp. 1–12.

132. Springer, T. E., Zawodzinski, T. A., and Gottesfeld, S. (1991). Polymer electrolyte fuel cell model, *J. Electrochem. Soc.*, **138**, pp. 2334–2342.

133. Stamenkovic, V., Mun, B. S., Mayrhofer, K. J. J., Ross, P. N., Markovic, N. M., Rossmeisl, J., Greeley, J., and Nørskov, J. K. (2006). Changing the activity of electrocatalysts for oxygen reduction by tuning the surface electronic structure, *Angew. Chem. Int. Ed.*, **45**, pp. 2897–2901.

134. Steinfeld, J. I., Francisco, J. S., and Hase, W. L. (1989). *Chemical Kinetics and Dynamics* (Prentice Hall, New Jersey).

135. Stern, O. Z. (1924). Zur Theorie der elektrolytischen Doppelschicht, *Z. Electrochem*, **30**, pp. 508–526.

136. Strasser, P., Koh, S., Anniyev, T., Greeley, J., More, K., Yu, C., Liu, Z., Kaya, S., Nordlund, D., Ogasawara, H., Toney, M. F., and Nilsson, A. (2010). Lattice-strain control of the activity in dealloyed core-shell fuel cell catalysts, *Nat. Chem.*, **2**, pp. 454–460.

137. Subbaraman, R., Strmcnik, D., Stamenkovic, V., and Markovic, N. M. (2010). Three phase interfaces at electrified metal-solid electrolyte systems 1. Study of the Pt(hkl)-Nafion interface, *J. Phys. Chem. C*, **114**, pp. 8414–8422.

138. Sundararaman, R., Guncerler, D., and Arias, T. A. (2014). Weighted-density functionals for cavity formation and dispersion energies in continuum solvation models, *J. Chem. Phys.*, **141**, pp. 134105.

139. Tada, M., Murata, S., Asakoka, T., Hiroshima, K., Okumura, K., Tanida, H., Uruga, T., Nakanishi, H., Matsumoto, S.-i., Inada, Y., Nomura, M., and Iwasawa, Y. (2007). In situ time-resolved dynamic surface events on the Pt/C cathode in a fuel cell under operando conditions, *Angew. Chem. Int. Ed.*, **46**, pp. 4310–4315.

140. Tapia, O. and Goscinski, O. (1975). Self-consistent reaction field theory of solvent effects, *Mol. Phys.*, **29**, pp. 1653–1661.

141. Taylor, C. D., Kelly, R. G., and Neurock, M. (2006). First-principles calculations of the electrochemical reactions of water at an immersed Ni(111)/H_2O interface, *J. Electrochem. Soc.*, **153**, pp. E207–E214.

142. Taylor, C. D., Wasileski, S. A., Filhol, J.-S., and Neurock, M. (2006). First principles reaction modeling of the electrochemical interface: consideration and calculation of a tunable surface potential from atomic and electronic structure, *Phys. Rev. B*, **73**, pp. 165402.

143. Teliska, M., Murthi, V. S., Mukerjee, S., and Ramaker, D. E. (2005). Correlation of water activation, surface properties, and oxygen reduction reactivity of supported Pt–M/C bimetallic electrocatalysts using XAS, *J. Electrochem. Soc.*, **152**, pp. A2159.

144. Teliska, M., O'Grady, W. E., and Ramaker, D. E. (2005). Determination of O and OH adsorption sites and coverage in situ on Pt electrodes from Pt L 23 X-ray absorption spectroscopy, *J. Phys. Chem. B*, **109**, pp. 8076–8084.

145. Tomasi, J., Mennucci, B., and Cammi, R. (2005). Quantum mechanical continuum solvation models, *Chem. Rev.*, **105**, pp. 2999–3093.

146. Toyoda, E., Jinnouchi, R., Ohsuna, T., Hatanaka, T., Aizawa, T., Otani, S., Kido, Y., and Morimoto, Y. (2013). Catalytic activity of Pt/TaB2(0001) for the oxygen reduction reaction, *Angew. Chem. Int. Ed.*, **52**, pp. 4137–4140.

147. Trasatti, S. (1991). Structure of the metal/electrolyte solution interface: new data for theory, *Electrochim. Acta*, **36**, pp. 1659–1667.

148. Völkening, S., Bedürftig, K., Jacobi, K., Wintterlin, J., and Ertl, G. (1999). Dual-path mechanism for catalytic oxidation of hydrogen on platinum surfaces, *Phys. Rev. Lett.*, **83**, pp. 2672–2675.

149. Vetter, K. J. and Schultze, J. W. (1972). The kinetics of the electrochemical formation and reduction of monomolecular oxide layers on platinum in 0.5M H2SO4, *J. Electroanal. Chem.*, **34**, pp. 141–158.

150. Viswanathan, V., Hansen, H. A., Rossmeisl, J., and Nørskov, J. K. (2012). Unifying the 2e– and 4e– reduction of oxygen on metal surfaces, *J. Phys. Chem. Lett.*, **3**, pp. 2948–2951.

151. Wakisaka, M., Asizawa, S., Uchida, H., and Watanabe, M. (2010). In situ STM observation of morphological changes of the Pt(111) electrode surface during potential cycling in 10 mM HF solution, *Phys. Chem. Chem. Phys.*, **12**, pp. 4184–4190.

152. Xin, H. L., Mundy, J. A., Liu, Z., Cabezas, R., Hovden, R., Kourkoutis, L. F., Zhang, J., Subramanian, N. P., Makharia, R., Wagner, F. T., and Muller, D. A. (2012). Atomic-resolution spectroscopic imaging of ensembles of nanocatalyst particles across the life of a fuel cell, *Nano Lett.*, **12**, pp. 490–497.

153. Yamamoto, M. and Kinoshita, M. (1997). Structure of the metal-liquid interface: self-consistent combination of the first-principles metal calculation and an integral equation method, *Chem. Phys. Lett.*, **274**, pp. 513–517.

154. Yu, L., Pan, X., Cao, X., Hu, P., and Bao, X. (2011). Oxygen reduction reaction mechanism on nitrogen-doped graphene: a density functional theory study, *J. Catal.*, **282**, pp. 183–190.

155. Zawodzinski, T. A., Derouin, C., Radzinski, S., Sherman, R. J., Smith, V. T., Springer, T. E., and Gottesfeld, S. (1993). Water uptake by and transport through Nafion 117 membranes, *J. Electrochem. Soc.*, **140**, pp. 1041–1047.

Chapter 4

Atomistic Modeling of Photoelectric Cells for Artificial Photosynthesis

Ryoji Asahi and Ryosuke Jinnouchi
Toyota Central R&D Laboratories, Inc., Nagakute, Aichi 480-1192, Japan
rasahi@mosk.tytlabs.co.jp

4.1 Introduction

In this chapter, we discuss computational methods and models used to evaluate physical quantities related to artificial photosynthesis. The photosynthesis process involves light absorption, electron transfer (ET), catalytic reactions, and so on. The design of artificial photosynthesis devices, therefore, requires considering many aspects of these complex, interrelated phenomena. Through this chapter, we hope to expose what could be the key factors of computational models, how to construct them, and how to evaluate them for improvement of the photosynthesis process.

Artificial photosynthesis is one of the promising technologies for sustainable and clean energy. In particular, photocatalytic CO_2 reduction into useful chemicals has attracted great interest,

Multiscale Simulations for Electrochemical Devices
Edited by Ryoji Asahi

ISBN 978-981-4800-71-6 (Hardcover), 978-0-429-29545-4 (eBook)
www.jennystanford.com

which has led to increasing studies for development of materials. Mechanisms of operation of materials that realize efficient conversion of solar energy are often similar to those of natural photosynthetic complexes. Both natural and artificial photosynthetic materials utilize multicomponent tandem structures known as Z-scheme. A typical structure of artificial photosynthesis utilizing semiconductors for the Z-scheme is schematically shown in Fig. 4.1. The Z-scheme is composed of at least two materials whose valence band (VB) and conduction band (CB) are offset to levels suitable for driving target half-cell redox reactions and whose energy gaps are designed to absorb light covering a wide range of the solar spectrum. The electrons and holes generated by solar light absorption transfer to the surface, where the electrons reduce CO_2 and the holes oxidize water, usually with the help of electrocatalysts. In contrast to photocatalysis for producing H_2, CO_2 reduction faces the additional difficulty of having to be selective to the H_2 evolution in the presence of water because the CO_2 molecule is highly stable. To this end, many efforts to explore combinations of semiconductors and electrocatalysts suitable for CO_2 reduction have been made in the last decades [11–16, 28, 40, 49, 50, 58, 67, 82–88, 90, 91, 93, 100–102, 104–106].

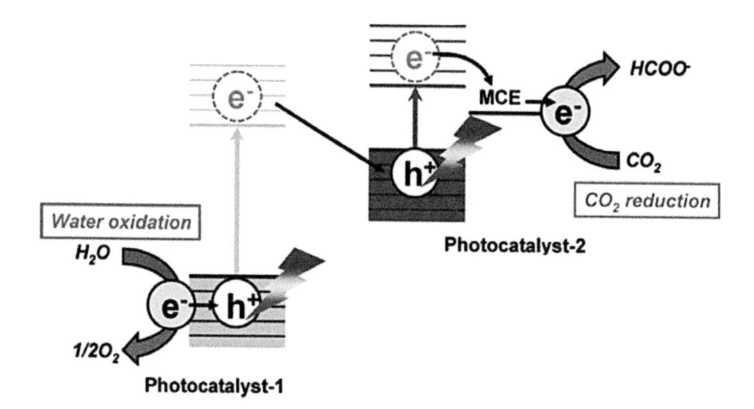

Figure 4.1 Example of the reaction process using the Z-scheme for artificial photosynthesis; here, a semiconductor (Photocatalyst-1) and a semiconductor/metal-complex (MCE) hybrid catalyst (Photocatalyst-2) are used for water oxidation and CO_2 reduction, respectively. Reprinted with permission from Ref. [84]. Copyright (2011) American Chemical Society.

One of the pioneering works on the use of semiconductors for artificial photosynthesis was done by Sato et al. [84]. They described a CO_2 reduction system composed of TiO_2 and N-doped Ta_2O_5, where water is oxidized by the photo-generated holes at the surface of a Pt-loaded TiO_2 photoanode, while electrons, excited in the Ta_2O_5, are injected into a Ru-complex catalyst loaded on the Ta_2O_5 surface. The injected electrons further participate in CO_2 reduction, producing HCOOH. Sato et al. [83] and Suzuki et al. [100] examined several Ru-complex/N-Ta_2O_5 composite systems with different anchors and found that those anchored by phosphonate exhibited high CO_2 photoreduction activity. As an extension of these studies, Arai et al. [14] demonstrated a monolithic tablet-shaped CO_2 photoreduction device that consisted of a triple junction of amorphous silicon-germanium as a light absorber, a Ru-complex polymer as a CO_2 reduction catalyst, and IrO_x as a water oxidation catalyst. The solar conversion efficiency for formate generation reached 4.6% in neutral pH aqueous media without any external electrical and chemical bias voltages. Further development of the CO_2 reduction reactions, such as the direct generation of CO and alcohol, has attracted great interest, making this a hot research field [96, 97].

The solar convergence efficiency of photosynthesis depends on many factors, including light absorption of the semiconductors, energy barriers along the reaction coordinate, ET rates inside semiconductors and at interfaces, and the effects of solvent. With these factors in mind, first-principles materials design has been intensely applied to the band engineering of semiconductors and interfaces so that the light absorption properties and energy band alignments can be evaluated and optimized [18, 20, 70, 78, 80, 109]. We should note that such computational design has become popular and accurate with the recent development of density functional theory (DFT) using sophisticated exchange-correlation functionals based on many-body perturbation theory, such as the hybrid exchange-correlation functional [62, 75] and the GW method [17, 94, 95, 115], along with recent computational hardware development. These methods successfully predict the bandgap and the band alignment once atomistic modeling is defined [26, 29, 44, 54, 71, 107].

In practice, computational modeling becomes much more complex and difficult to construct in cases where surfaces and defects are involved. Defects in semiconductors change light absorption properties and the mobility of electrons and holes, as in the case of photocatalysts [20, 24, 53, 121]. In addition, atomistic inhomogeneity and modifications induced around the surface affect electrostatic potentials and thus band alignments for photosynthesis. An example related to the surface modifications is metal-complex catalysts or metal-nanoparticle cocatalysts loaded on the semiconductor and another is the oxygen vacancies introduced in the semiconductor surface region [120, 122]. Importantly, such surface modifications or surface defects strongly depend on materials synthesis processing. Therefore, understanding defect formations leads to an optimization of synthesis processing [19].

While the static properties mentioned above provide the fundamental factors for understanding and designing the photosynthesis process, evaluations of ET are also desired to determine the rate-limiting steps of the entire photosynthesis process. In particular, interfacial ET is sensitive to electronic coupling between donor and acceptor states, as shown by studies of atomistic details of materials [3, 32, 79, 112]. Coupling strength can vary notably depending on the way the adsorbate is attached to the surface, including the length [10, 25, 118] and nature of linkers [9, 39, 52, 59, 63, 117], as well as the anchor-binding mode. Quantitative calculations of the ET rate, however, require significant computations, and therefore, several approximate calculations have been employed. Following Fermi's golden rule, one can expect that ET rates depend on the density of acceptor states, as extensively studied in a number of photovoltaic systems [1, 21, 36]. The Marcus theory predicts a strong dependence of the rates on the free energy difference, ΔG, of the ET process [61]. On the other hand, methodological development of atomistic non-adiabatic molecular dynamics (NA-MD) allows us to perform direct simulations of the electron dynamics. Efficient non-adiabatic ET simulations have been implemented by means of a trajectory surface hopping (SH) NA-MD algorithm [4, 5, 31, 37, 38, 113].

In the following sections of this chapter, we focus on two major issues involved in atomistic modeling of artificial photosynthesis. In Section 4.2, we discuss surface modification of semiconductors. An electrostatic effect of metal-nanoparticles loaded on TiO_2 and the resulting band bending are explained [89]. Furthermore, a detailed analysis of surface-defect formations in Ta_2O_5 induced by N-doping, which affects ΔG at the N-doped Ta_2O_5/Ru-complex interface, is provided [51]. In Section 4.3, a study of the ET rate from N-doped Ta_2O_5 to Ru-complex catalysts using NA-MD is presented [6, 7]. These issues cover many crucial factors to be considered for designing artificial photosynthesis, as noted in Section 4.4.

4.2 Surface Modification of Semiconductors

As stated in the introduction, surface modification of semiconductors is one of the fundamental issues that significantly affect the performance of photosynthesis. Here we present two examples, namely, metal-nanoparticles loaded on TiO_2 [89] and defect formations of N-doped Ta_2O_5 [51], which are used for photocatalysis and artificial photosynthesis, respectively. We focus on computational modeling and results, so for the detailed methodology, one should refer to the literature cited.

4.2.1 Metal-Nanoparticles Loaded on TiO_2

Photocatalytic activity is often improved by the surface adsorption of metal or metal-oxide cocatalysts [20, 43, 45, 73]. For example, TiO_2 with a surface loading of nanoscale Pt decomposed volatile organic compounds more efficiently than without Pt. One of the reasons for this effect is that fast transfer of the photoexcited electrons to the surface Pt enhances the electron-hole charge separation, which prevents carrier recombination [35]. Such ET should be promoted by electrostatic effects at the vicinity of the metal-nanoparticle/semiconductor interface; however, atomistic details were not well understood. In fact, there were additional complexities and physical insights involved [89], as explained in this section.

To illustrate the electrostatic effects, the binding energies of Ti $2p_{3/2}$ in TiO_2 were calculated using the DFT method. The TiO_2 surface was modeled as a three atomic-layer, rutile-type $TiO_2(110)$ slab with 4 × 2 two-dimensional periodicity, with Pt particles or thin layers composed of 1–120 Pt atoms placed on top of the TiO_2 surface, as illustrated in Fig. 4.2. Similarly, Rh- and Au-loaded TiO_2 models were set up for the calculations. The geometry of each model was optimized until forces acting on all atoms converged within 0.05

eV/Å. The binding energies were evaluated [89, 108] by introducing a core hole of a 2p orbital at a Ti atom and averaged over 16 Ti atoms in the middle layer of the slab. The transferred charge values, ΔQ, were also calculated from gross charges in a Mulliken population analysis [69]. All calculations employed a generalized gradient approximation based on a Perdue–Burke–Ernzerhof function [76, 77].

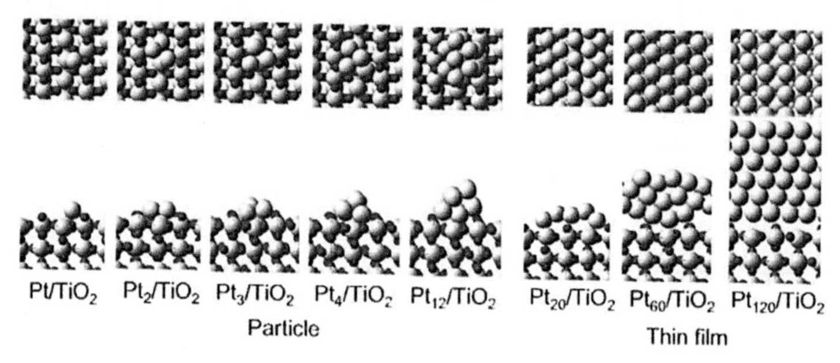

Figure 4.2 The TiO_2(110) slab models with Pt particles on the surface, which are optimized by DFT calculations. The upper and lower parts illustrate the top and side views of each model, respectively. Reprinted from Ref. [89] with permission from the PCCP Owner Societies.

The calculated Ti $2p_{3/2}$ binding energies for Pt particles or thin films on TiO_2 are presented in Fig. 4.3. The Ti $2p_{3/2}$ binding energies are shifted negatively with these Pt loadings. The amount of the shift increases with increasing nanoparticle size, which is much larger than that in thin film models for the same number of Pt atoms. These results indicate that the binding-energy shift depends on both size and morphology of the Pt loading. When Pt is loaded on TiO_2, the charge redistribution forms a metal-semiconductor bonding at the interface. The origin of the binding energy shift is thus presumably from the formation of a dipole as a result of the charge redistribution. Given the same amount of Pt loading per unit surface area, the Pt nanoparticles can effectively make a larger dipole than the uniform Pt thin film.

In Fig. 4.4, the binding energies for Rh and Au are also plotted with those for Pt. Among them, Rh is predicted to give the largest shift at the same morphology of loading. On the contrary, when the binding energy shifts are plotted as a function of the transferred charge, these metals fall on a single line, as shown in Fig. 4.5. This means that the negative Ti $2p_{3/2}$ peak shift is determined by the amount of charge redistribution from the nanoscale metal to the TiO_2. At a glance, such a charge redistribution seems to depend on the electron negativities (or the ionization energies in kJ/mol) among the metals, which are 1.54 (659) for Ti, 2.54 (890) for Au, 2.28 (720) for Rh, and 2.28 (870) for Pt. However, as seen clearly in the calculation results, the shifts are quantitatively determined by the detailed quantum mechanical effects, which are considered in the present atomistic simulations.

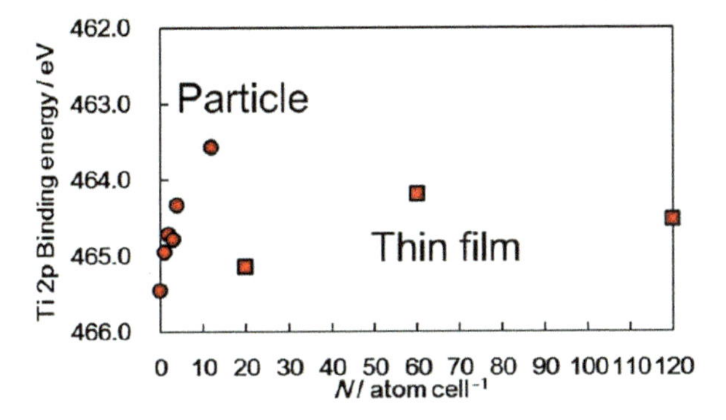

Figure 4.3 Calculated Ti $2p_{3/2}$ binding energies for Pt particles (closed circles) or thin films (closed squares) on TiO_2. Reprinted from Ref. [89] with permission from the PCCP Owner Societies.

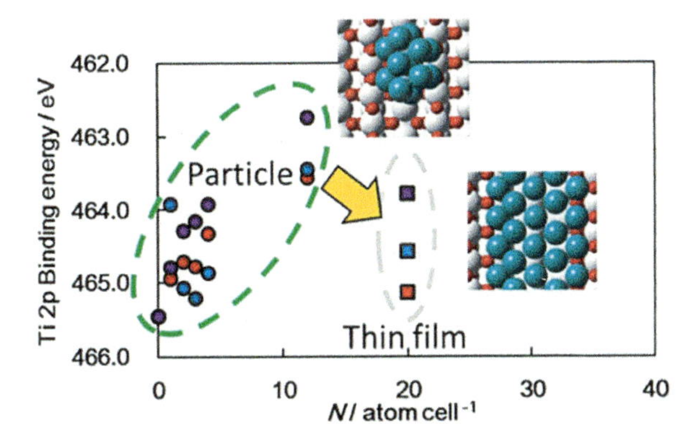

Figure 4.4 Calculated Ti $2p_{3/2}$ binding energies for metal particles (circles) or thin films (squares) on TiO_2: red, blue, and purple represent Pt, Au, and Rh, respectively. Reprinted from Ref. [89] with permission from the PCCP Owner Societies.

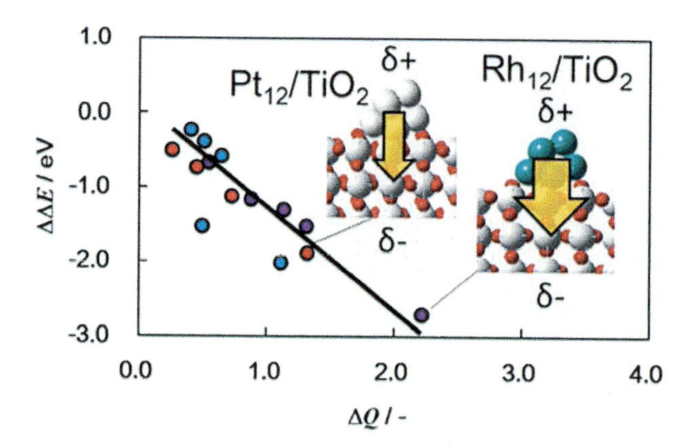

Figure 4.5 Ti $2p_{3/2}$ binding energy shifts ($\Delta\Delta E$) plotted as a function of the transferred charge (ΔQ). The red, blue, and purple circles represent Pt, Au, and Rh particles on TiO_2, respectively. Reprinted from Ref. [89] with permission from the PCCP Owner Societies.

As an experimental verification, the binding energies of the Ti $2p_{3/2}$ peak have been measured by angular-resolved hard X-ray photoemission spectroscopy (HAXPES) and are shown in Fig. 4.6. The samples were prepared with TiO_2 films where Pt, Au, and Rh were loaded with a nominal thickness of 1 nm. A large shift of binding energies is observed for Rh and Pt, in good agreement with the theoretical results. However, a small shift is seen for Au loading. In fact, the morphologies of these metal loadings are different, as shown in Fig. 4.7. In particular, Au forms a large aggregation on the TiO_2 surface while Rh and Pt form uniformly distributed nanoparticles. Such an aggregation of Au can produce a rather small amount of interface charge redistribution and thus form a small dipole. The angular dependencies of the HAXPES measurement shown in Fig. 4.6 indicate band bending at the metal/semiconductor interfaces with a large depth extension (the take-off angle of 65° represents a depth of approximately 30 nm) from the surface (the zero take-off angle). This band bending is explained qualitatively using the Mott–Schottky model, originating from the large work functions of Pt, Au, and Rh compared with TiO_2 [65, 114]. The total electrostatic effects combining the dipole formation and the Mott–Schottky effect are important to understanding and improving photocatalytic activity because they affect the charge separation of the photoexcited electrons and holes, the charge transfer, and ΔG in the photocatalytic activity on the semiconductor surface.

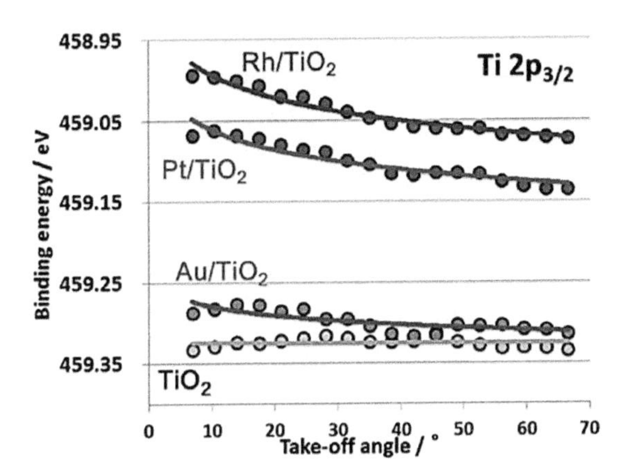

Figure 4.6 Experimentally measured Ti $2p_{3/2}$ binding energies for TiO_2, Au/TiO_2, Pt/TiO_2, and Rh/TiO_2, obtained by angular-resolved hard X-ray photoemission spectroscopy (HAXPES). Reprinted from Ref. [89] with permission from the PCCP Owner Societies.

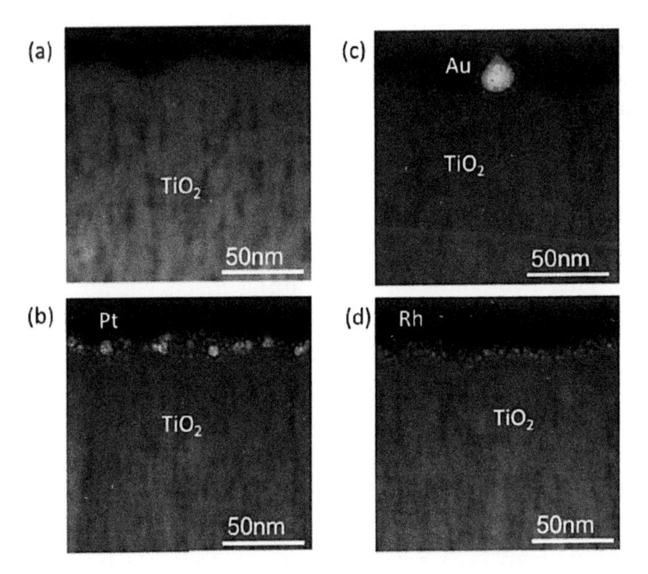

Figure 4.7 Transmission electron microscopy (TEM) images of (a) TiO_2, (b) Pt/TiO_2, and (c) Au/TiO_2. Reprinted from Ref. [89] with permission from the PCCP Owner Societies.

4.2.2 Defect Formations of N-doped Ta_2O_5

As described in the introduction, an N-doped Ta_2O_5/Ru complex was proposed as an active cathode electrode for CO_2 reduction photosynthesis [66, 83, 101, 103]. The excited electrons are injected from the CB of N-doped Ta_2O_5 into the Ru-complex, followed by the CO_2 reduction reaction to generate formate on the Ru complex, namely:

$$2CO_2 + 4H^+ + 4e^- \rightarrow 2HCOOH \quad (4.1)$$

The virtue of using the Ru complex is that Reaction 4.1 is promoted highly selectively on the complex rather than generating H_2 as soon as the electrons are injected. The efficient electron injection is driven by the suitable energy alignment between N-doped Ta_2O_5 and the Ru complex. Firstly, the lowest unoccupied molecular orbital (LUMO) level of the complex must be higher than the redox level of the CO_2 reduction, that is, –4.4 eV on a vacuum scale for the case of Reaction 4.1 [23]. Secondly, the energy level of the conduction band minimum (CBM) of the N-doped Ta_2O_5 must be even higher than the LUMO level to make the electron injections possible [83, 120]. It should be also noted that the bandgap of the semiconductors must be narrow enough to make the most of the solar light available for electron excitations [11, 66, 83].

However, little was known about the mechanisms dominating the energy alignments. In particular, several factors add complexity to the energy alignments, such as N-doping and defect formations introduced in Ta_2O_5 as well as anchoring effects at the interface between N-doped Ta_2O_5 and the Ru complex. In this section, we focus on the former; the latter anchoring effects will be discussed in the next section.

Redox levels of Ru complexes such as $[Ru(bpy)_2(CO)_2]^{2+}$ (bpy: 2,2′-bipyridine), $[Ru(dcbpy)(bpy)(CO)_2]^{2+}$ (dcbpy: 4,4′-dicarboxy-2,2′bipyridine), $[Ru(dcpby)_2(CO)_2]^{2+}$, and $[Ru(dpbpy)(Cl)_2(CO)_2]$ (dpbpy: 4,4′-diphosphonate-2,2′bipyridine), are experimentally measured as –3.8 to –3.5 eV on a vacuum scale [84], while the CBM of nondoped Ta_2O_5 is –4.03 eV [30]. Therefore, the electrons on the CB of the nondoped Ta_2O_5 cannot be injected to the LUMO levels of Ru complexes, which is consistent with the absence of the photocurrent on the nondoped Ta_2O_5 samples in experiments [84]. When 8 at.% of N-doped Ta_2O_5 is used, a photocurrent is generated. Electrochemical measurements and photoelectron spectroscopy in air indicate that the CBM of Ta_2O_5 is raised to −3.3 eV by N-doping [66, 83]. These experimental results indicate that the upward shift by N-doping plays a key role in achieving electron injections from Ta_2O_5 into the Ru complexes. From the fact that the CBM consists of dominant Ta d bands, we may expect that anion doping for oxides does not affect CBM as much. This is true for CBMs of TaON and Ta_3N_5, in good agreement with that of nondoped Ta_2O_5 [30]. The mechanism of the upward shift of the CBM by N-doping is thus an intriguing issue and disclosed by the detailed investigation, including defects, as shown below.

In this study, DFT calculations are executed on modeled Ta_2O_5 surfaces with and without N-doping. The CBM and valence band maximum (VBM) of the semiconductor surfaces are calculated and compared with the theoretically obtained redox levels of Ru complexes as well as experimentally obtained energy alignments.

The bulk and surface models of Ta_2O_5 are prepared on the basis of the λ model in the orthorhombic phase [57]. It is well known that metal oxides could involve oxygen defects and hydrogen impurities [33, 92, 119] depending on synthesis conditions. In the present study oxygen vacancy (O_v), interstitial hydrogen impurity (H_{int}), and their combination (H_{int} + O_v) are considered. For N-doping, the substitutional nitrogen atom (N_{sub}) or interstitial one (N_{int}) is introduced to the site where the DFT calculations determine it is most stable. The concentration of nitrogen atoms corresponds to about 7 at.% for the present study, which is comparable to an experimental range of 2–9 at.%.

For the surface model, the surface terminations still retain arbitrariness, which can be sensitive to the environment. Similar to the surfaces of other transition metal oxides, such as TiO_2 [99, 116], under slightly humidified conditions as in actual experimental conditions [83], the metal-oxide surface is likely terminated by hydroxyl groups. As seen in the DFT results [51], an OH-terminated Ta_2O_5 surface is shown to be more stable thermodynamically than Ta_2O_5 surfaces with other terminations, such as H, O, and Ta terminations. Hereafter, we focus our discussion on the OH-terminated Ta_2O_5 (001) surface slab model using a four atomic layer. Generating the surface model from the λ-Ta_2O_5 bulk model, the formal oxidation states have +5 and −2 for Ta and O atoms, respectively. In the case of N-doping, charge neutrality could stabilize OH vacancies (OH_v) on the surface, thus including the combined model, $OH_v + N_{sub}$. Inhomogeneity of the N and defects distributions should be important and taken into consideration. To this end, the homogeneously doped nitrogen atoms, denoted as $N^h{}_{sub}$ and $N^h{}_{int}$, and the nitrogen atoms segregated on the surfaces, denoted as $N^s{}_{sub}$ and $N^s{}_{int}$, are modeled. The surface models considered are schematically summarized in Fig. 4.8.

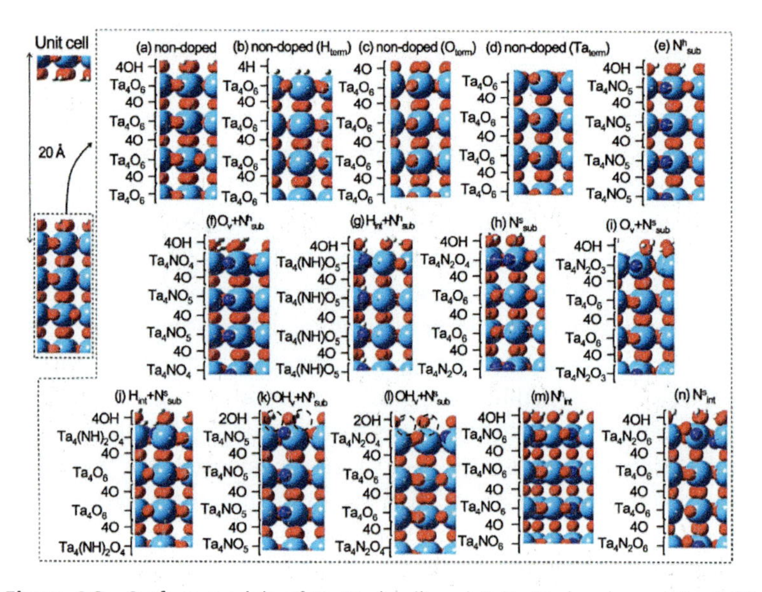

Figure 4.8 Surface models of Ta_2O_5 (a–d) and N-Ta_2O_5 (e–n) used for DFT calculations. The composition of each plane is shown, and the locations of OH vacancies (OH_v) are shown as dashed circles. Reprinted with permission from Ref. [51]. Copyright (2015) American Chemical Society.

In experiments, an ammonia treatment or sputtering in a N_2/air atmosphere is adopted to dope nitrogen atoms into Ta_2O_5 [66, 83]. Under the former conditions, N-Ta_2O_5 is likely formed through the following reaction:

$$Ta_2O_5 + xNH_3 \rightarrow N_xH_yTa_2O_{(5-z)} + zH_2O + [(3/2)x - (y/2) - z]H_2 \quad (4.2)$$

This reaction defines the reaction free energy as

$$\Delta G = G[N_xH_yTa_2O_{(5-z)}] + zG[H_2O] + [(3/2)x - (y/2) - z]G[H_2] - G[Ta_2O_5] - xG[NH_3], \quad (4.3)$$

where G indicates the Gibbs free energy, including temperature-dependent enthalpy and entropy contributions on the basis of the ideal gas model and the harmonic oscillator model [98]. Furthermore, the Gibbs free energy of the gaseous molecules, such as NH_3, H_2, and H_2O, depends on the partial pressure p with respect to p^0 as follows:

$$\Delta G = k_B T \ln p/p^0, \tag{4.4}$$

where p^0 was set as 0.1 MPa in the present study. Equations 4.3 and 4.4 connect to the environmental conditions and determine the phase diagram mapping of stable bulk and surface phases.

The DFT calculations were performed using the Vienna ab initio Simulation Package code [55, 56] implemented with the projector-augmented wave method [27]. Structural optimizations were performed using the GGA[a]-PBE[b] functional [76], and bandgaps, optical properties, and energy alignments were obtained by the HSE06 hybrid functional [41, 42, 75]. Vacuum layers with a thickness of 20 Å were placed between the slabs to avoid interactions among the repeated slabs, and dipole corrections [60, 72] were applied to eliminate the long-range electrostatic interactions among the repeated slabs. The redox levels of the isolated Ru complexes were evaluated by the DFT calculations using the Gaussian 09 package [34]. The HSE06 functional was used to compare to the CBM of the N-Ta_2O_5 models. In the calculations, the solution was modeled by a polarizable continuum model [110] with parameters for acetonitrile [83].

From the calculated equilibrium N-Ta_2O_5 bulk phases, (O_v + N_{sub}) and (N_{int}) become stable at H_2O-poor conditions and H_2-poor conditions, respectively, while the nondoped Ta_2O_5 bulk phase is stable in a wide free energy range. Given normal experimental conditions, (O_v + N_{sub}) is concluded to be the most plausible model for N-Ta_2O_5. The calculated bandgaps and lattice constants for bulk N-Ta_2O_5 models in comparison with the experiment are summarized in Table 4.1. The theoretical bandgap and lattice constants of the nondoped Ta_2O_5 agree well with the experimental ones. The bandgap is narrowed significantly when N_{sub} is introduced into Ta_2O_5. When O_v is introduced with N_{sub}, the bandgap is widened but still smaller than that of nondoped Ta_2O_5.

---

[a]Generalized gradient approximation
[b]Perdew–Burke–Ernzerhof

Table 4.1 Experimental and theoretical lattice constants (Å) and bandgaps (eV) of nondoped and N-doped Ta_2O_5 bulks

	Material	Lattice constants	Bandgap
DFT (HSE)	Nondoped Ta_2O_5	$a = 6.17, b = 40.07, c = 3.77$	3.7
	N-Ta_2O_5 (N_{sub})	$a = 6.17, b = 40.46, c = 3.77$	2.0
	N-Ta_2O_5 (O_v+N_{sub})	$a = 6.23, b = 40.51, c = 3.73$	3.2
Exp.	Nondoped Ta_2O_5	$a = 6.16, b = 40.30, c = 3.89$	3.8
	N-Ta_2O_5	$a = 6.20, b = 40.26, c = 3.88$	2.4

Note: Experimental data are taken from Refs. [83] and [66].

The theoretically obtained phase diagrams of nondoped and N-doped Ta_2O_5 surfaces with and without the surface segregation of the doped N atoms are summarized in Fig. 4.9. The symbols (a–m) correspond to the models shown in Fig. 4.8. Importantly, the OH-terminated nondoped Ta_2O_5 surface model (Fig. 4.8a) is stable in a wide free energy range, as expected. N-doped Ta_2O_5 surfaces are stable in the more negative regions of $\Delta G\,[H_2O]$ or $\Delta G\,[H_2]$, similar to the bulk systems; $OH_v + N^h{}_{sub}$ (k) and $OH_v + N^s{}_{sub}$ (l) are stable in the negative $\Delta G\,[H_2O]$ region; and $N^h{}_{int}$ (m) is stable in a very negative $\Delta G\,[H_2]$ region, where normal experiment hardly reaches.

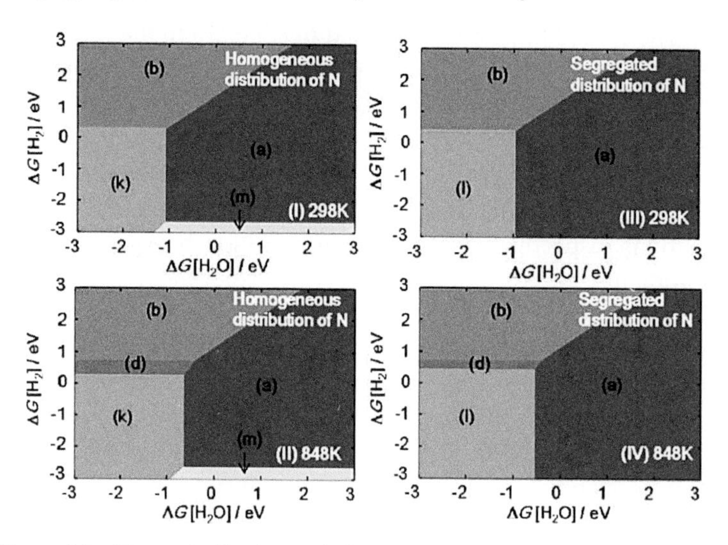

Figure 4.9 Theoretically obtained phase diagrams of nondoped and N-doped Ta_2O_5 surfaces with and without the surface segregation of the doped N atoms. Here $\Delta G[NH_3]$ was fixed at 0 eV. The symbols (a–m) indicate the models shown in Fig. 4.8. Reprinted with permission from Ref. [51]. Copyright (2015) American Chemical Society.

The calculated CBMs and VBMs of the surface models are summarized in Table 4.2. The calculated CBM and VBM of the nondoped Ta_2O_5 with OH surface adsorbates agree very well with the experimental results [30]. More importantly, the experimentally observed upward shifts in CBM and VBM by N-doping [66, 83] are reproduced well by the N-Ta_2O_5 surface models (f) and (k) (O_v+ N^h_{sub} and OH_v+ N^h_{sub}, respectively), where oxygen and hydroxyl defects, respectively, are introduced into the surface planes and nitrogen atoms are homogeneously doped. As shown in Table 4.3, the redox levels of Ru complexes are also reproduced well by the theory. Consequently, the energy alignments of Ta_2O_5/Ru complex are drastically changed by the N-doping and defect formations so that the electron injections from the Ta_2O_5 surface to the Ru complexes are indeed plausible as illustrated in Fig. 4.10.

Table 4.2 Theoretical and experimental valence band maximums ε_{VBM} and conduction band minimums ε_{CBM} of Ta_2O_5 and N-Ta_2O_5 surfaces shown in Fig. 4.8

	Surface model	ε_{VBM}	ε_{CBM}	$\Delta\varepsilon_{VBM}$	$\Delta\varepsilon_{CBM}$
Exp.	Nondoped Ta_2O_5	−7.9	−4.0	–	–
	N-Ta_2O_5	−5.7	−3.3	+2.2	+0.7
Cal.	Nondoped Ta_2O_5 (a)	−7.99	−4.25	–	–
	N^h_{sub} (e)	−7.89	−5.75	+0.10	−1.50
	O_v+ N^h_{sub} (f)	−5.51	−3.67	+2.48	+0.58
	H_{int} + N^h_{sub} (g)	−7.95	−5.41	+0.04	−1.16
	N^s_{sub} (h)	−7.65	−5.52	+0.34	−1.27
	O_v + N^s_{sub} (i)	−7.71	−4.35	+0.28	−0.10
	H_{int}+ N^s_{sub} (j)	−7.33	−4.01	+0.66	+0.24
	OH_v+ N^h_{sub} (k)	−5.74	−3.74	+2.25	+0.51
	OH_v + N^s_{sub} (l)	−7.36	−4.52	+0.63	−0.27
	N^h_{int} (m)	−7.17	−4.61	+0.82	−0.36
	N^s_{int} (n)	−7.49	−5.21	+0.82	−0.95

Note: $\Delta\varepsilon_{VBM}$ and $\Delta\varepsilon_{CBM}$ indicate the relative shift caused by the N-doping and defect formations. Experimental data are taken from Refs. [83] and [30].

Table 4.3 Experimental and theoretical redox levels (eV on a vacuum scale) of four kinds of Ru complexes, (r–u)

Molecule	Cal.	Exp.
$[Ru(dpbpy)(CO)_2(Cl)_2]$ (r)	-3.34 (−1.26)	−3.5 (−1.1)
$[Ru(bpy)_2(CO)_2]^{2+}$ (s)	-3.46 (−1.14)	−3.6 (−1.0)
$[Ru(dcbpy)(bpy)(CO)_2]^{2+}$ (t)	-3.89 (−0.71)	−3.7 (−0.9)
$[Ru(dcbpy)_2(CO)_2]^{2+}$ (u)	-3.96 (−0.64)	−3.8 (−0.8)

Note: Values in parenthesis are the redox potentials scaled in SHE. Experimental data are taken from Ref. [83].

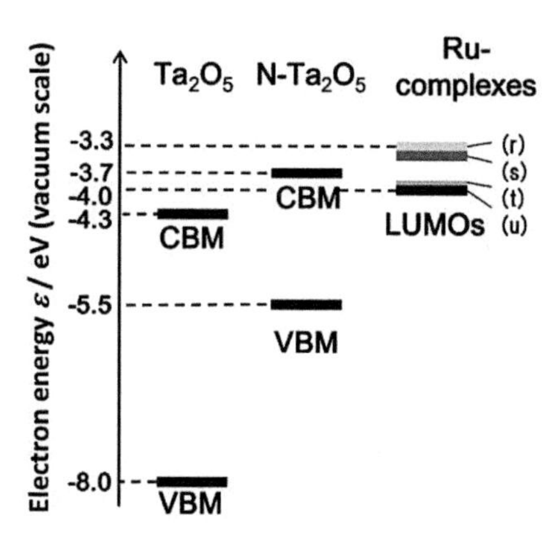

Figure 4.10 Theoretically obtained energy alignments among nondoped Ta_2O_5, N-Ta_2O_5 of the surface model (f), and Ru complexes (r–u) indicated in Table 4.3. Reprinted with permission from Ref. [51]. Copyright (2015) American Chemical Society.

It is worth understanding why the surface models (f) and (k) result in the upward shifts of CBM. Figure 4.11 shows the calculated charge distributions for (f) and (k) surface models compared to those for the other surface models as well as the nondoped surface model. The excess Bader charges are evaluated for each layer in these surface slab models. It is clear that only the two surface models (f) and (k) have prominent dipole moments normal to the surfaces at the first and second Ta_2O_3 planes. These dipole moments should raise the electrostatic potential at the surface of N-Ta_2O_5. To depict this effect, the local potentials across the surface are calculated and shown in Fig. 4.12. The upward shifts in the CBM and VBM are clearly shown as the result of the surface dipole formation in the surface models (f) and (k). The amount of the shift is about 1 eV.

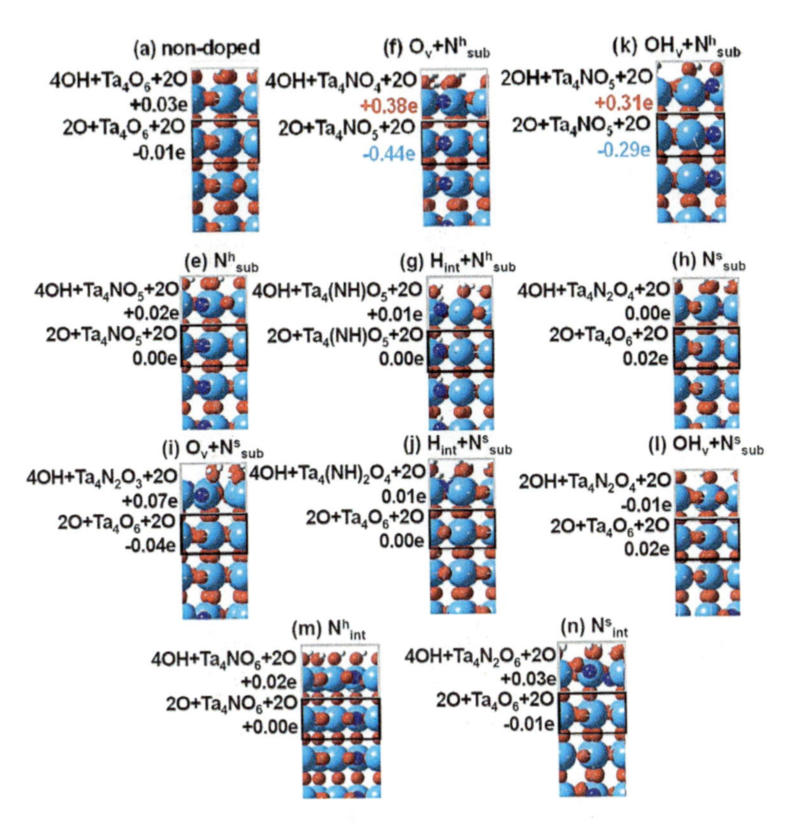

Figure 4.11 Calculated excess Bader charges for the models indicated in Fig. 4.8. Compared to the nondoped Ta_2O_5 surface (a), N-doped Ta_2O_5 surface models (f) and (k) clearly show charge redistribution forming the surface dipole. Reprinted with permission from Ref. [51]. Copyright (2015) American Chemical Society.

With the detailed atomistic analyses, the mechanism of the upward shifts in the CBM can be understood as follows. In the process of nitrification with NH_3, the substitutional nitrogen induces O_V or OH_V at the surface of the Ta_2O_5, stabilizing the system through charge compensation, as realized in the surface models (f) and (k). In such concurrent formations of N-doping and defects, positive charges are induced near the defect sites and negative charges are induced near the nitrogen atoms. Because the concentrations of the doped nitrogen atoms in the surface layers of the N-Ta_2O_5 models (f) and (k) are not high enough to fully accept the unpaired electrons from the Ta atoms next to the defect sites, an excess positive charge near the defect sites in the surface layers and an excess negative charge induced near the nitrogen in the inner layer form a dipole normal to the surface. In contrast, in other models, because the defects are not created, or concentrations of the doped nitrogen atoms in the surface layers are high enough, a dipole moment normal to the surface is not generated.

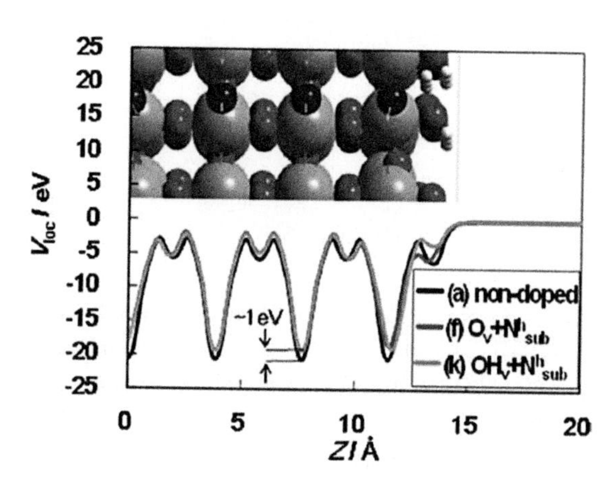

Figure 4.12 Local potentials in the nondoped (a) and N-doped (f, k) Ta_2O_5 surface models. Reprinted with permission from Ref. [51]. Copyright (2015) American Chemical Society.

As illustrated in Fig. 4.10, the band alignment cannot be determined only by considering each isolated and/or defect-free system. Subtle interface effects, including charge redistributions and inhomogeneous defect formations, significantly modify the band alignments. A synthesis process, such as the partial pressures of H_2 and H_2O, can control the N-doping level and defect formations, which are essential for realizing the desired energy alignments for accelerating the CO_2 reduction reactions.

4.3 Electron Transfer Dynamics in Semiconductor/Metal-Complex for CO_2 Reduction

In the previous section, we discussed the band alignment of a semiconductor/metal-complex photosynthesis system. The band alignment should be sufficient to have a driving force, ΔG, large enough to promote the CO_2 reduction according to the Marcus theory. In this context, the ET from nondoped Ta_2O_5 to Ru complex is not allowed, while N-doped Ta_2O_5 possesses a positive ΔG for ET to Ru complex [51, 84]. However, the ET rate is determined not only by ΔG but also by the detailed bonding nature at the interface typically characterized by the anchor group that attaches the Ru complex to the N-Ta_2O_5 surface. In fact, experiments have reported that a series of Ru complexes with different anchor groups, such as $[Ru(bpy)_2(CO)_2]^{2+}$, $[Ru(dcbpy)_2(CO)_2]^{2+}$, $[Ru(dcbpy)(bpy)(CO)_2]^{2+}$, and $[Ru(dpbpy)(bpy)(CO)_2]^{2+}$, where bpy, dcbpy, and dpbpy refer to 2,2′-bipyridyne, 4,4′-dicarboxy-2,2′-bipyridyne, and 4,4′-diphosphonate-2,2′-bipyridyne ligands, respectively, presented substantially different reduction efficiencies [84, 100]. One can explain that the efficiencies correlate with the number of COOH groups that act as anchors, which can naturally be attributed to enhancing the charge transfer. On the contrary, the last two systems, $[Ru(dcbpy)(bpy)(CO)_2]^{2+}$ and $[Ru(dpbpy)(bpy)(CO)_2]^{2+}$, differ only in the substituent groups in disubstituted bipyridyne ligands. The experimentally measured driving forces ΔG for the two systems are comparable to each other, and the apparent number of anchor groups per Ru complex is the same for the two systems. Yet, the observed CO_2 reduction rates are strikingly different; they are much larger for the catalyst attached with the PO_3H_2 anchor than with the COOH anchor. These experimental results clearly indicate that anchoring groups play a crucial role in ET. However, the fundamental reasons behind these experimental observations are unclear, motivating the need for explicit atomistic computational studies. In the next section, we employ atomistic NA-MD, to investigate the detailed time-domain properties of ET in the N–Ta_2O_5/Ru complex, and in particular, to examine the anchor effects on ET [6, 7].

4.3.1 Methodology

To carry out non-adiabatic ET simulations, we utilize the PYXAID code [4, 5] designed for modeling NA-MD in nanoscale systems. In the NA-MD simulations, we first solve the time-dependent Schrödinger equation (TD-SE),

$$i\hbar\frac{\partial\Psi}{\partial t} = H\Psi, \tag{4.5}$$

where $H = T_{\text{nucl}} + H_{\text{el}}$ is the total Hamiltonian of the system. The overall electron-nuclear wave function, Ψ, is represented as a superposition of electronic basis states, $|i\rangle$, weighted by the time-dependent coefficients, $c_i(t)$: $\Psi = \sum_i c_i(t)|i\rangle$. The TD-SE in the electronic basis states reduces to the semiclassical TD-SE:

$$i\hbar\frac{\partial c_i(t)}{\partial t} = \sum_j (H_{\text{vib}})_{ij} c_j(t), \tag{4.6}$$

where H_{vib} is the vibronic (electron-nuclear) Hamiltonian, defined in terms of state energies, E_i, and the non-adiabatic coupling (NAC) for pairs of states, $D_{ij} = \langle i|\partial/\partial t|j\rangle$, as in,

$$(H_{\text{vib}})_{ij} = E_i\delta_{ij} - i\hbar D_{ij}. \tag{4.7}$$

The energies and NAC needed to construct the vibronic Hamiltonian are computed in MD simulations. The solution of the semiclassical TD-SE does not account for detailed balance, allowing barrier-less access to high-energy states from any arbitrary initial state. The electron-vibrational energy relaxation taking place during the ET process is thus poorly treated, and thermodynamic equilibrium cannot be achieved in the long time limit. To remedy this problem, we use the obtained coefficients, $c_i(t)$, only as auxiliary quantities for the fewest-switches surface-hopping (FSSH) NA-MD algorithm [3, 113]. In the FSSH, a swarm of trajectories is propagated, and each trajectory may diffuse in the Hilbert space of electronic basis states included in the model. The hopping probabilities depend on the semiclassical TD-SE amplitudes, $c_i(t)$, as well as on the NAC between source and target states, $D_{\text{source,target}}$. For a hopping transition to a new state, the nuclear velocities are rescaled to conserve the total electron-nuclear, quantum-classical energy. If such rescaling is not possible, the transition is rejected.

In this way, FSSH is designed to generate different trajectories for each distinct electronic state and properly describe the quantum-classical equilibrium and energy exchange. Averaging all stochastic realizations of the FSSH trajectories and all initial conditions (different starting configurations) yields SH populations of different states,

$$P_i(t) = \left\langle \frac{N_i(t)}{N_{\text{tot}}} \right\rangle, \tag{4.8}$$

where <> denotes thermal averaging.

The calculation models used are depicted in Fig. 4.13. The electroneutral Ru(di-X-bpy) $(CO)_2Cl_2$ complexes, with X = COOH and PO_3H_3, are attached to N-Ta_2O_5. The Ru complexes are bound via two anchor groups, COOH and PO_3H_3. The hydroxyl groups of the N-Ta_2O_5 surface that bind the anchors are chosen to minimize distortion of the adsorbate. The 8.9% substitutional nitrogen doping for oxygen sites is introduced in the Ta_2O_5 cluster model. Because the present molecular models are rather too large to perform NA-MD simulations, approximated calculations for MD and electronic structures are employed as explained below.

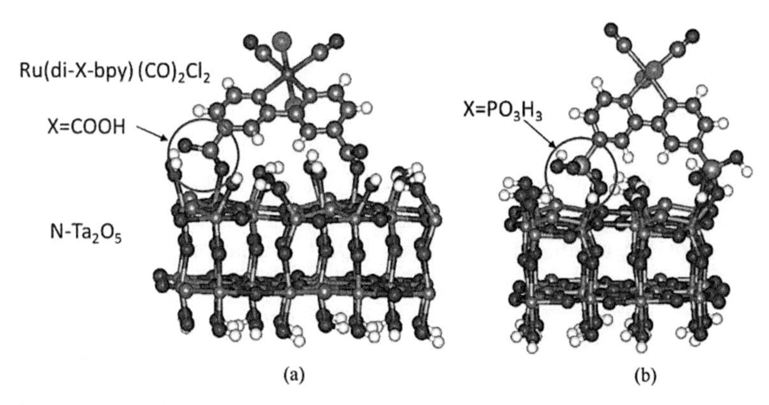

Figure 4.13 Calculation models of Ru complexes attached to N-Ta_2O_5 clusters via COOH (a) and PO_3H_3 (b) anchors (red circles).

The MD trajectories are obtained using rigid-body molecular dynamics (RB-MD) [2]. In the RB-MD, one constrains arbitrary degrees of freedom by grouping specific atoms in united fragments. The nuclear dynamics of the system are then described in terms of translations and rotations of those rigid fragments. Currently, we focus solely on the role of the anchor and, therefore, treat the entire Ru complex and the cluster representing the N-Ta_2O_5 surface as two distinct rigid fragments. However, the anchor groups are fully flexible at the atomistic level. Hence, the Ru complex/N-Ta_2O_5 system with the COOH is composed of 8 fragments, 6 of which are the atoms of two –COO groups, and the PO_3H_2 system is composed of 12 fragments, 10 of which are the atoms of two –PO_3H groups. The geometry of the N-Ta_2O_5 cluster and Ru complexes is taken from DFT calculations performed using the hybrid functional of HSE06 [41, 42]. The interactions between all rigid bodies are computed on the basis of the explicit-atom universal force field (UFF) [81]. While the UFF would predict the structures of the Ru complex and N-Ta_2O_5 to be different from the DFT-optimized structures, the RB approach helps us to avoid any spurious effects due to structural differences. At the same time, the all-atomic description of the anchor groups allows us to focus on the ET dynamics through the anchors and to reveal the factors that stem only from their nature.

The electronic structure calculations are performed using an implementation of the semiempirical extended Hückel theory (EHT) [7, 46]. EHT utilizes a minimal basis set of atomic s, p, and d orbitals. The use of EHT is essential for efficient computations with the systems of interest; the N-Ta_2O_5 cluster contains many Ta atoms, each contributing many electrons and orbitals. Thus, the electronic structure calculations quickly become expensive even for small clusters. Keeping in mind that such calculations must be done for many configurations along the MD trajectory and that NAC matrices must be computed as well, the computational expenses become unfeasible for standard DFT methods. At the same time, the tight-binding theories, including EHT, are well known for capturing essential physics [8, 47], which makes them reliable and transparent for qualitative analysis of chemical phenomena. EHT usually lacks explicit electrostatic effects. Therefore, in the present work, effective parameterization is employed to reproduce the DFT or experimental band alignment between N-Ta_2O_5 and Ru complexes with different anchor groups, as described in Section 4.2.2. EHT calculations reproduce a monotonic increase of the energy levels in a series COOH $<$ PO_3H_2 $<$ OH, which is also observed for the DFT values. In addition, the CBM energy of the N-Ta_2O_5 cluster is higher than the

LUMO (and also LUMO+1, for COOH) energy levels of isolated Ru complexes. Proper energy-level alignment, reproduced by EHT, is essential for making interfacial ET possible.

The computed projected density of state (pDOS) for each system is shown in Fig. 4.14. We project the DOS on the sets of atoms that constitute the Ru complex together with the anchor and the N-Ta_2O_5 substrate. One can observe that there are two unoccupied acceptor levels (LUMO and LUMO+1) below the N-Ta_2O_5 CBM in the COOH system (Fig. 4.14a) and there is only one such level (LUMO) in the PO_3H_2 system (Fig. 4.14b). The energy-level alignment in the bound systems is consistent with that in the unbound components, suggesting that the present parametrization is suitable for modeling the ET dynamics in the bound system.

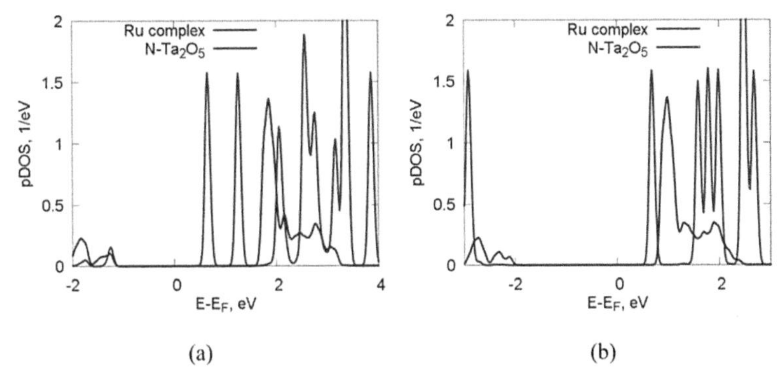

Figure 4.14 Computed projected density of states (pDOS) for COOH (a) and PO_3H_2 (b) systems. The pDOS of the N-Ta_2O_5 is decreased by a factor of 50 for clarity. Reprinted with permission from Ref. [6]. Copyright (2015) American Chemical Society.

4.3.2 Results and Discussion

After the system is thermalized by running an MD simulation for 25 ps in the NVT ensemble with the target temperature of 278 K, the temperature is maintained for a 5 ps production run using the Nose–Hoover chain thermostat [48, 74] with a length of 3. The trajectories obtained are used to compute the coordinate-dependent vibronic Hamiltonians needed for NA-MD simulations. NA-MD simulation provides the total survival probability for a system to remain in any of the donor states after a time delay, $P_{\mathrm{donor}}(t)$. The computed probability is averaged over initial excitations to all considered donor states.

The results of NA-MD calculations are presented in Fig. 4.15, showing the total population transfer dynamics of a donor in N-Ta_2O_5. The initial relaxation shows weak Gaussian character, which is typical of quantum processes. The Gaussian component decays when quantum dynamics develops to encompass multiple states. The rest of the relaxation follows the exponential-decay law. We determine the ET time to be 530 fs for the COOH system and 5.50 ps for the PO_3H_2 system. The ET occurs on a much shorter time scale in the COOH system.

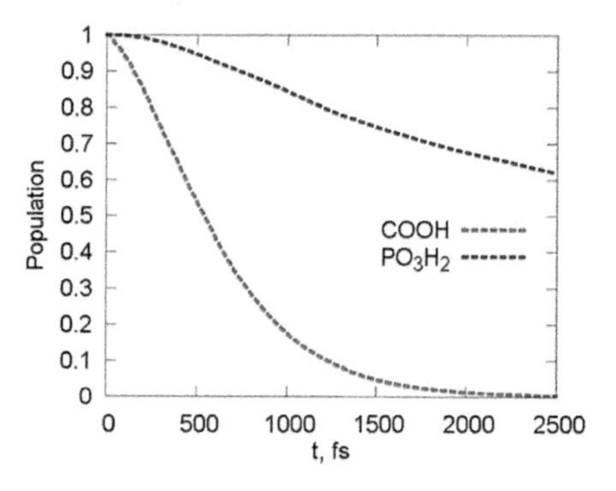

Figure 4.15 Decays of the total population of donor states in N-Ta_2O_5 for COOH and PO_3H_2 anchors. Reprinted with permission from Ref. [6]. Copyright (2015) American Chemical Society.

In the FSSH description, population transfer is governed by NAC and the energy gap between donor and acceptor states. The energy gap magnitudes are very similar in the COOH and PO_3H_2 systems, that is, 0.78 eV and 0.72 eV, respectively, evaluated by the DFT. We, therefore, assume that the major reason for very distinct efficiencies of the ET in these systems is the disparity in the NAC magnitudes. Figure 4.16 shows the time-averaged NAC magnitudes for all pairs of orbitals as in the two-dimensional maps, which are computed by

$$\left\langle \mathrm{Im}(H_{\mathrm{vib},ij}) \right\rangle = \frac{\hbar}{T} \sum_{t=0}^{T-1} \left| D_{ij}(t) \right|, \tag{4.9}$$

where T is the number of MD steps for the time averaging. The COOH system (Fig. 4.16a) exhibits strong coupling of the LUMO+1 and LUMO+2 levels, with a magnitude of approximately 300 meV. This large coupling realizes an efficient population transfer from the lowest donor (LUMO+2) state to the upper acceptor (LUMO+1) level. One can also observe that there is a secondary channel for exciton relaxation via direct coupling between the LUMO+2 state and the LUMO+4–LUMO+6 states. This channel promotes faster electron relaxation to the lowest donor state (LUMO+2) and therefore facilitates interfacial ET. The situation is different for the PO_3H_2 system (Fig. 4.16b). Here, we observe only a single ET channel, between the LUMO acceptor state and the LUMO+1 donor state. Most importantly, the average magnitudes of the couplings are almost an order of magnitude smaller than those in the COOH system. The coupling between LUMO and LUMO+1 is on the order of only 20–30 meV.

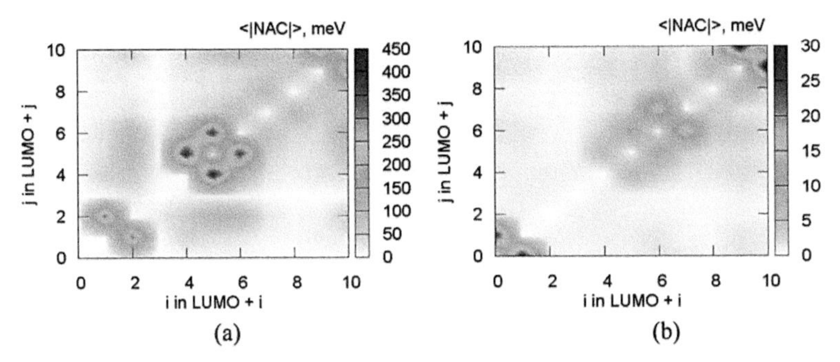

Figure 4.16 Visualization of the non-adiabatic coupling (NAC) between unoccupied orbitals of N-Ta_2O_5 and the Ru complex attached via either COOH (a) or PO_3H_2 (b) anchors. Reprinted with permission from Ref. [6]. Copyright (2015) American Chemical Society.

It is important to explain the fundamental reasons leading to large NACs in the COOH system compared to those in PO_3H_2. Following its definition, NAC characterizes the wave function sensitivity to nuclear motion. One may expect larger couplings for the PO_3H_2 anchor because the group is more flexible, due to sp^3 hybridization of the central P atom, whereas COOH is more rigid, due to sp^2 hybridization of the C atom. However, nuclear motion is not the determining factor. The wave function isosurfaces of the acceptor and lowest donor states in the COOH and PO_3H_2 system are shown in Fig. 4.17. One can observe a distinct difference in the localization of the orbitals, consistent with an intuitive assumption about the nature of donor and acceptor states. As seen in Fig. 4.17 along with Fig. 4.14, the pDOS of the COOH system shows that both LUMO and LUMO+1 contain notable contributions from the central C atom of the COOH anchor because the anchor is conjugated to the ligand π-system. Thus, the motion of the COOH group strongly affects the acceptor wave functions and the donor–acceptor NAC. On the contrary, the pDOS of the PO_3H_2 system shows little contribution of anchor orbitals to the acceptor state (LUMO). Therefore, motions of the PO_3H_2 group do not affect the acceptor wave function notably, leading to a smaller NAC.

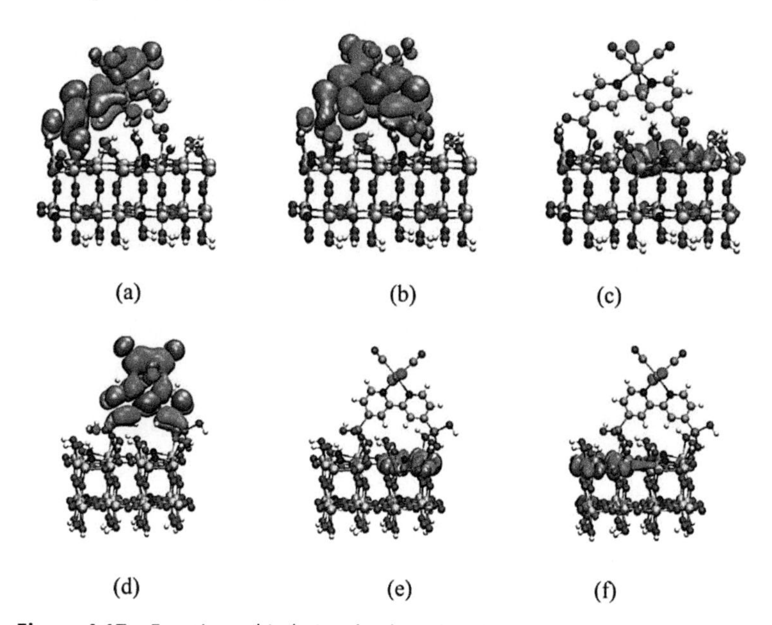

Figure 4.17 Frontier orbitals involved in the charge transfer: LUMO (a, d), LUMO+1 (b, e), and LUMO+2 (c, f). The top row and the bottom row correspond to orbitals of the Ru complex with COOH and PO_3H_2 anchors, respectively. The isosurface magnitude is 0.01 bohr^{-1}. Reprinted with permission from Ref. [6]. Copyright (2015) American Chemical Society.

The calculations predict the COOH anchor to be more efficient than PO_3H_2, whereas the experimental efficiency is larger for the PO_3H_2 system [100]. Such a discrepancy can be explained by differences in the computational modeling and the experiment. The most important structural difference is the use of electroneutral

$Ru(di\text{-}X\text{-}bpy)(CO)_2Cl_2$ complexes in the computational models whereas the experiment uses $[Ru(dcbpy)\ (bpy)\ (CO)_2]^{2+}$ and $[Ru(dpbpy)\ (bpy)\ (CO)_2]^{2+}$. Such a structural difference can result in different electronic structures and thus influence the magnitude of the NAC. It is also known from other studies that molecules attached via the COOH anchor are very sensitive to conditions such as pH and solvent [22, 64]. There is another factor included in experiments, which makes it easier for the system with the COOH anchor to detach from the substrate while no such stabilization problem is observed for the PO_3H_2 anchor. The PO_3H_2 system is more stable [22, 100, 111], leading to a larger surface coverage and, hence, higher efficiency. There are some other examples regarding the COOH and PO_3H_2 anchors. An electron injection from chalcogenorhodamine dyes anchored to TiO_2 is twice as efficient via carboxylate linkages as via phosphonate linkages [68]. Alternatively, $(AgIn)_{0.22}Zn_{1.56}S_2$ or Ni (0.2 mol%)-doped ZnS combined with a neutral Ru complex incorporating the phosphonate ligand $[Ru(4,4'\text{diphosphonate-}2,2'\text{-bipyridine})(CO)_2Cl_2]$ shows higher CO_2 photoreduction activity than the carboxyl ligand $[Ru(4,4'\text{-dicarboxy-}2,2'\text{-bipyridine})(CO)_2Cl_2]$ [104]. Thus, as we have seen in this section, the ET rate is quite sensitive to the selection of Ru complexes and semiconductors because the magnitude of NAC is attributed to the detailed bonding nature at the interface.

4.4 Summary and Future Scope

In this chapter, we discussed some of the fundamental factors to understand and design artificial photosynthesis systems. While the energy alignments are crucial for ET and photosynthesis efficiency, they are highly sensitive to surface modifications and defects introduced into bulks and interfaces. As seen in Section 4.2, inhomogeneous defects and charge redistributions give rise to significant modification of the electrostatic potential. We showed these effects in metal-nanoparticles loaded on TiO_2 and in defect formations in N-doped Ta_2O_5, generating a surface dipole that is attributed to a shift of the CBM and VBM of the semiconductor. We also presented in Section 4.3 that the ET rate is directly evaluated by employing atomistic NA-MD. The results successfully yielded the microscopic origin of the ET efficiency in relation to the kind of anchors used for N-Ta_2O_5/Ru-complex interfaces. We emphasized that NAC is an important contribution to the ET.

Through the discussions in this chapter, we arrive at an important conclusion. Experimentally observed photocatalytic efficiencies include too many factors to identify the critical few factors needed for improvement. On the other hand, atomistic models, because of computational limitations, can focus only on a few of the factors under consideration. The key to make efficient progress is thus an interplay between theory and experiment, converging to the most fundamental and controllable factors for further improvement. To this end, a simulation methodology that can handle realistic larger-scale modeling with a reasonable computational resource needs to be developed. In particular, because the photosynthesis process involves photoinduced excitations, a theoretical approach going beyond the ground states, such as many-body perturbation theory, should participate in this challenge, along with its efficient computational implementation.

References

1. Ai, X., Anderson, N. A., Guo, J. C., and Lian, T. Q. (2005). Electron injection dynamics of Ru polypyridyl complexes on SnO_2 nanocrystalline thin films, *J. Phys. Chem. B*, **109**, pp. 7088–7094.
2. Akimov, A. V. and Kolomeisky, A. B. (2011). Recursive Taylor series expansion method for rigid-body molecular dynamics, *J. Chem. Theory Comput.*, **7**, pp. 3062–3071.
3. Akimov, A. V., Neukirch, A. J., and Prezhdo, O. V. (2013). Theoretical insights into photoinduced charge transfer and catalysis at oxide interfaces, *Chem. Rev.*, **113**, pp. 4496–4565.
4. Akimov, A. V. and Prezhdo, O. V. (2013). The PYXAID program for non-adiabatic molecular dynamics in condensed matter systems, *J. Chem. Theory Comput.*, **9**, pp. 4959–4972.
5. Akimov, A. V. and Prezhdo, O. V. (2014). Advanced capabilities of the PYXAID program: integration schemes, decoherence effects, multiexcitonic states, and field-matter interaction, *J. Chem. Theory Comput.*, **10**, pp. 789–804.
6. Akimov, A. V., Asahi, R., Jinnouchi, R., and Prezhdo, O. V. (2015). What makes the photocatalytic CO_2 reduction on N-doped Ta_2O_5 efficient:

insights from nonadiabatic molecular dynamics, *J. Am. Chem. Soc.*, **137**, pp. 11517–11525.

7. Akimov, A. V., Jinnouchi, R., Shirai, S., Asahi, R., and Prezhdo, O. V. (2015). Theoretical insights into the impact of Ru catalyst anchors on the efficiency of photocatalytic CO_2 reduction on Ta_2O_5, *J. Phys. Chem. B*, **119**, pp. 7186–7197.
8. Akimov, A. V. and Prezhdo, O. V. (2015). Analysis of self-consistent extended Hückel theory (SC-EHT): a new look at the old method, *J. Math. Chem.*, **53**, pp. 528–550.
9. Ambrosio, F., Martsinovich, N., and Troisi, A. (2012). What is the best anchoring group for a dye in a dye-sensitized solar cell?, *J. Phys. Chem. Lett.*, **3**, pp. 1531–1535.
10. Anderson, N. A., Ai, X., Chen, D. T., Mohler, D. L., and Lian, T. Q. (2003). Bridge-assisted ultrafast interfacial electron transfer to nanocrystalline SnO_2 thin films, *J. Phys. Chem. B*, **107**, pp. 14231–14239.
11. Arai, T., Sato, S., Uemura, K., Morikawa, T., Kajino, T., and Motohiro, T. (2010). Photoelectrochemical reduction of CO_2 in water under visible-light irradiation by a p-type InP photocathode modified with an electropolymerized ruthenium complex, *Chem. Commun.*, **46**, pp. 6944–6946.
12. Arai, T., Tajima, S., Sato, S., Uemura, K., Morikawa, T., and Kajino, T. (2011). Selective CO_2 conversion to formate in water using a CZTS photocathode modified with a ruthenium complex polymer, *Chem. Commun.*, **47**, pp. 12664–12666.
13. Arai, T., Sato, S., Kajino, T., and Morikawa, T. (2013). Solar CO_2 reduction using H_2O by a semiconductor/metal-complex hybrid photocatalyst: enhanced efficiency and demonstration of a wireless system using $SrTiO_3$ photoanodes, *Energy Environ. Sci.*, **6**, pp. 1274–1282.
14. Arai, T., Sato, S., and Morikawa, T. (2015). A monolithic device for CO_2 photoreduction to generate liquid organic substances in a single-compartment reactor, *Energy Environ. Sci.*, **8**, pp. 1998–2002.
15. Arai, T., Sato, S., and Morikawa, T. (2016). Aminoalkylsilane-modified silver cathodes for electrochemical CO_2 reduction, *Chem. Lett.*, **45**, pp. 1362–1364.
16. Arai, T., Sato, S., Sekizawa, K., Suzuki, T. M., and Morikawa, T. (2019). Solar-driven CO_2 to CO reduction utilizing H_2O as an electron donor by earth-abundant Mn-bipyridine complex and Ni-modified Fe-oxyhydroxide catalysts activated in a single-compartment reactor, *Chem. Commun.*, **55**, pp. 237–240.

17. Aryasetiawan, F. and Gunnarsson, O. (1998). The GW method, *Rep. Prog. Phys.*, **61**, pp. 237–312.

18. Asahi, R., Morikawa, T., Ohwaki, T., Aoki, K., and Taga, Y. (2001). Visible-light photocatalysis in nitrogen-doped titanium oxides, *Science*, **293**, pp. 269–271.

19. Asahi, R. and Morikawa, T. (2007). Nitrogen complex species and its chemical nature in TiO_2 for visible-light sensitized photocatalysis, *Chem. Phys.*, **339**, pp. 57–63.

20. Asahi, R., Morikawa, T., Irie, H., and Ohwaki, T. (2014). Nitrogen-doped titanium dioxide as visible-light-sensitive photocatalyst: designs, developments, and prospects, *Chem. Rev.*, **114**, pp. 9824–9852.

21. Asbury, J. B., Hao, E., Wang, Y. Q., Ghosh, H. N., and Lian, T. Q. (2001). Ultrafast electron transfer dynamics from molecular adsorbates to semiconductor nanocrystalline thin films, *J. Phys. Chem. B*, **105**, pp. 4545–4557.

22. Bae, E. Y., Choi, W. Y., Park, J. W., Shin, H. S., Kim, S. B., and Lee, J. S. (2004). Effects of surface anchoring groups (carboxylate vs phosphonate) in ruthenium-complex-sensitized TiO_2 on visible light reactivity in aqueous suspensions, *J. Phys. Chem. B*, **108**, pp. 14093–14101.

23. Bard, A. J., Parsons, R., and Jordan, J. (1985). *Standard Potentials in Aqueous Solution* (Marcel Dekker, New York).

24. Batzill, M., Morales, E. H., and Diebold, U. (2006). Influence of nitrogen doping on the defect formation and surface properties of TiO_2 rutile and anatase, *Phys. Rev. Lett.*, **96**, pp. 026103.

25. Bauer, C., Teuscher, J., Pelet, S., Wenger, B., Bonhote, P., Nazeeruddin, M. K., Zakeeruddin, S. M., Comte, P., Grätzel, M., and Moser, J.-E. (2010). Ultrafast charge transfer through p-oligo(phenylene) bridges: effect of nonequilibrium vibrations, *Curr. Sci.*, **99**, pp. 343–352.

26. Biller, A., Tamblyn, I., Neaton, J. B., and Kronik, L. (2011). Electronic level alignment at a metal-molecule interface from a short-range hybrid functional, *J. Chem. Phys.*, **135**, pp. 164706.

27. Blöchl, P. E. (1994). Projector augmented-wave method, *Phys. Rev. B*, **50**, pp. 17953–17979.

28. Chardon-Noblat, S., Deronzier, A., Ziessel, R., and Zsoldos, D. (1998). Electroreduction of CO, catalyzed by polymeric $[Ru(bpy)(CO)_2]_n$ films in aqueous media: parameters influencing the reaction selectivity, *J. Electroanal. Chem.*, **444**, pp. 253–260.

29. Chen, W. and Pasquarello, A. (2012). Band-edge levels in semiconductors and insulators: hybrid density functional theory versus many-body perturbation theory, *Phys. Rev. B*, **86**, pp. 035134.

30. Chun, W. J., Ishikawa, A., Fujisawa, H., Takata, T., Kondo, J. N., Hara, M., Kawai, M., Matsumoto, Y., and Domen, K. (2003). Conduction and valence band positions of Ta_2O_5, TaON, and Ta_3N_5 by UPS and electrochemical methods, *J. Phys. Chem. B*, **107**, pp. 1798–1803.

31. Drukker, K. (1999). Basics of surface hopping in mixed quantum/classical simulations, *J. Comput. Phys.*, **153**, pp. 225–272.

32. Duncan, W. R. and Prezhdo, O. V. (2007). Theoretical studies of photoinduced electron transfer in dye-sensitized TiO_2, *Annu. Rev. Phys. Chem.*, **58**, pp. 143–184.

33. Finazzi, E., Di Valentin, C., Pacchioni, G., and Selloni, A. (2008). Excess electron states in reduced bulk anatase TiO_2: comparison of standard GGA, GGA plus U, and hybrid DFT calculations, *J. Chem. Phys.*, **129**, pp. 154113.

34. Frisch, M. J., Trucks, G. W., Schlegel, H. B., Scuseria, G. E., Robb, M. A., Cheeseman, J. R., Scalmani, G., Barone, V., Petersson, G. A., Nakatsuji, H., Li, X., Caricato, M., Marenich, A., Bloino, J., Janesko, B. G., Gomperts, R., Mennucci, B., Hratchian, H. P., Ortiz, J. V., Izmaylov, A. F., Sonnenberg, J. L., Williams-Young, D., Ding, F., Lipparini, F., Egidi, F., Goings, J., Peng, B., Petrone, A., Henderson, T., Ranasinghe, D., Zakrzewski, V. G., Gao, J., Rega, N., Zheng, G., Liang, W., Hada, M., Ehara, M., Toyota, K., Fukuda, R., Hasegawa, J., Ishida, M., Nakajima, T., Honda, Y., Kitao, O., Nakai, H., Vreven, T., Throssell, K., J. A. Montgomery, J., Peralta, J. E., Ogliaro, F., Bearpark, M., Heyd, J. J., Brothers, E., Kudin, K. N., Staroverov, V. N., Keith, T., Kobayashi, R., Normand, J., Raghavachari, K., Rendell, A., Burant, J. C., Iyengar, S. S., Tomasi, J., Cossi, M., Millam, J. M., Klene, M., Adamo, C., Cammi, R., Ochterski, J. W., Martin, R. L., Morokuma, K., Farkas, O., Foresman, J. B., and Fox, D. J., *Gaussian 09 Rev. D.01*. 2009, Gaussian, Inc., Wallingford CT, 2016: Wallingford, CT.

35. Furube, A., Asahi, T., Masuhara, H., Yamashita, H., and Anpo, M. (1999). Charge carrier dynamics of standard TiO_2 catalysts revealed by femtosecond diffuse reflectance spectroscopy, *J. Phys. Chem. B*, **103**, pp. 3120–3127.

36. Guo, J. C., Stockwell, D., Ai, X., She, C. X., Anderson, N. A., and Lian, T. Q. (2006). Electron-transfer dynamics from Ru polypyridyl complexes to In_2O_3 nanocrystalline thin films, *J. Phys. Chem. B*, **110**, pp. 5238–5244.

37. Hack, M. D. and Truhlar, D. G. (2000). Nonadiabatic trajectories at an exhibition, *J. Phys. Chem. A*, **104**, pp. 7917–7926.

38. Hack, M. D., Wensmann, A. M., Truhlar, D. G., Ben-Nun, M., and Martinez, T. J. (2001). Comparison of full multiple spawning, trajectory surface hopping, and converged quantum mechanics for electronically nonadiabatic dynamics, *J. Chem. Phys.*, **115**, pp. 1172–1186.

39. Hammarstrom, L. and Styring, S. (2011). Proton-coupled electron transfer of tyrosines in Photosystem II and model systems for artificial photosynthesis: the role of a redox-active link between catalyst and photosensitizer, *Energy Environ. Sci.*, **4**, pp. 2379–2388.

40. Hawecker, J., Lehn, J. M., and Ziessel, R. (1986). Photochemical and electrochemical reduction of carbon dioxide to carbon monoxide mediated by (2,2-bipyridine) tricarbonylchlororhenium(I) and related complexes as homogeneous catalysts, *Helv. Chim. Acta*, **69**, pp. 1990–2012.

41. Heyd, J., Scuseria, G. E., and Ernzerhof, M. (2003). Hybrid functionals based on a screened Coulomb potential, *J. Chem. Phys.*, **118**, pp. 8207–8215.

42. Heyd, J., Scuseria, G. E., and Ernzerhof, M. (2006). Hybrid functionals based on a screened Coulomb potential (vol 118, pg 8207, 2003), *J. Chem. Phys.*, **124**, pp. 219906.

43. Higashimoto, S., Tanihata, W., Nakagawa, Y., Azuma, M., Ohue, H., and Sakata, Y. (2008). Effective photocatalytic decomposition of VOC under visible-light irradiation on N-doped TiO_2 modified by vanadium species, *Appl. Catal., A*, **340**, pp. 98–104.

44. Hinuma, Y., Gruneis, A., Kresse, G., and Oba, F. (2014). Band alignment of semiconductors from density-functional theory and many-body perturbation theory, *Phys. Rev. B*, **90**, pp. 155405.

45. Hoffmann, M. R., Martin, S. T., Choi, W. Y., and Bahnemann, D. W. (1995). Environmental applications of semiconductor photocatalysis. *Chem. Rev.*, **95**, pp. 69–96.

46. Hoffmann, R. (1963). An extended Hückel theory. I. Hydrocarbons, *J. Chem. Phys.*, **39**, pp. 1397–1412.

47. Hoffmann, R. (1988). A chemical and theoretical way to look at bonding on surfaces, *Rev. Mod. Phys.*, **60**, pp. 601–628.

48. Hoover, W. G. (2007). Nose-Hoover nonequilibrium dynamics and statistical mechanics, *Mol. Simul.*, **33**, pp. 13–19.

49. Hori, H., Johnson, F. P. A., Koike, K., Ishitani, O., and Ibusuki, T. (1996). Efficient photocatalytic CO_2 reduction using $[Re(bpy)(CO)_3\{P(OEt)_3\}]^+$, *J. Photochem. Photobiol., A*, **96**, pp. 171–174.

50. Ishida, H., Tanaka, K., and Tanaka, T. (1987). Electrochemical CO_2 reduction catalyzed by ruthenium complexes $[Ru(bpy)_2(CO)_2]^{2+}$ and $[Ru(bpy)_2(CO)Cl]^{+}$. Effect of pH on the formation of CO and $HCOO^-$, *Organometallics*, **6**, pp. 181–186.

51. Jinnouchi, R., Akimov, A. V., Shirai, S., Asahi, R., and Prezhdo, O. V. (2015). Upward shift in conduction band of Ta_2O_5 due to surface dipoles induced by N-doping, *J. Phys. Chem. C*, **119**, pp. 26925–26936.

52. Jones, D. R. and Troisi, A. (2010). A method to rapidly predict the charge injection rate in dye sensitized solar cells, *Phys. Chem. Chem. Phys.*, **12**, pp. 4625–4634.

53. Katoh, R., Furube, A., Yamanaka, K., and Morikawa, T. (2010). Charge separation and trapping in N-doped TiO_2 photocatalysts: a time-resolved microwave conductivity study, *J. Phys. Chem. Lett.*, **1**, pp. 3261–3265.

54. Kharche, N., Muckerman, J. T., and Hybertsen, M. S. (2014). First-principles approach to calculating energy level alignment at aqueous semiconductor interfaces, *Phys. Rev. Lett.*, **113**, pp. 176802.

55. Kresse, G. and Furthmüller, J. (1996). Efficient iterative schemes for ab initio total-energy calculations using a plane-wave basis set, *Phys. Rev. B*, **54**, pp. 11169–11185.

56. Kresse, G. and Furthmüller, J. (1996). Efficiency of ab-initio total energy calculations for metals and semiconductors using a plane-wave basis set, *Comput. Mater. Sci.*, **6**, pp. 15–50.

57. Lee, S.-H., Kim, J., Kim, S.-J., Kim, S., and Park, G.-S. (2013). Hidden structural order in orthorhombic Ta_2O_5, *Phys. Rev. Lett.*, **110**, pp. 235502.

58. Leung, K., Nielsen, I. M. B., Sai, N., Medforth, C., and Shelnutt, J. A. (2010). Cobalt-porphyrin catalyzed electrochemical reduction of carbon dioxide in water. 2. Mechanism from first principles, *J. Phys. Chem. A*, **114**, pp. 10174–10184.

59. Li, J., Nilsing, M., Kondov, I., Wang, H., Persson, P., Lunell, S., and Thoss, M. (2008). Dynamical simulation of photoinduced electron transfer reactions in dye-semiconductor systems with different anchor groups, *J. Phys. Chem. C*, **112**, pp. 12326–12333.

60. Makov, G. and Payne, M. C. (1995). Periodic boundary conditions in ab initio calculations *Phys. Rev. B*, **51**, pp. 4014–4022.

61. Marcus, R. A. and Sutin, N. (1985). Electron transfers in chemistry and biology, *Biochim. Biophys. Acta*, **811**, pp. 265–322.

62. Marsman, M., Paier, J., Stroppa, A., and Kresse, G. (2008). Hybrid functionals applied to extended systems, *J. Phys.: Condens. Matter*, **20**.

63. Martsinovich, N. and Troisi, A. (2011). High-throughput computational screening of chromophores for dye-sensitized solar cells, *J. Phys. Chem. C*, **115**, pp. 11781–11792.

64. McNamara, W. R., Milot, R. L., Song, H.-E., Snoeberger, R. C., III, Batista, V. S., Schmuttenmaer, C. A., Brudvig, G. W., and Crabtree, R. H. (2010). Water-stable, hydroxamate anchors for functionalization of TiO_2 surfaces with ultrafast interfacial electron transfer, *Energy Environ. Sci.*, **3**, pp. 917–923.

65. Mori, T., Kozawa, T., Ohwaki, T., Taga, Y., Nagai, S., Yamasaki, S., Asami, S., Shibata, N., and Koike, M. (1996). Schottky barriers and contact resistances on p-type GaN, *Appl. Phys. Lett.*, **69**, pp. 3537–3539.

66. Morikawa, T., Saeki, S., Suzuki, T., Kajino, T., and Motohiro, T. (2010). Dual functional modification by N doping of Ta_2O_5: p-type conduction in visible-light-activated N-doped Ta_2O_5, *Appl. Phys. Lett.*, **96**, pp. 142111.

67. Morris, A. J., Meyer, G. J., and Fujita, E. (2009). Molecular approaches to the photocatalytic reduction of carbon dioxide for solar fuels, *Acc. Chem. Res.*, **42**, pp. 1983–1994.

68. Mulhern, K. R., Detty, M. R., and Watson, D. F. (2013). Effects of surface-anchoring mode and aggregation state on electron injection from chalcogenorhodamine dyes to titanium dioxide, *J. Photochem. Photobiol., A*, **264**, pp. 18–25.

69. Mulliken, R. S. (1955). Electronic population analysis on LCAO–MO molecular wave functions. I, *J. Chem. Phys.*, **23**, pp. 1833–1840.

70. Murugesan, S., Huda, M. N., Yan, Y. F., Al-Jassim, M. M., and Subramanian, V. (2010). Band-engineered bismuth titanate pyrochlores for visible light photocatalysis, *J. Phys. Chem. C*, **114**, pp. 10598–10605.

71. Nagoya, A., Asahi, R., and Kresse, G. (2011). First-principles study of Cu_2ZnSnS_4 and the related band offsets for photovoltaic applications, *J. Phys.: Condens. Matter*, **23**.

72. Neugebauer, J. and Scheffler, M. (1992). Adsorbate-substrate and adsorbate-adsorbate interactions of Na and K adlayers on Al(111), *Phys. Rev. B*, **46**, pp. 16067–16080.

73. Nosaka, Y., Norimatsu, K., and Miyama, H. (1984). The function of metals in metal-compounded semiconductor photocatalysts, *Chem. Phys. Lett.*, **106**, pp. 128–131.

74. Nose, S. (1984). A unified formulation of the constant temperature molecular dynamics methods, *J. Chem. Phys.*, **81**, pp. 511–519.

75. Paier, J., Marsman, M., Hummer, K., Kresse, G., Gerber, I. C., and Angyan, J. G. (2006). Screened hybrid density functionals applied to solids, *J. Chem. Phys.*, **124**.

76. Perdew, J. P., Burke, K., and Ernzerhof, M. (1996). Generalized gradient approximation made simple, *Phys. Rev. Lett.*, **77**, pp. 3865–3868.

77. Perdew, J. P., Burke, K., and Ernzerhof, M. (1997). Generalized gradient approximation made simple (vol 77, pg 3865, 1996), *Phys. Rev. Lett.*, **78**, pp. 1396–1396.

78. Pozun, Z. D. and Henkelman, G. (2011). Hybrid density functional theory band structure engineering in hematite, *J. Chem. Phys.*, **134**.

79. Prezhdo, O. V., Duncan, W. R., and Prezhdo, V. V. (2009). Photoinduced electron dynamics at the chromophore-semiconductor interface: a time-domain ab initio perspective, *Prog. Surf. Sci.*, **84**, pp. 30–68.

80. Rahimi, N., Pax, R. A., and Gray, E. M. (2016). Review of functional titanium oxides. I: TiO_2 and its modifications, *Prog. Solid State Chem.*, **44**, pp. 86–105.

81. Rappe, A. K., Casewit, C. J., Colwell, K. S., Goddard, W. A., and Skiff, W. M. (1992). UFF, a full periodic table force field for molecular mechanics and molecular dynamics simulations, *J. Am. Chem. Soc.*, **114**, pp. 10024–10035.

82. Sahara, G., Abe, R., Higashi, M., Morikawa, T., Maeda, K., Ueda, Y., and Ishitani, O. (2015). Photoelectrochemical CO_2 reduction using a Ru(II)-Re(I) multinuclear metal complex on a p-type semiconducting NiO electrode, *Chem. Commun.*, **51**, pp. 10722–10725.

83. Sato, S., Morikawa, T., Saeki, S., Kajino, T., and Motohiro, T. (2010). Visible-light-induced selective CO_2 reduction utilizing a ruthenium complex electrocatalyst linked to a p-type nitrogen-doped Ta_2O_5 semiconductor, *Angew. Chem. Int. Ed.*, **49**, pp. 5101–5105.

84. Sato, S., Arai, T., Morikawa, T., Uemura, K., Suzuki, T. M., Tanaka, H., and Kajino, T. (2011). Selective CO_2 conversion to formate conjugated with H_2O oxidation utilizing semiconductor/complex hybrid photocatalysts, *J. Am. Chem. Soc.*, **133**, pp. 15240–15243.

85. Sato, S., Morikawa, T., Kajino, T., and Ishitani, O. (2013). A highly efficient mononuclear iridium complex photocatalyst for CO_2 reduction under visible light, *Angew. Chem. Int. Ed.*, **52**, pp. 988–992.

86. Sato, S., Arai, T., and Morikawa, T. (2015). Toward solar-driven photocatalytic CO_2 reduction using water as an electron donor, *Inorg. Chem.*, **54**, pp. 5105–5113.

87. Sato, S., Arai, T., and Morikawa, T. (2016). Carbon microfiber layer as noble metal-catalyst support for selective CO_2 photoconversion in phosphate solution: toward artificial photosynthesis in a single-compartment reactor, *J. Photochem. Photobiol., A*, **327**, pp. 1–5.

88. Sato, S., Arai, T., and Morikawa, T. (2018). Electrocatalytic CO_2 reduction near the theoretical potential in water using Ru complex supported on carbon nanotubes, *Nanotechnology*, **29**.

89. Sato, S., Kataoka, K., Jinnouchi, R., Takahashi, N., Sekizawa, K., Kitazumi, K., Ikenaga, E., Asahi, R., and Morikawa, T. (2018). Band bending and dipole effect at interface of metal-nanoparticles and TiO_2 directly observed by angular-resolved hard X-ray photoemission spectroscopy, *Phys. Chem. Chem. Phys.*, **20**, pp. 11342–11346.

90. Sato, S., Saita, K., Sekizawa, K., Maeda, S., and Morikawa, T. (2018). Low-energy electrocatalytic CO_2 reduction in water over Mn-complex catalyst electrode aided by a nanocarbon support and K^+ cations, *ACS Catal.*, **8**, pp. 4452–4458.

91. Saveant, J. M. (2008). Molecular catalysis of electrochemical reactions. Mechanistic aspects, *Chem. Rev.*, **108**, pp. 2348–2378.

92. Sawada, H. and Kawakami, K. (1999). Electronic structure of oxygen vacancy in Ta_2O_5, *J. Appl. Phys.*, **86**, pp. 956–959.

93. Sekizawa, K., Sato, S., Arai, T., and Morikawa, T. (2018). Solar-driven photocatalytic CO_2 reduction in water utilizing a ruthenium complex catalyst on p-type Fe_2O_3 with a multiheterojunction, *ACS Catal.*, **8**, pp. 1405–1416.

94. Shishkin, M. and Kresse, G. (2006). Implementation and performance of the frequency-dependent GW method within the PAW framework, *Phys. Rev. B*, **74**, pp. 035101.

95. Shishkin, M. and Kresse, G. (2007). Self-consistent GW calculations for semiconductors and insulators, *Phys. Rev. B*, **75**, pp. 235102.

96. Singh, M. R. and Bell, A. T. (2016). Design of an artificial photosynthetic system for production of alcohols in high concentration from CO_2, *Energy Environ. Sci.*, **9**, pp. 193–199.

97. Singh, V., Beltran, I. J. C., Ribot, J. C., and Nagpal, P. (2014). Photocatalysis deconstructed: design of a new selective catalyst for artificial photosynthesis, *Nano Lett.*, **14**, pp. 597–603.

98. Steinfeld, J. I., Francisco, J. S., and Hase, W. L. (1989). *Chemical Kinetics and Dynamics* (Prentice Hall, New Jersey).

99. Sumita, M., Hu, C., and Tateyama, Y. (2010). Interface water on TiO_2 anatase (101) and (001) surfaces: first-principles study with TiO_2 slabs dipped in bulk water, *J. Phys. Chem. C*, **114**, pp. 18529–18537.

100. Suzuki, T. M., Tanaka, H., Morikawa, T., Iwaki, M., Sato, S., Saeki, S., Inoue, M., Kajino, T., and Motohiro, T. (2011). Direct assembly synthesis of metal complex-semiconductor hybrid photocatalysts anchored by phosphonate for highly efficient CO_2 reduction, *Chem. Commun.*, **47**, pp. 8673–8675.

101. Suzuki, T. M., Nakamura, T., Saeki, S., Matsuoka, Y., Tanaka, H., Yano, K., Kajino, T., and Morikawa, T. (2012). Visible light-sensitive mesoporous N-doped Ta_2O_5 spheres: synthesis and photocatalytic activity for hydrogen evolution and CO_2 reduction, *J. Mater. Chem.*, **22**, pp. 24584–24590.

102. Suzuki, T. M., Iwase, A., Tanaka, H., Sato, S., Kudo, A., and Morikawa, T. (2015). Z-scheme water splitting under visible light irradiation over powdered metal-complex/semiconductor hybrid photocatalysts mediated by reduced graphene oxide, *J. Mater. Chem. A*, **3**, pp. 13283–13290.

103. Suzuki, T. M., Saeki, S., Sekizawa, K., Kitazumi, K., Takahashi, N., and Morikawa, T. (2017). Photoelectrochemical hydrogen production by water splitting over dual-functionally modified oxide: p-type N-doped Ta_2O_5 photocathode active under visible light irradiation, *Appl. Catal., B*, **202**, pp. 597–604.

104. Suzuki, T. M., Takayama, T., Sato, S., Iwase, A., Kudo, A., and Morikawa, T. (2018). Enhancement of CO_2 reduction activity under visible light irradiation over Zn-based metal sulfides by combination with Ru-complex catalysts, *Appl. Catal., B*, **224**, pp. 572–578.

105. Suzuki, T. M., Yoshino, S., Takayama, T., Iwase, A., Kudo, A., and Morikawa, T. (2018). Z-Schematic and visible-light-driven CO_2 reduction using H_2O as an electron donor by a particulate mixture of a Ru-complex/$(CuGa)_{1-x}Zn_{2x}S_2$ hybrid catalyst, $BiVO_4$ and an electron mediator, *Chem. Commun.*, **54**, pp. 10199–10202.

106. Takeda, H., Koike, K., Inoue, H., and Ishitani, O. (2008). Development of an efficient photocatalytic system for CO_2 reduction using rhenium(I) complexes based on mechanistic studies, *J. Am. Chem. Soc.*, **130**, pp. 2023–2031.

107. Tamblyn, I., Darancet, P., Quek, S. Y., Bonev, S. A., and Neaton, J. B. (2011). Electronic energy level alignment at metal-molecule interfaces with a GW approach, *Phys. Rev. B*, **84**, pp. 201402.

108. Teruyasu, M., Isao, T., Shang-Peng, G., and Chris, J. P. (2009). First-principles calculation of spectral features, chemical shift and absolute threshold of ELNES and XANES using a plane wave pseudopotential method, *J. Phys.: Condens. Matter*, **21**, pp. 104204.

109. Thimsen, E., Biswas, S., Lo, C. S., and Biswas, P. (2009). Predicting the band structure of mixed transition metal oxides: theory and experiment, *J. Phys. Chem. C*, **113**, pp. 2014–2021.

110. Tomasi, J., Mennucci, B., and Cammi, R. (2005). Quantum mechanical continuum solvation models, *Chem. Rev.*, **105**, pp. 2999–3093.

111. Trammell, S. A., Moss, J. A., Yang, J. C., Nakhle, B. M., Slate, C. A., Odobel, F., Sykora, M., Erickson, B. W., and Meyer, T. J. (1999). Sensitization of TiO_2 by phosphonate-derivatized proline assemblies, *Inorg. Chem.*, **38**, pp. 3665–3669.

112. Tsuchiya, T. and Jakubikova, E. (2012). Role of Noncoplanar conformation in facilitating ground state hole transfer in oxidized porphyrin dyads, *J. Phys. Chem. A*, **116**, pp. 10107–10114.

113. Tully, J. C. (1990). Molecular dynamics with electronic transitions, *J. Chem. Phys.*, **93**, pp. 1061–1071.

114. Tung, R. T. (2001). Recent advances in Schottky barrier concepts, *Mater. Sci. Eng., R*, **35**, pp. 1–138.

115. van Schilfgaarde, M., Kotani, T., and Faleev, S. (2006). Quasiparticle self-consistent GW theory, *Phys. Rev. Lett.*, **96**, pp. 226402.

116. Vittadini, A., Selloni, A., Rotzinger, F. P., and Grätzel, M. (1998). Structure and energetics of water adsorbed at TiO_2 anatase (101) and (001) surfaces, *Phys. Rev. Lett.*, **81**, pp. 2954–2957.

117. Wang, L. X., Ernstorfer, R., Willig, F., and May, V. (2005). Absorption spectra related to heterogeneous electron transfer reactions: the perylene TiO_2 system, *J. Phys. Chem. B*, **109**, pp. 9589–9595.

118. Wenger, B., Gratzel, M., and Moser, J. E. (2005). Rationale for kinetic heterogeneity of ultrafast light-induced electron transfer from Ru(II) complex sensitizers to nanocrystalline TiO_2, *J. Am. Chem. Soc.*, **127**, pp. 12150–12151.

119. Wu, Y.-N., Li, L., and Cheng, H.-P. (2011). First-principles studies of Ta_2O_5 polymorphs, *Phys. Rev. B*, **83**, pp. 144105.

120. Yamanaka, K., Sato, S., Iwaki, M., Kajino, T., and Morikawa, T. (2011). Photoinduced electron transfer from nitrogen-doped tantalum oxide

to adsorbed ruthenium complex, *J. Phys. Chem. C*, **115**, pp. 18348–18353.

121. Yamanaka, K. and Morikawa, T. (2012). Charge-carrier dynamics in nitrogen-doped TiO_2 powder studied by femtosecond time-resolved diffuse reflectance spectroscopy, *J. Phys. Chem. C*, **116**, pp. 1286–1292.

122. Yamanaka, K., Ohwaki, T., and Morikawa, T. (2013). Charge-carrier dynamics in Cu- or Fe-loaded nitrogen-doped TiO_2 powder studied by femtosecond diffuse reflectance spectroscopy, *J. Phys. Chem. C*, **117**, pp. 16448–16456.

Chapter 5

Large-Scale Simulations I: Methods and Applications for a Li-Ion Battery

Nobuko Ohba[a] and Shuji Ogata[b]
[a]*Toyota Central R&D Laboratories, Inc., Nagakute, Aichi 480-1192, Japan*
[b]*Graduate School of Engineering, Nagoya Institute of Technology, Gokiso-cho, Nagoya, Aichi 466-8555, Japan*
e4606@mosk.tytlabs.co.jp; ogata@nitech.ac.jp

5.1 Introduction

From the viewpoint of suppressing global warming, the development of green or environmentally friendly vehicles such as hybrid cars, plug-in hybrids, electric cars, and fuel cell cars has evolved rapidly in recent years. Lithium (Li)-ion rechargeable batteries [1] are indispensable for these vehicles because they have a high energy density. In a Li-ion battery (LIB), there are various key reactions and processes. On application of an electric voltage, Li ions traverse from positive to negative electrodes through a nonaqueous electrolyte and separator. The migration of the Li ions in both the electrode and the electrolyte materials is directly related to the speed of charging and discharging of the LIB. Reducing the resistance of the charge-

Multiscale Simulations for Electrochemical Devices
Edited by Ryoji Asahi

ISBN 978-981-4800-71-6 (Hardcover), 978-0-429-29545-4 (eBook)
www.jennystanford.com

transfer reaction at the electrolyte/electrode interfaces leads to improvements in the input/output characteristics of the battery. The control of the internal structural changes of the electrodes accompanying the insertion/deinsertion of Li ions improves the durability of these types of batteries.

To elucidate the various reaction mechanisms in such an LIB system, atomistic- and electronic-scale simulations such as electronic state calculations and molecular dynamics (MD) calculations have been conducted in a number of studies [6, 45, 46, 49, 51]. In practice, realistic modeling of, for example, the electrode structural change, the interface phenomenon, and the potential influence due to Li-ion movement requires the development and use of a large-scale calculation method linking electronic-level microscopic processes and macroscopic dynamics processes. Various simulations have been conducted in order to analyze the transport properties of Li ions in the electrode and electrolyte materials and their interface in LIBs. Borodin and coworkers [4, 6, 7] investigated the structural and transport properties of Li ions and their solvated states by a combined simulation of quantum chemical calculations and classical MD simulations for the organic solvent used in an electrolyte liquid. A classical MD simulation method with a polarizable force field [3] was developed previously, and the ion self-diffusion in ionic liquids at several temperatures has been validated [5]. A long-time-scale (~1 μs) study [38] of the transport properties of Li ions in solid/electrolyte interphase (SEI) films, which formed during charging and discharging cycles in LIBs (see next paragraph for more details), was conducted using the GROMACS[a] MD simulation package [73]. Although a nonpolarizable force field was used in this study, the simulation results for structural and transport properties of Li ions in a model SEI corresponded with those of a prior polarizable-force-field study [7]. In others, classical MD simulation was performed for the local environment of Li ions and their transport properties in high-molecular-weight poly(trimethylene carbonate) [66] of solid polymer electrolytes in LIBs.

Various chemical reactions accompany the movement of Li ions in an LIB. For example, in an organic electrolyte liquid based on ethylene carbonate (EC), the electrolyte undergoes a reductive

[a]GROningen MAchine for Chemical Simulations

decomposition reaction in the potential region in the vicinity of 1 V during the initial stage of charging [2, 33, 79]. As a result, a passive film forms on the surface of the negative electrode, which suppresses further decomposition of the electrolyte. Among such passive films, those that are permeable to Li ions are known as SEI films. Although the SEI affects the migration of Li ions at the electrode interface, its components and features are not fully understood. To identify the elementary formation mechanism of the SEI films, computational studies such as the ab initio MD (AIMD) method have been performed [21, 32, 61]. AIMD simulation for EC/graphite interfaces demonstrated that the EC breakdown was affected by the termination of the edge of the graphite electrode [32]. AIMD simulation results performed by Ganesh et al. [21] investigated the place where the SEI were formed. Li_2CO_3, which is the primary component in the SEI film, forms near the anode surface. On the other hand, LiF, which is formed from the decomposition of a Li salt such as $LiPF_6$, agglomerates at the anode interfaces. The hybrid Monte Carlo (MC)/MD reaction method revealed the atomistic differences between the EC- and propylene carbonate (PC)-based SEI films on the negative graphite electrode in LIBs [67]. This hybrid MC/MD method was applied for both LIBs and sodium-ion batteries to understand the role of the electrolyte additives in the formation of the SEI film [68]. Moreover, the molecular mechanism of a highly concentrated electrolyte system is also investigated [69]. In this method, a list of elementary reaction processes must be prepared in advance.

In this chapter, we describe two other types of atomistic calculation methods for the large-scale simulation of Li-ion batteries. First, the order-N divide-and-conquer-type real-space grid density functional theory (DFT) code (DC-RGDFT) was developed to reduce the calculation cost of DFT for a large-scale model that is divided into subspaces, without degrading the accuracy. Next, a hybrid quantum-classical (QM-CL) simulation method has been implemented to couple the electronic state and CL region described using a force field in a condensed solid. As an application for Li-ion batteries, microscopic mechanisms of Li-ion transfer through the boundary between the organic SEI formed on the graphite anode and liquid electrolyte are investigated theoretically. In addition, an application of the hybrid QM-CL simulation to elucidate the thermal diffusion process of Li ions in the graphite anode is presented, which demonstrates how such a large-scale simulation provides physical insights into a realistic materials model.

5.2 Method

5.2.1 Real-Space Grid Kohn–Sham DFT (RGDFT) Method [48]

In this section we summarize the RGDFT method [10, 11, 13, 14, 25, 31, 62, 75] before we advance to the DC type. Let us consider an atomic cluster composed of N_{ion} ions with charge numbers $\{Z_i\}$ at positions $\{\vec{R}_i\}(i=1,2,\ldots,N_{\text{ion}})$ and the valence electrons. For simplicity, the charge-neutral system is assumed though the formulation is applicable to non-neutral systems also; the total number of electrons $N_e = \Sigma_i Z_i$.

The well-known Kohn–Sham (KS) equation for the eigen orbital $\phi_n(\vec{r})$ with the eigen energy ε_n is

$$\left[-\frac{1}{2}\nabla^2 + v_{\text{H}}(\vec{r}) + \sum_{i=1}^{N_{\text{ion}}} v_{\text{ion},i}(\vec{r}) + v_{\text{XC}}(\vec{r})\right]\phi_n(\vec{r}) = \varepsilon_n \phi_n(\vec{r}) \tag{5.1}$$

in the atomic unit (i.e., $m_e = \hbar = e = 1$, where m_e, $\hbar$, and e, indicate electron rest mass, reduced Planck constant, and elementary charge, respectively). The Hartree potential in Eq. 5.1 is

$$v_{\text{H}}(\vec{r}) = \int d\vec{r}' \frac{\rho(\vec{r}')}{\left|\vec{r}-\vec{r}'\right|} \tag{5.2}$$

or

$$\nabla^2 v_{\text{H}}(\vec{r}) = -4\pi\rho(\vec{r}) \tag{5.3}$$

with the density of electrons

$$\rho(\vec{r}) = \sum_{n=1}^{N_e/2} 2\left|\phi_n(\vec{r})\right|^2 \tag{5.4}$$

in the spin neutral case.

The pseudopotential of ion i, $v_{\text{ion},i}(\vec{r})$ for $\phi_n(\vec{r})$ acts as

$$v_{\text{ion},i}(\vec{r})\phi_n(\vec{r}) = v_{\text{L},i}(\vec{r})\phi_n(\vec{r}) + v_{\text{NL},i}(\vec{r})\,|\,\phi_n\rangle \tag{5.5}$$

with

$$v_{\mathrm{NL},i}(\vec{r})|\phi_n\rangle = \sum_{l=0}^{l_{\max}} \sum_{m=-l}^{l} \frac{\phi_{lm,i}^{\mathrm{PS}}(\vec{r})\Delta v_{l,i}(\vec{r}) \int d\vec{r}' \phi_{lm,i}^{\mathrm{PS}*}(\vec{r}')\Delta v_{l,i}(\vec{r})\phi_n(\vec{r}')}{\int d\vec{r} \phi_{lm,i}^{\mathrm{PS}*}(\vec{r})\Delta v_{l,i}(\vec{r})\phi_{lm,i}^{\mathrm{PS}}(\vec{r})}. \tag{5.6}$$

Here the Kleinman–Bylander form [28] is adopted for the treatment of the nonlocal pseudopotential. The norm-conserving pseudopotentials [72] are used. The $\phi_{lm,i}^{\mathrm{ps}}(\vec{r})$ in Eq. 5.6 is the pseudo eigen orbital for a free atom i at the angular state (l,m). The pseudopotential at a chosen angular state $l = l_{\mathrm{loc}}$ (often the maximum of l) is regarded as the local pseudopotential and the deviation of the pseudopotential from the local one as the nonlocal pseudopotential:

$$v_{\mathrm{L},i}(\vec{r}) \equiv v_{\mathrm{loc},i}(\vec{r}) \quad \text{and} \quad \Delta v_{l,i}(\vec{r}) \equiv v_{l,i}(\vec{r}) - v_{\mathrm{loc},i}(\vec{r}) \tag{5.7}$$

with the pseudopotential $v_{l,i}(\vec{r})$ or a free ion i at the angular state l. The $v_{\mathrm{L},i}(\vec{r})$ contains the long-ranged Coulomb potential, while the $\Delta v_{l,i}(\vec{r})$ short ranged.

The $v_{\mathrm{xc}}(\vec{r})$ in Eq. 5.1 is the exchange-correlation potential defined as the functional derivative of the exchange-correlation energy:

$$v_{\mathrm{xc}}(\vec{r}) = \frac{\delta E_{\mathrm{xc}}(\rho)}{\delta \rho(\vec{r})}. \tag{5.8}$$

Various approximation formulas of $E_{\mathrm{xc}}(\rho)$ are given in literatures. For simplicity, we use the local density approximation (LDA) formula in Ref. [56] .

The eigen orbitals in Eq. 5.1 are solved numerically through the self-consistent field (SCF) iteration [12, 30, 54] under the orthonormalization constraint:

$$\int d\vec{r} \phi_i^*(\vec{r})\phi_j(\vec{r}) = \delta_{i,j} \equiv \begin{cases} 1, \text{for } i = j, \\ 0, \text{for } i \neq j. \end{cases} \tag{5.9}$$

The number of SCF iterations required to reach the convergence, which is independent of the target system size, is typically 20. In the conventional planewave-based KS-DFT method, eigen orbitals are represented using the planewaves under the periodic boundary conditions. And the SCF iteration procedure contains the local iterations for all the energy levels considered. In sweeping the orbitals for a given $\rho(\vec{r})$ in the local iteration procedure, orbitals are updated one by one from the lowest to the highest energy levels by the conjugate gradient (CG) method with the Gram–Schmidt orthonormalization [58] to the orbitals at lower energy levels.

In the RGDFT method, we set the Cartesian grid points in 3D with the grid size h to describe the eigen orbitals and the potentials. The grid size h in the unit of the Bohr radius $a_B \approx 0.529$ Å corresponds to the cutoff energy $0.5(\pi/h)^2$ (au) (1 au of energy ≈ 27.2 eV) in the conventional planewave-based KS-DFT method. The overall shape of the grid points is spherical, with radius r_{max}, which is determined to enclose all the ions with a few angstrom vacuum width so that $\rho(\vec{r}) = 0$ at $r = r_{max}$. The second derivative operations in the three directions in the KS and Poisson equations are calculated by the high-order (fourth or more) finite difference method [13, 25, 58, 75] using the data on multiple grid points in both plus and minus sides. For the ion with a relatively deep pseudopotential as oxygen, a smaller grid size of $h/3$ is used at around the ion only to represent the pseudopotential accurately [53]. The RGDFT method is well suited to the parallel computation environment. It is free from the fast Fourier transform method, which becomes inefficient for massively parallel machines. The idea of spatial decomposition of the grid points works well for parallel machines. In addition, the RGDFT method has a unique feature of numerical stability, which helps to realize high computation performance of the Gram–Schmidt orthonormalization of the orbitals as explained below. While the Gram–Schmidt orthonormalization needs to be performed orbital by orbital for stability reasons in the conventional planewave-based method, it can be performed for all the orbitals together in the RGDFT method after the orbital sweep in the local iteration procedure [25]. This rearrangement of the Gram–Schmidt orthonormalization procedure improves the computation performance [25] on a parallel machine by employing a highly tuned linear-algebra library.

5.2.2 Divide-and-Conquer-Type RGDFT Method [48]

For the present formulation, we divide a target system into the total of N_d basic domains arranged three dimensionally, as depicted in Fig. 5.1a for a 2D case. The boundary surface between the neighboring basic domains is denoted as $\mathbf{S}_{bdry}$. To each basic domain, we add a surface layer taken from the neighboring basic domains with the cutoff depth d_c from $\mathbf{S}_{\text{bdry}}$ to define the domain that overlaps with neighboring domains. For a domain, we call the ions located in the basic domain as the real ions, as depicted in Fig. 5.1b. The ions located in the additional surface layer are called buffer ions. Ions that are neither real nor buffer are the external ions. The Coulomb potential due to the external ion should be considered in the domain, while it will be modified in the vicinity of the ion so that the valence electrons become unbound from it. The weight function $W_I(\vec{r})$ for domain I should obey the sum rule of $\sum_{I=0}^{N_d-1} W_I(\vec{r})=1$ at any $\vec{r}$ For simplicity we assume the stepwise form for $W_I(\vec{r})$ that is, $W_I(\vec{r})=1$ in basic domain I, while $W_I(\vec{r})=0$ outside the basic domain.

For domain I, the density $\rho_I(\vec{r})$ is obtained through the KS-type equation explained below, with Fermi energy common to all the domains. The grid points form a sphere of radius r_{max}, with its center set at the averaged position of the real and buffer ions. The value of r_{max} is determined so that the real and buffer ions are enclosed with sufficient vacuum width so that $\rho_I(\vec{r})=0$ at $r = r_{\text{max}}$. The total density defined as

$$\rho_{\text{tot}}(\vec{r}) \equiv \sum_{I=0}^{N_d-1} W_I(\vec{r})\rho_I(\vec{r}) \tag{5.10}$$

is thereby calculated, which corresponds to $\rho(\vec{r})$ in the RGDFT method. We call $W_I(\vec{r})\rho_I(\vec{r})$ the real density and $[1-W_I(\vec{r})]\rho_I(\vec{r})$ the shadow density of domain I.

The KS-type equation in the DC-RGDFT method for domain I is

$$\left[-\frac{1}{2}\nabla^2 + v_{\text{H}}^{\text{DC}}(\vec{r}) + \sum_{i=1}^{N_{\text{ion}}} v_{\text{ion},i}^{\text{DC}}(\vec{r}) + v_{\text{XC}}^{\text{DC}}(\vec{r}) + w_{\text{DT}}(\vec{r})v_{\text{DT}}(\vec{r}) + \left(1-w_{\text{DT}}(\vec{r})\right)v_{\text{emb}}(\vec{r})\right]\phi_n(\vec{r}) = \varepsilon_n\phi_n(\vec{r}). \tag{5.11}$$

Here the kinetic energy term (i.e., the second derivative term) is the same as in the standard KS equation (see Eq. 5.1). The exchange-correlation potential is

$$v_{\text{XC}}^{\text{DC}}(\vec{r}) = \frac{\delta E_{\text{xc}}(\rho_{\text{tot}})}{\delta\rho_{\text{tot}}(\vec{r})}. \tag{5.12}$$

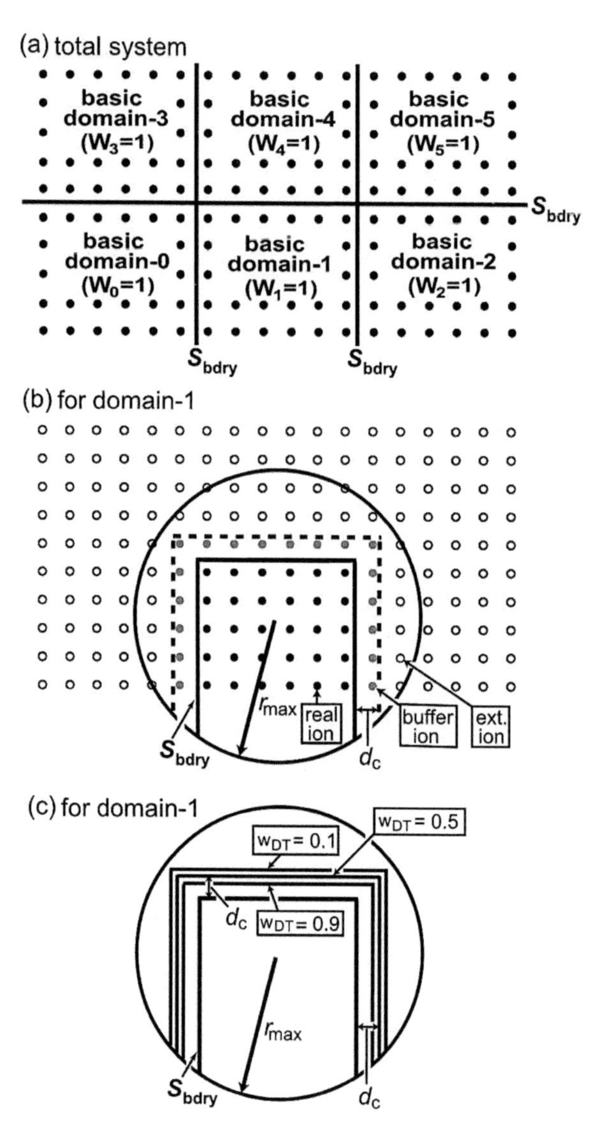

Figure 5.1 (a) Schematic 2D view of the division of a target system into basic domains in the DC-RGDFT method. The dots represent the ions. The boundary surfaces $\mathbf{S}_{\text{bdry}}$ are indicated by the lines. (b) Identification of all the ions as real, buffer, or external ones for domain-1 in (a). The cutoff depth d_c and the maximum radius r_{max} of the spherical grid points for domain-1 are depicted by the boundary surfaces. (c) The contour map of the support function, $w_{DT}(\vec{r})$ for domain-1 in (b). Reprinted from Ref. [48], Copyright (2012), with permission from Elsevier.

The density $\rho_I(\vec{r})$ is calculated as

$$\rho_I(\vec{r}) = \sum_{n=1}^{N_{e,I}/2} 2|\phi_n(\vec{r})|^2 \tag{5.13}$$

with either integer or noninteger number of electrons $N_{e,I}$ for domain I. The global Fermi level is determined so that the total number of electrons becomes

$$N_e = \sum_{I=0}^{N_d-1} N_{e,I}.$$

The Hartree potential in Eq. 5.11 is defined as

$$v_H^{DC}(\vec{r}) = \int d\vec{r}' \frac{\rho_{tot}(\vec{r}')}{|\vec{r}-\vec{r}'|}. \tag{5.14}$$

It includes the embedding effect of the electronic Coulomb potential from other domains. Note that $v_H^{DC}(\vec{r})$ is felt by both real and shadow densities.

As for the pseudopotentials $v_{ion,i}^{DC}(\vec{r})$ in Eq. 5.11, those terms relating to the real and buffer ions are the same as that in $v_{ion,i}(\vec{r})$ in Eq. 5.1. On the other hand, the $v_{ion,i}^{DC}(\vec{r})$ relating to each external ion is modified at small r so that no bound orbital exists on it. That is, while a relatively short cutoff radius r_L^c is used to calculate the local pseudopotential $v_{L,i}(\vec{r})$ of an external ion at $\vec{r}$ if the $\vec{r}$ corresponds to the angle within plus and minus 60° from the direction orienting from the external ion to the nearest-neighbor buffer ion, a longer cutoff distance than r_L^c is used if the condition is not met. The nonlocal pseudopotentials by the external ions are assumed to be zero.

Continuation of $\rho_{tot}(\vec{r})$ at the boundary $\mathbf{S}_{bdry}$ is a principal property that should be held in the DC-RGDFT method. To realize this, the existing method [63, 64] used a soft weight function with the embedding potential in a unique form. In the method, however, artificial density decrease of $\rho_{tot}(\vec{r})$ occurs at around $\mathbf{S}_{bdry}$ even if the unweighted density $\rho_I(\vec{r})$ is accurate at the location, as explained later. We, therefore, take a different route by introducing a potential in the KS-type equation to minimize the possible difference in density at the boundary. Firstly, the density-template potential defined as

$$v_{\mathrm{DT}}(\vec{r}) = \frac{\rho_I(\vec{r}) - \rho_{\mathrm{tot}}(\vec{r})}{\alpha} \tag{5.15}$$

is considered for domain I with the adjustable parameter α (>0). If $\rho_I(\vec{r}) < \rho_{\mathrm{tot}}(\vec{r})$, the $v_{\mathrm{DT}}(\vec{r})$ lowers the base of the potential in the KS-type equation and hence acts to increase $\rho_I(\vec{r})$, and vice versa. Since the $v_{\mathrm{DT}}(\vec{r})$ should work at around $\mathbf{S}_{\mathrm{bdry}}$ only, we introduce the support function $w_{\mathrm{DT}}(\vec{r})$ to define a finite-depth layer just outside the $\mathbf{S}_{\mathrm{bdry}}$ as depicted in Fig. 5.1c:

$$w_{\mathrm{DT}}(\vec{r}) = \frac{1}{1 + \exp[r - d_c/b]}, \tag{5.16}$$

is considered for where $b = d_c/2$ and r is measured from $\mathbf{S}_{\mathrm{bdry}}$ toward the outside. The supported density-template potential $w_{\mathrm{DT}}(\vec{r})v_{\mathrm{DT}}(\vec{r})$ is finally added to the KS-type equation for domain I (see Eq. 5.11). Note that the $w_{\mathrm{DT}}(\vec{r})v_{\mathrm{DT}}(\vec{r})$ works on both real and shadow densities. We will demonstrate in the next section that the supported density-template potential is effective in stabilizing the shadow density.

The $\rho_I(\vec{r})$ is assumed to decrease to zero as the $\vec{r}$ approaches r_{max}. Such abrupt decrease in $\rho_I(\vec{r})$ at peripheral grid points, which is an artifact of introducing the domains, may modify the eigen orbitals and energies substantially. To minimize the modification, we consider the quantum embedding effects [22] of the kinetic and exchange-correlation energies of electrons by defining the embedding potential

$$v_{\mathrm{emb}}(\vec{r}) = \frac{\delta T_s(\rho_{\mathrm{tot}})}{\delta \rho_{\mathrm{tot}}(\vec{r})} - \frac{\delta T_s(\rho_I)}{\delta \rho_I(\vec{r})} + \frac{\delta E_{\mathrm{xc}}(\rho_{\mathrm{tot}})}{\delta \rho_{\mathrm{tot}}(\vec{r})} - \frac{\delta E_{\mathrm{xc}}(\rho_I)}{\delta \rho_I(\vec{r})}, \tag{5.17}$$

where $T_s = \sum_{n=1}^{N_{e,I}/2} \int d\vec{r} \phi_n^*(\vec{r})(-\nabla^2)\phi_n(\vec{r})$ and E_{xc} is the exchange-correlation energy. The LDA is applied to T_s and E_{xc}. When $\rho_I(\vec{r})$ differs from $\rho_{\mathrm{tot}}(\vec{r})$, the $v_{\mathrm{emb}}(\vec{r})$ acts to shift up or down the base of the potential to take into account the many-body quantum effects of electrons. In a high-density situation with $r_s \equiv \frac{(3/4\pi\rho_{\mathrm{tot}})^{\frac{1}{3}}}{a_{\mathrm{B}}} < 3$, the Fermi degeneracy effect dominates and thereby shifts up the averaged electron energy; at a low density with $r_s > 3$, the exchange-correlation effect dominates and shifts down the averaged electron energy. Since the $v_{\text{emb}}(\vec{r})$ should work only at the artificially decreasing tail of $\rho_I(\vec{r})$, the supported embedding potential defined as $\left[1-w_{\text{DT}}(\vec{r})\right]v_{\text{emb}}(\vec{r})$ is added in the KS-type equation for domain I (see Eq. 5.11).

Let us compare the present method with similar existing methods. The atomic orbital (AO)-based KS-DFT (AODFT) method uses the precomputed AO basis set to describe the eigen orbitals. To advance the AODFT method to the DC type, only the classical embedding effects of the Hartree and ionic Coulomb potentials are considered with soft weight functions [80–83]. Hence the DC-AODFT method requires relatively thick buffer layers to obtain accurate results. As a side effect of using the AO basis set, the shadow density relating to the buffer ions fluctuates little in the DC-AODFT method. Therefore, the density template potential is not considered in the DC-AODFT method.

The difference between the existing method in Ref. [63] and the present one is clarified below. Both methods use the RGDFT method for each domain. In Ref. [63], the quantum embedding effect of the kinetic energy is considered for a domain in addition to the classical embedding effect of the Hartree and ionic Coulomb potentials, while the density-template potential is not considered. Hence the KS-type equation for domain I in Ref. [63] contains the embedding potential

$$\bar{v}_{\text{emb}}(\vec{r})=\frac{\delta T_{\text{s}}(\rho_{\text{tot}})}{\delta\rho_{\text{tot}}(\vec{r})}-\left(\frac{\delta T_{\text{s}}(\rho)}{\delta\rho}\right)_{\rho=W_I(\vec{r})\rho_I(\vec{r})} \tag{5.18}$$

with a soft weight function $W_I(\vec{r})$. The method ignores the shadow density and regards the real density as embedded in $\rho_{\text{tot}}(\vec{r})$. The method may cause the following problem: Let us consider the case of applying the method to a homogeneous system with a soft weight function. In the case, the $\rho_I(\vec{r})$ should be nearly equal to $\rho_{\text{tot}}(\vec{r})$ at around $\mathbf{S}_{\text{bdry}}$. Since the inequality $W_I(\vec{r})\rho_I(\vec{r})<\rho_{\text{tot}}(\vec{r})$ holds for a thick region (relating to the softness of the weight function) at around $\mathbf{S}_{\text{bdry}}$, the embedding potential takes on positive values, that is, $\bar{v}_{\text{emb}}(\vec{r})>0$, in the region. It means that $\rho_I(\vec{r})$ will be suppressed artificially in the region by $\bar{v}_{\text{emb}}(\vec{r})$ despite the intrinsic homogeneity of the system, resulting in inhomogeneity of $\rho_{\text{tot}}(\vec{r})$. In our formulation for the embedding potential, no such unphysical situation is expected to occur.

In Refs. [18, 19], both density-template and embedding potentials are ignored in the KS-type equation for a domain. As in the present method, the eigen orbitals are assumed to vanish outside the spherical grid points. The radii, that is, (r_{max}) and locations of the domains are optimized adaptively for a target system during the calculation. Inclusion of both potentials in the KS-type equation makes the calculated density and atomic forces have similar high accuracies irrespective of target systems without such an optimization procedure [48] .

In the DC-RGDFT method we calculate the forces on atoms as follows: For an ion i with charge number Z_i at position $\vec{R}_i$, we firstly identify the basic domain I to which the ion belongs. Using the eigen orbitals and $\rho_I(\vec{r})$ for domain I, we then calculate the force $\vec{F}_i$ acting on atom i based on the Hellmann–Feynman formula [54]:

$$\vec{F}_i = -\int d\vec{r}\, v_{\text{xc}}\left(\rho_I(\vec{r}) + \rho_{\text{c},I}(\vec{r})\right)\frac{\partial \rho_{\text{c},I}(\vec{r})}{\partial \vec{R}_i} - \int d\vec{r}\, v_{\text{xc}} \frac{\partial v_{\text{L},i}(\vec{r})}{\partial \vec{R}_i}\rho_{\text{tot}}(\vec{r})$$

$$-\sum_{n=1}^{N_{\text{e},I}/2}\left\langle 2\phi_n \left| \frac{\partial v_{\text{NL},i}(\vec{r})}{\partial \vec{R}_i} \right| \phi_n \right\rangle + \sum_{j}^{\neq i} \frac{Z_i Z_j (\vec{R}_i - \vec{R}_j)}{\left|\vec{R}_i - \vec{R}_j\right|^3}, \qquad (5.19)$$

where $\rho_{\text{c},I}(\vec{r})$ is the partial core-charge density [35] of the real ions. The atomic force, $\vec{F}_i$, is explicitly independent of the weight function. We expect the combination of the density-template and embedding potentials makes the total density in the DC-RGDFT method comparable to that in the RGDFT method, resulting in high accuracy in atomic forces.

The KS-type equation for each domain (Eq. 5.11) is treated on the real-space grid points that form a sphere. The grid size h is common to all the domains. The KS-type equations are solved simultaneously for all the domains by repeating the following SCF iteration procedure. As shown in Fig. 5.2, the procedure for a given set of densities $\{\rho_I(\vec{r})\}$ is composed of the following three major steps:

1. Local iteration in each domain about the orbitals by the CG method [54, 58] with their mutual orthonormalization. The global Fermi energy is determined.
2. Transfer of data on the overlapping grid points between the domains with respect to the weighted densities, $\{W_I(\vec{r})\rho_I(\vec{r})\}$, and the corresponding Hartree potentials.

3. Update of the set of densities $\{\rho_I(\vec{r})\}$ using the Pulay mixing method [59, 60].

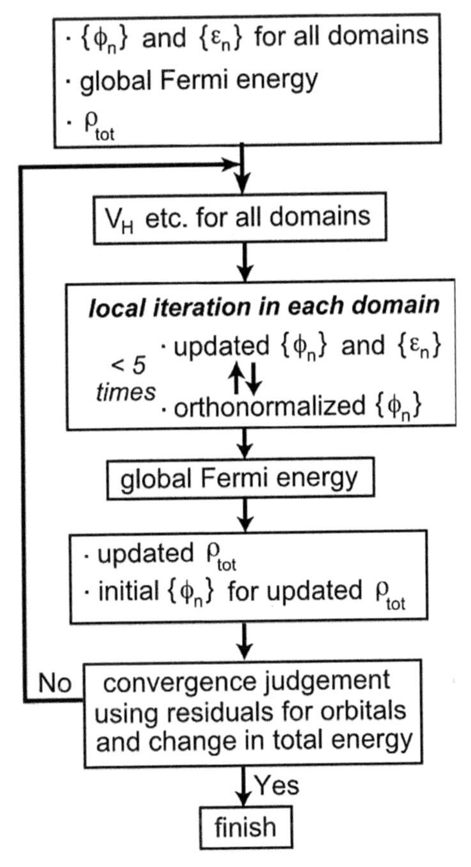

Figure 5.2 Brief flowchart of the SCF iteration procedure in the DC-RGDFT method. $\{\phi_n\}$ and $\{\varepsilon_n\}$ are the eigen orbitals and energies for the KS-type equations, respectively. ρ_{tot} is the total density of electrons. Reprinted from Ref. [48], Copyright (2012), with permission from Elsevier.

For Step 1 we exploit the RGDFT code [40, 41, 62] that is parallelized following the spatial decomposition strategy to treat the grid points by multiple compute nodes. The local CG iteration is repeated three to five times per orbital. It is composed of the updates of the orbitals and of their Gram–Schmidt orthonormalization all together. The preconditioning (or smoothing) of the gradients of the orbitals using the six nearest-neighbor grid point data (two for each direction) with the relative weight 0.1 with respect to the central point is important for numerical stability. The fourth-order finite difference method that uses nine data points for each direction is adopted to evaluate the second derivative term; the possible error is order-h^{10}. Details of the local CG iteration have been explained in Refs. [25, 40, 41, 62, 75].

As for Step 2, the communicator (or the group of compute nodes) for interdomain communication is prepared for the message passing interface (MPI) standard [24], in addition to the communicator for each domain for intradomain communication. For a given cutoff depth d_c, the buffer ions are selected and then the maximum radius r_{max} for the spherical grid points is determined for each domain. The overlaps of the spheres are precomputed and saved for transferring data on the overlapping grid points between the domains. To compute the Hartree potential relating to $\rho_{tot}(\vec{r})$, the Poisson equation for the weighted density $W_I(\vec{r})\rho_I(\vec{r})$ is solved for each domain. If a grid point of a domain overlaps to the point of a neighboring domain, the Hartree potential on the point obtained by the Poisson solution for the domain is transferred to the overlapping domain. To the nonoverlapping (or far) domains, the multipole data (up to the 8th order in the spherical harmonics) [23, 42] of the weighted density are transferred instead of the Hartree potentials on the grid points. The total Hartree potential is thereby constructed by summing those contributions in each domain.

In Step 3, the set of densities $\{\rho_I(\vec{r})\}$ is updated all together by the Pulay method [59, 60]. A maximum of 10 previous sets are used to get the updated densities. For the next iteration, the eigen orbitals relating to the updated density are evaluated in the subspace spanned by the orbitals obtained at the last iteration. Thereafter the local iteration using the updated potentials in the KS-type equations is repeated. Such SCF iteration repeats 20–40 times until convergence that is judged using the residuals for orbitals or the change of total energy. If the residuals are not sufficiently small, we have the option to fix the $\{N_{e,I}\}$ and the Hartree potential due to the other domains to the averaged values in each domain and to perform additional iterations several times until sufficiently small residuals are obtained in each domain.

The DC-RGDFT code is parallelized in four points: (i) domain decomposition of a simulation system, (ii) spatial decomposition of the grid points in each domain, (iii) parallel updating of orbitals in each domain, and (iv) the OpenMP parallelization of various do-loops. Through a comparison of the computation time [47] between the DC-RGDFT and RGDFT, we will use the RGDFT for a system with $N_{atom} \leq 300$ while the DC-RGDFT for $N_{atom} > 300$ in the present simulation. In using the DC-RGDFT, we will set 8 domains with d_c = 7 a_B in the present simulation because the maximum deviation of the ion forces due to the domain decomposition becomes satisfactorily small (averaged relative accuracy is smaller than 10^{-3}. The spatial decomposition of each domain will be set to 4 × 4 × 4 = 64. The parallelization degree of the orbital calculation will be 4 for each domain. In total 8 × 64 × 4 = 2048 MPI processes will be executed on a parallel machine.

5.2.3 Hybrid Quantum-Classical Simulation Method

The hybrid QM-CL simulation is expected to be one of the calculation methods that aim for both large-scale and high accuracy [43, 44]. In this method, the reaction region, where the electronic structure is treated by a highly accurate calculation technique such as the DFT, is embedded in a classical dynamics system of atoms based on an empirical interaction model. In the hybrid QM-CL simulation, the atomic bond is cut at the QM-CL boundary and a dangling bond forms. As for this dangling bond, its influence on the electronic state or bonding distance of the atoms in the QM region should be removed. The link-atom (or handshake atom) method [41] that uses hydrogen atoms for the termination of the QM atoms is usually applied to couple the QM and CL regions. In the link-atom method, however, the influence of the surface reconstruction with the relaxation of the boundary atom extends to the atoms of the whole system and a large distortion from the original stable structure is produced. In this study, we adopt the buffered cluster method (BCM) [43], which requires no link atoms and is more precise and a general-purpose model. In the following section we present an outline of the BCM.

5.2.3.1 Buffered cluster method

Figure 5.3 shows the schematic views of the hybrid QM-CL simulation with the BCM. In the BCM, additional atoms, called buffer atoms, are put to terminate the dangling bond of the QM atoms at

QM-CL boundaries. The positions of the buffer atoms are adjusted so as to minimize the potential energy under the constraint of fixing the position of the QM atoms for the CL calculation of the QM cluster region. In the QM calculation, the positions of the buffer atoms are not relaxed. Therefore, various surface reconstructions of the QM cluster region are suppressed in the BCM. Demonstrations of the simulation of the stress corrosion clacking mechanism of silicon and the alumina make this technique practical [43].

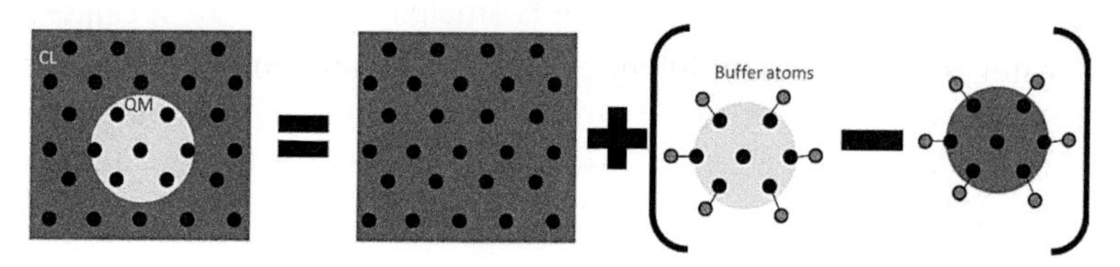

Figure 5.3 Schematic views of hybrid quantum-classical (QM-CL) simulation with the buffered cluster method.

In the target system, we assume a region dealt with by the QM calculation as the cluster region. The subscript CL denotes the physical properties calculated from the CL calculation, and QM denotes ones computed from the QM calculation.

In the hybrid QM-CL simulation, the Hamiltonian of the system is defined as follows:

$$H(R_{\text{all}}, P_{\text{all}}) = H_{\text{CL}}^{\text{system}}(R_{\text{all}}, P_{\text{all}}) + \sum_{\text{cluster}} (E_{\text{QM}}^{\text{cluster}} - E_{\text{CL}}^{\text{cluster}}) \quad (5.20)$$

Here, R_{all} and P_{all} represent the positions and momenta of all the atoms, respectively. Hamiltonian $H_{\text{CL}}^{\text{system}}$ is obtained by applying the CL calculation to the whole system: $H_{\text{CL}}^{\text{system}} = E_{\text{kin}}(P_{\text{all}}) + E_{\text{CL}}^{\text{system}}(R_{\text{all}})$. The last two terms in Eq. 5.20 are energies given from the QM and CL calculations in the cluster region, respectively.

We assume the atomic position in the cluster region to be $\{r_{\text{cluster}}\}$. The energy term in the cluster region is given by the function depending on only $\{r_{\text{cluster}}\}$ in the BCM:

$$E_{\text{QM}}^{\text{cluster}} = E_{\text{QM}}^{\text{cluster}}(\{\mathbf{r}_{\text{cluster}}\}) \quad (5.21)$$

and

$$E_{\text{CL}}^{\text{cluster}} = E_{\text{CL}}^{\text{cluster}}(\{\mathbf{r}_{\text{cluster}}\}) \quad (5.22)$$

In the QM and CL calculations in the cluster region, additional atoms are set on the dangling bonds in order to terminate the bonds of the atoms in the cluster that are cut on the boundary or surface of the region. This additional atom is called the "buffer atom."

The constituent atomic element of the system is used as the buffer atom to mimic the original bonding in the CL calculation in the cluster region. The position of this buffer atom is adjusted under the condition of the atomic position r_{cluster} in the cluster region to minimize potential energy $E_{\text{CL}}^{\text{cluster}}$ In the QM calculation in the cluster region, the constituent atom or hydrogen atom is chosen as the buffer atom corresponding to the coordination number of the original atom. Its position is provided with reference to a position of the buffer atom determined by the CL calculation in the cluster region and not relaxed when the QM calculation is conducted. If buffer atoms are relaxed in the QM calculation, there is the possibility of various surface reconstructions. When a hydrogen atom is chosen as the buffer atom, its position $\{r_{\text{b}}\}$ is set to be $\mathbf{r}_{\text{b}} = \beta \mathbf{r}_{\text{CL}}^{\text{buffer}} + (1-\beta)\mathbf{r}_{\text{cluster}}$, with a scaling factor β. On the other hand, if the constituent atom is put as the buffer atom, it becomes $\mathbf{r}_{\text{b}} = \beta \mathbf{r}_{\text{CL}}^{\text{buffer}}$.

5.3 Applications

5.3.1 Li-Ion Transfer through the Boundary between Solid/Electrolyte Interphase and Liquid Electrolyte [45]

The SEI film is formed during charging and discharging cycles in an LIB and plays an essential role in the currently used LIBs. Li ions can get transferred through the SEI; on the other hand, the electron current is blocked. So, the SEI prevents further decomposition of the electrolyte [76–78]. The SEI is basically composed of inorganic solids (Li_2O, Li_2CO_3, LiF, etc.) and organic solids like di-lithium ethylene di-carbonate (Li_2EDC) [15, 39, 55, 70]. Li-ion diffusion in the SEI is very important because it is related to battery resistance directly. Li ions are expected to diffuse through the relatively soft organic region. Classical MD simulations for Li-ion diffusion in bulk Li_2EDC, with EC as the electrolyte, were performed [4, 6, 7]. The dynamics of Li ions through the boundary between Li_2EDC and EC are more important and complex than those in the bulk system because you need to consider de-bonding of Li ions from EDC^{2-} and bonding (or solvation) of Li ions to EC. Then we perform the first-principles molecular dynamics (FPMD) simulation runs using a highly parallelized DFT code (DC-RGDFT) to observe the Li-ion transfer dynamics from the SEI to the liquid electrolyte of the EC with or without $LiPF_6$, where $LiPF_6$ is commonly added as the salt in the EC liquid.

An amorphous Li_2EDC system for the SEI and the liquid EC without salt is separately prepared. The Li_2EDC system consists of 72 molecules, the EC system consists of 120 EC molecules, and each density is set to 1.69 g/cm^3 and 1.2 g/cm^3 (10% smaller than the experimental values) [77]. Relaxation calculations are performed with the confining potential corresponding to the averaged pressure of 1 GPa. We then connect the 72 Li_2EDC and 120 EC systems in the x direction to have the total system that models the SEI-electrolyte boundary; the small x and large x region are, respectively, the Li_2EDC and EC subsystems. The total system is set with sizes (L_x, L_y, L_z) = (24.8 Å, 30.9 Å, 30.9 Å). Figure 5.4 shows the atomic configuration of the system. Additionally we prepare the system with the salt, $LiPF_6$, included by the molar concentration 1 M in the EC region. Similar to the previous reports, the dissociation behavior of $LiPF_6$ was observed during the relaxation run.

We have prepared three charge-neutral systems as the initial configurations of the present FPMD runs: (i) 72 Li_2EDC and 120 EC at T = 825 K; (ii) 72 Li_2EDC, 112 EC, and 8 $LiPF_6$ at T = 825 K; (iii) 72 Li_2 EDC and 120 EC at T = 1100 K. The total numbers of atoms are 2352 for (i) and (iii) and 2324 for (ii). In a present run, we insert 36 Li ions at time t = 0, which form a square lattice on an x plane at x = −2 Å. Every 0.3 ps we remove those Li ions located within 10% of L_x from $x = L_x$. Irrespective of such insertion and removal of Li ions, the total number of electrons is fixed. Since the total number of Li ions in the Li_2EDC region is about (72 × 2) – 10 = 134 before starting the present run, the sudden insertion of 36 Li ions corresponds to the increase in the number of Li ions by 27%. We simulate the transfer of Li ions through the EDC^{2-}-EC boundary to understand their dynamics and the effects of $LiPF_6$ salt and temperature on the dynamics.

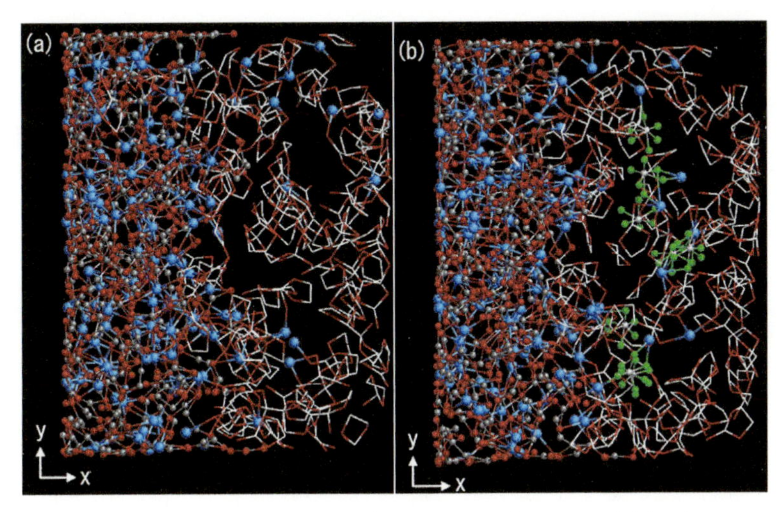

Figure 5.4 Atomic configuration before inserting the Li ions: (a) for the run at $T = 825$ K without salt and (b), at $T = 825$ K with 1 M $LiPF_6$. The EDC^{2-} are drawn by red (O) and gray (C) spheres with bonds; EC, by bonds only; Li ions, by blue spheres; PF_6^- by gray (P) and green (F) spheres with bonds. The H atoms are hidden. Reprinted with permission from Ref. [45]. Copyright (2013) American Chemical Society.

5.3.1.1 Time Evolution of Distribution of Li Ions

Figure 5.5 depicts the time evolution of the distribution of the Li ions along the x direction in the present simulation: panel (a) for the run at $T = 825$ K without salt, (b) at $T = 825$ K with 1 M $LiPF_6$, and (c) at $T = 1100$ K without salt. The location of the EDC^{2-}-EC boundary corresponds to $x = 9 - 12$ Å depending on the y,z position. Each data point in each panel represents the x width of 2 Å averaged over 0.25 ps. In the run at $T = 825$ K without salt, depicted in Fig. 5.5a, the Li-ion density in the range $x = 2.5$–7.5 Å decreases as the run progresses. The growth in the magnitude of the x gradient of the Li density at around $x = 11$ Å suggests that the Li ions are impeded to some extent when passing through the EDC^{2-}-EC boundary. In the run at $T = 825$ K with the salt, depicted in Fig. 5.5b, the Li-ion density in the range $x = 2.5$–7.5 Å decreases as the run progresses. Note in Fig. 5.5b that the magnitude of the x gradient of the Li-ion density at around $x = 11$ Å decreases in time, which suggests enhancement of the Li-ion transfer due to the salt. The hump in the Li-ion density observed in Fig. 5.5b at x = 16–22 Å relates to partial bindings of the Li ions to the 8 PF_6^- molecules. In the run at T = 1100 K without salt, depicted in Fig. 5.5c, the Li-ion density is nearly uniform in the range x = 2.5–7.5 Å during 1.5–1.75 ps, which reflects faster diffusion of Li ions in the EDC^{2-} region as compared to that at the boundary. We also find a substantial growth in the magnitude of the x gradient of the Li density at around x = 11 Å in a similar way as in the run at T = 825 K without salt.

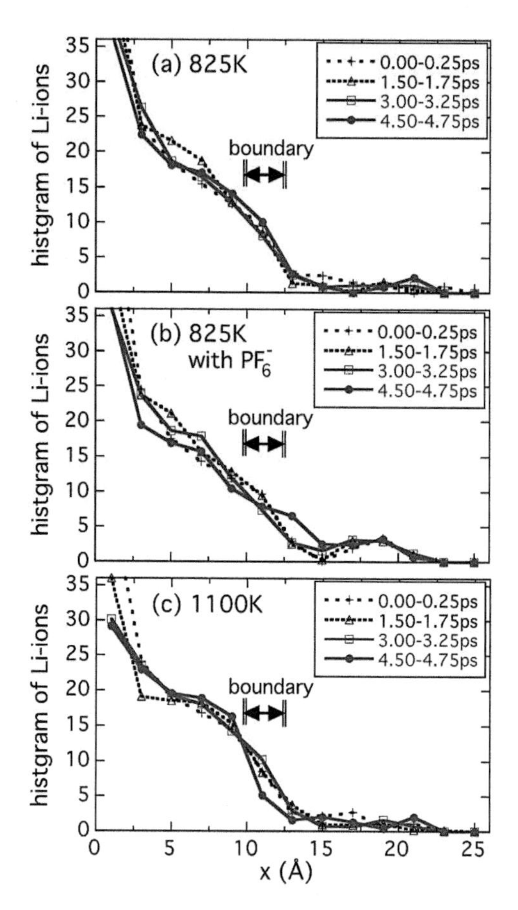

Figure 5.5 Time evolution of x distribution of Li ions (x bin size is 2 Å): (a) for the run at T = 825 K without salt, (b) at T = 825 K with PF_6^-, and (c) at T = 1100 K without salt. Reprinted with permission from Ref. [45]. Copyright (2013) American Chemical Society.

We observe in the present runs that a Li ion hops between the carbonates of EDC^{2-} molecules. As the diffusion constant of the Li ion in the EDC^{2-} region is $D_{Li}(EDC) \approx 10^{-8}$ m^2/s in the present runs, it requires a few picoseconds for a Li ion to move by 4 Å, which is the averaged distance between the nearest-neighbor carbonates of EDC^{2-}. Once a Li ion reaches a site at which both EDC^{2-} and EC bind to the Li ion, that is, the EDC^{2-}-EC boundary, the Li ion becomes energetically more stable (refer to the next section for details). To evaluate the degree of resistance that such a stabilized Li ion experiences in its migration, we exploit the x distribution of the Li ions at a later period of the run. From the Li distributions during t = 4.5–4.75 ps in Figs. 5.5a and 5.5c, we estimate that the site with the characteristic size of 4 Å in reality corresponds to 12–20 Å of bulk EDC^{2-} at T = 825 K and 1100 K if no salt is included. In other words, the EDC^{2-}-EC boundary can be regarded as a virtual material with several times higher resistance for the Li-ion transfer than bulk EDC^{2-}. As for the case at T = 825 K with 1 M $LiPF_6$, the EDC^{2-}-EC boundary is nearly barrierless or its effective depth is nearly zero, as seen in Fig. 5.5b. We will explain later the physical reasons for the resistance lowering due to the salt.

Figure 5.6 shows the time evolution of the accumulated number of Li ions that crossed the boundary from the EDC^{2-} region to the EC region in the present run; Li passing in the reverse direction is taken into consideration. In the analyses we regard a Li ion as having passed the boundary if its x position becomes larger than x = 15.5 Å. The time evolution of the accumulated number of Li ions removed at the large x end is depicted also in Fig. 5.6. At T = 825 K without salt, about 3.2 Li ions pass the boundary in 4.8 ps, as depicted in Fig. 5.6a. We also find in Fig. 5.6a that the number of removed Li ions is larger than the number of passed ones for most of the time during the run. It means that the boundary passing of the Li ions is the rate-limiting step in the whole Li-ion transfer process. Significantly different behaviors of the Li-ion transfer are observed at T = 825 K with the 1 M salt. In the case, as depicted in Fig. 5.6b, the rate of boundary passing of the Li ions is quite high for 1 ps and decreases by about 30% after that. On average 6.5 Li ions cross the boundary in 4.8 ps. That is, the 1 M salt $LiPF_6$ enhances the Li-passing rate by a factor of 2. Here we note that the accumulated number of passed Li ions is larger than or equal to that of removed ones, as seen in

Fig. 5.6b. Contrary to the case without salt at the same temperature (see Fig. 5.6a), the Li-ion diffusion in the EC region is the rate-limiting step in the whole Li transfer process in the case at T = 825 K with the salt.

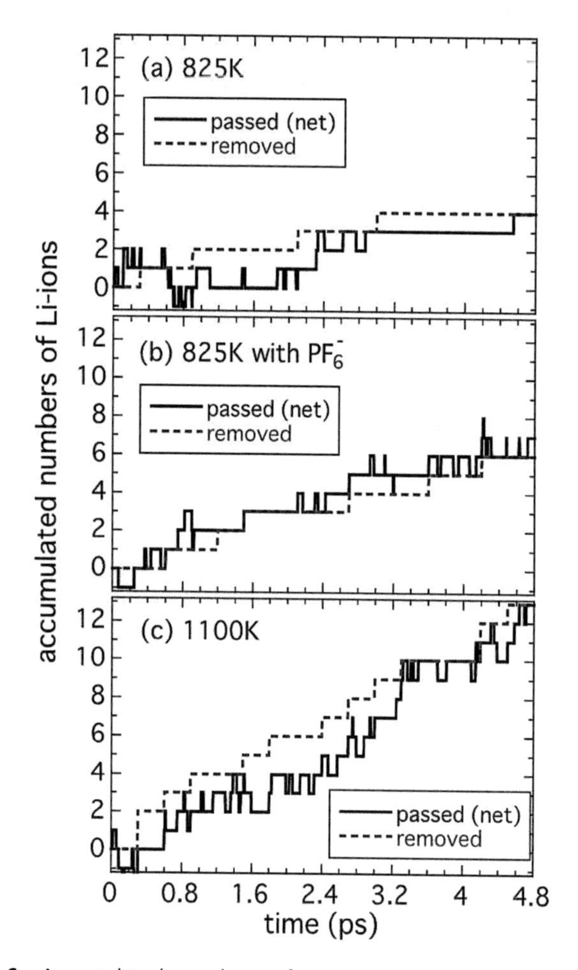

Figure 5.6 Accumulated numbers of Li ions that passed the EDC^{2-}–EC boundary and that are removed at the end of the EC region: (a) for the run at T = 825 K without salt, (b) at T = 825 K with PF_6^-, and (c) at T = 1100 K without salt. Reprinted with permission from Ref. [45]. Copyright (2013) American Chemical Society.

In the case at T = 1100 K without salt, about 11 Li ions pass the EDC^{2-}-EC boundary in 4.8 ps, as seen in Fig. 5.6c. In a similar way to the case at T = 825 K without salt, the accumulated number of removed Li ions is larger than or equal to that of passed ones; that is, the boundary passing is the rate-limiting step. In fact, the Li transfer rate increases by a factor of 11/3.2 = 3.4 while the Li diffusion in the EC region increases by a similar factor of 3.6 when the system temperature increases by a factor of 1100/825 = 1.33. From the temperature dependence of the Li transfer rate, we estimate that about 60 ps would be needed for a single Li ion to pass the EDC^{2-}-EC boundary if a run is performed at T = 350 K without salt.

5.3.1.2 Microscopic mechanisms of Li-Ion transfer through the SEI boundary

We investigate microscopic mechanisms of the resistance of the EDC^{2-}-EC boundary to Li transfer. Figure 5.7 shows the averaged density of oxygen atoms n_O at a distance $r_{Li\text{-}O}$ from a Li ion in either the EDC^{2-} or the EC region: panel (a) for the run at T = 825 K without salt and (b) at T = 1100 K without salt. While n_O in the EDC^{2-} region differs little between the T = 825 K and T = 1100 K cases, the height of the first peak of n_O in the EC region decreases substantially as T increases. In both runs, the location of the first bottom of n_O is 2.8 Å, irrespective of region.

By defining the Li–O bonds as $r_{Li\text{-}O} < 2.8$ Å, we calculate in Table 5.1 the statistics of the binding molecules per Li ion and the residence times of the Li ions. In Table 5.1, we find that the average number of binding EC molecules per Li ion is 2.89 (2.60) at T = 825 K (1100 K) in the EC region. It is less than the room temperature value of 4 [6] because the present runs are performed at higher temperatures. It is particularly remarkable in Table 5.1 that the number of binding molecules per Li ion is substantially larger at the EDC^{2-}-EC boundary than that in either the EDC^{2-} or the EC region. A combination of the smallness of EC molecule compared to EDC^{2-} and the sticking-out shape of EDC^{2-} on the boundary toward the EC region makes EC molecules bind to the Li ion at the boundary in addition to EDC^{2-} molecules. An increase in the total number of binding molecules per Li ion lowers the energy state of the Li ion, resulting in enhanced stability of the Li ion at the boundary. The residence time of a Li ion at the boundary, defined as the lasting period for a Li ion to bind to the same set of molecules, does not get longer because of frequent exchange of binding EC molecules, as seen in Table 5.1.

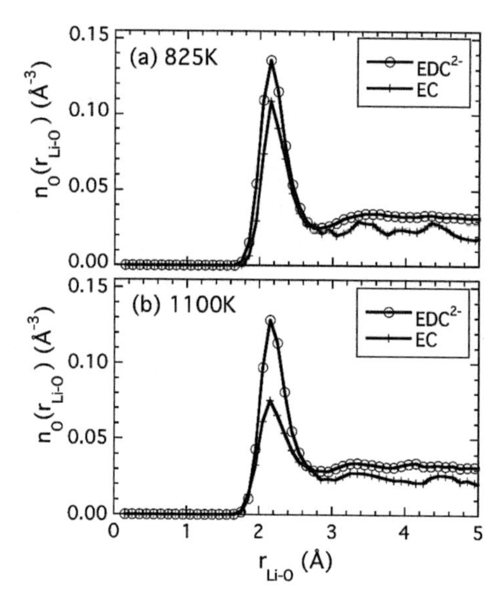

Figure 5.7 The density n_O of O atoms at a distance r_{Li-O} from the Li ion in the EDC^{2-} or EC region: (a) for the run at T = 825 K without salt and (b) at T = 1100 K without salt. Reprinted with permission from Ref. [45]. Copyright (2013) American Chemical Society.

Table 5.1 Averaged numbers of binding molecules per Li ion and residence times of Li ions in the runs without salt

Region	Number of molecules per Li ion	Residence time (ps)
Run at T = 825 K without salt		
EDC^{2-}	2.79	0.32
EDC^{2-}–EC boundary	3.30 (1.79 EDC^{2-}, 1.51 EC)	0.29
EC	2.89	0.27
Run at T = 1100 K without salt		
EDC^{2-}	2.79	0.25
EDC^{2-}–EC boundary	3.32 (1.80 EDC^{2-}, 1.52 EC)	0.20
EC	2.60	0.16

Source: Adapted with permission from Ref. [45]. Copyright (2013) American Chemical Society.

In fact, the probabilities of a Li ion at the EDC^{2-}-EC boundary going to (EC, EDC^{2-}, boundary) regions when the set of binding molecules changes are (0.079, 0.127, 0.794) in the run at T = 825 K without salt. In the run at T = 1100 K without salt, the probabilities are (0.112, 0.164, 0.724). In both runs, the Li ion at the EDC^{2-}-EC boundary has a high probability of staying at the boundary. Hence, we state that the enhanced stability of the Li ion at the EDC^{2-}-EC boundary acts to restrict the Li-diffusion direction to some extent along the boundary.

A typical event of the Li-ion transfer, observed in the run at T = 825 K without salt proceeds as follows. During the period t = 0.95–2.04 ps, a Li ion diffuses along the boundary by changing both EDC^{2-} and EC molecules (numbers of binding molecules are 1 or 2 for EDC^{2-}, while 2 or 1 for EC). The Li ion binds to 1 EDC^{2-} and 2 EC just before t = 2.04 ps. At t = 2.04 ps the Li ion detaches from the boundary, accompanied by additional binding of 1 EC. Hence the Li ion detaches from the boundary without changing the total number of binding EDC^{2-} and EC molecules.

We recall two facts: (i) the Li ion at the EDC^{2-}-EC boundary has the probability of staying at the boundary several times as high as that of going to the EDC^{2-} region and (ii) the Li residence time at the EDC^{2-}-EC boundary differs little from that in the EDC^{2-} region, as shown in Table 5.1. Those facts in combination give the physical reason of our observation that the contact site between EDC^{2-} and EC can be regarded as a virtual material of several times higher resistance for the Li-ion transfer than bulk EDC^{2-}, which is mentioned above relating to Figs. 5.5a and 5.5c.

Figure 5.8 depicts the time evolution of the number of Li ions bonding to PF_6^- in the run at T = 825 K with the 1 M $LiPF_6$. For the analyses we set the cutoff distance of 2.3 Å to define the Li–F bonds, which is 21% longer than the Li–F distance of $LiPF_6$ in vacuum in the ground state. The number of bonding Li ions fluctuates in time significantly between 2 and 12, which relates to the flow of Li ions in the EC region. When the number is less than 8, the 8 PF_6^- molecules assume partially dissociated states. For 4–5 ps the number is about 8, which means the saturation of Li ions around PF_6^- molecules due to combined effects of a relatively high Li-ion transfer rate through the boundary and a relatively slow Li-ion diffusion in the EC region.

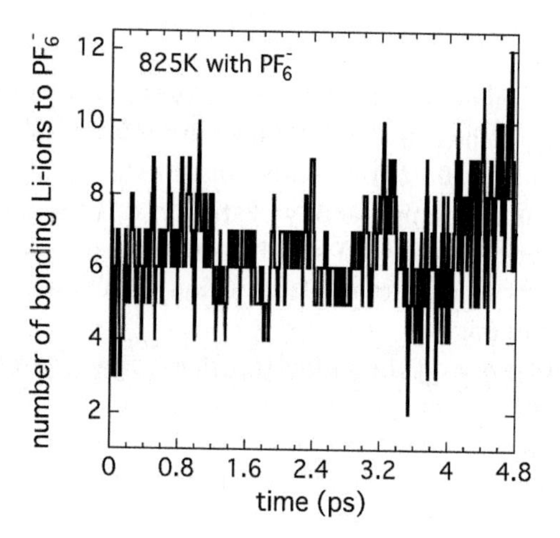

Figure 5.8 Time evolution of the number of bonding Li ions to PF_6^- in the run at T = 825 K with PF_6^-. Reprinted with permission from Ref. [45]. Copyright (2013) American Chemical Society.

We explain the physical reasons for the enhanced Li-ion transfer rate due to PF_6^-. A detailed examination of the run at T = 825 K with the 1 M salt clarifies that a Li ion binds to 1 or 2 EDC^{2-}, 1 EC, and 1 PF_6^- just before its detachment from EDC^{2-} at the boundary. We pick up an atomic configuration of only those atoms of 1 $LiEDC^-$, 1 Li^+, 1 EC, and 1 PF_6^- that are involved in such a detachment event. Figure 5.9c depicts the ground state configuration of the system in vacuum. Separately, we replace PF_6^- in the system by EC to model the case without salt. We obtain the ground state of it as shown in Fig. 5.9a. In the case without salt (Fig. 5.9a), a Li ion binds to two O atoms of EDC^{2-} with distances 1.96 Å and 2.09 Å and to an O of an EC with 2.36 Å; the distance $r_{\mathrm{Li-C_c}}$ between the Li ion and the C of carbonate of EDC^{2-} is $r^{\mathrm{eq}}_{\mathrm{Li-C_c}} = 2.16$ Å. Note that the Li ion does not bind to another EC. Figure 5.9 (bottom) shows the system energy calculated with the RGDFT as a function of $r_{\mathrm{Li-C_c}}$ to give the Li detaching energy 1.7 eV. The configuration at $r_{\mathrm{Li-C_c}} = r^{\mathrm{eq}}_{\mathrm{Li-C_c}} + 1.5$ Å is depicted in Fig. 5.9b, in which the Li ion binds to an O of an EC with the distance 2.04 Å and to an O of another EC with the distance 2.07 Å.

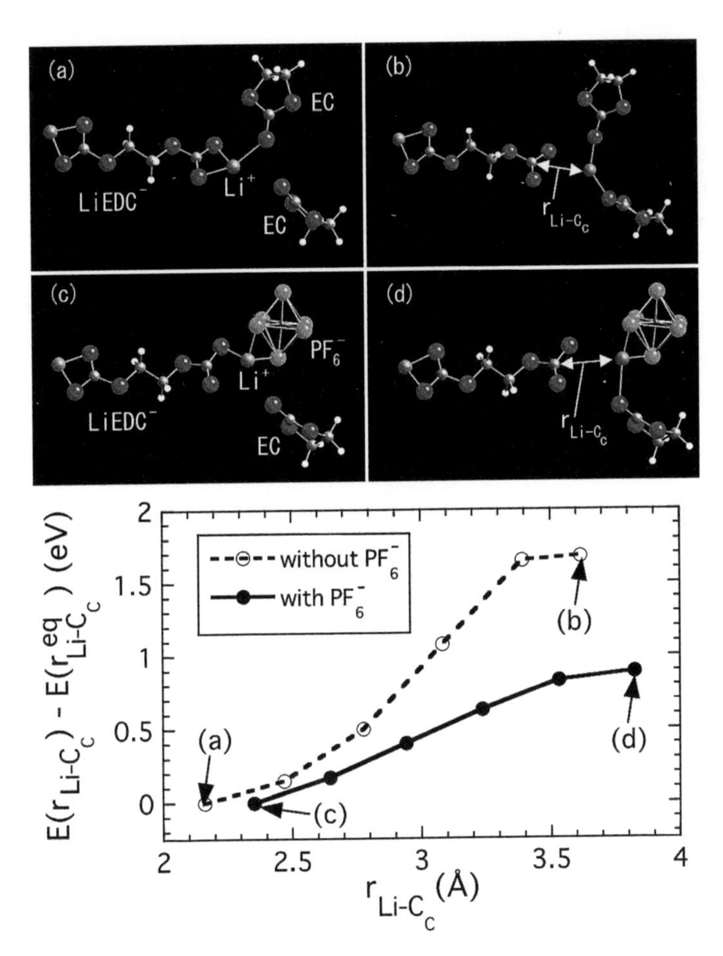

Figure 5.9 (a) Atomic configuration of $LiEDC^-$, Li^+, and 2 EC in the ground state ($r^{eq}_{Li-C_c} = 2.16$ Å). (b) Same as (a) but $r_{Li-C_c} = r^{eq}_{Li-C_c} + 1.5$ Å . (c) Atomic configuration of $LiEDC^-$, Li^+, EC, and PF_6^- in the ground state ($r^{eq}_{Li-C_c} = 2.35$ Å). (d) Same as (c) but $r_{Li-C_c} = r^{eq}_{Li-C_c} + 1.5$ Å . (bottom) The system energies as functions of r_{Li-C_c} for the two systems depicted in (a, b) and (c, d). Reprinted with permission from Ref. [45]. Copyright (2013) American Chemical Society.

As for the case with PF_6^- in Fig. 5.9c, a Li ion binds to an O of EDC^{2-} with the distance 1.95 Å and to two F atoms of PF_6^- with distances 2.01 Å and 1.96 Å; $r^{eq}_{Li-C_c}$ = 2.35 Å. The Li ion does not form a bond with an O of EC; the distance is 2.83 Å. The Li detaching energy is 0.9 eV, as seen in Fig. 5.9 (bottom). When $r_{Li-C_c} = r^{eq}_{Li-C_c}$ + 1.5 Å, as shown in Fig. 5.9d, the Li ion binds to 2 F atoms with distances 1.99 Å and 1.92 Å and to an O of EC with a distance 2.48 Å. The Li detaching energy decreases significantly because the strong binding of PF_6^- to the Li ion weakens the interaction between the Li ion and EDC^{2-}, as manifested in the existence of only a single Li–O bond in Fig. 5.9c as compared to two Li–O bonds in Fig. 5.9a. After detaching of the Li ion from the boundary in the combined form of $Li^+ - PF_6^-$, the Li ion has a high probability of dissociation (breaking of Li–F bonds) to be solvated by EC. After the dissociation, the PF_6^- diffuses and binds to another Li ion at the EDC^{2-}-EC boundary. By this way, the effective barrier for the Li-ion transfer through the EDC^{2-}-EC boundary is lowered by PF_6^-.

5.3.2 Hybrid Quantum-Classical Simulation on the Diffusivity of Lithium in Graphite [51]

5.3.2.1 Intraplane correlation between Li ions

In this section we investigate diffusivity of multiple Li ions inserted into the same interlayer space. The lithium-graphite intercalation compound (Li-GIC) is put to practical use as the negative electrode of the LIBs. In the LIB, Li ions move from the negative electrode to the positive one during discharge and back during charging. The transport rate of Li ions in graphite is related to the output power and fast charging of the LIBs. Improving LIBs requires theoretical understanding of the Li-ion dynamics in realistic settings with high accuracies. In the Li-GIC, a Li ion creates a long-ranged stress field around itself by expanding the interlayer distance by about 10% [46, 49]. It is, therefore, necessary to take into account such distortion of C layers of graphite in simulating the diffusivity of Li ions in graphite. Since the weak interlayer interaction of graphite originates from the van der Waals interaction, conventional DFT with LDA or generalized-gradient approximations cannot treat it accurately. In the hybrid QM-CL simulation, the van der Waals interlayer interaction in graphite can be taken into account through a classical interatomic potential for both QM and CL regions. Moreover, the expansion of the interlayer space by about 10% increases the potential energy of the graphite. Such vertical deformation spreads horizontally to about 5 Å, corresponding to the size of a six-membered C ring. When multiple Li ions exist in the same interlayer space of graphite, they gather together, with each Li ion sitting in the center of two 6-membered C rings (i.e., C_6Li structure) in the ground state; in other words, Li ions are trapped in a C cage. It is because the insertion energy of another Li ion to already expanded C layers is lower than that without expansion as the positive charge of a Li ion is screened well by the neighboring C atoms. In the present study we will investigate possible formation of the C cage for Li ions even at an elevated temperature and its effects on thermal diffusivity of Li ions. Figure 5.10 depicts a snapshot of a simulation run at temperature T = 443 K. The total system contains nine C layers composed of 12,096 C atoms. In total, seven Li atoms are inserted into the same interlayer space. The two C layers sandwiching the Li ions are set to AA-stacking (i.e., the C atoms on the two layers have identical *x*-*y* positions), while others are set to AB-stacking. As the stacking sequence of graphite is known experimentally for two extreme cases only (AB-stacking for pure graphite and AA-stacking for the maximally Li-intercalated graphite at C_6Li stoichiometry), we start the simulation with AA-stacking for the two sandwiching C layers and then let the stacking sequence relax thermally. As we see in Fig. 5.10 and Fig. 5.18b, the stacking sequence of the two sandwiching C layers fluctuates and has substantial chances of assuming both AA- and AB-stackings. A region around Li ions is set as the QM one. The periodic boundary conditions are applied to *x*, *y*, and *z* directions. The *z* axis is perpendicular to the C layers. The system size is set to (L_x, L_y, L_z) = (59.57, 58.96, 30.58) Å. The atomic configuration in Fig. 5.10 is obtained after relaxation for 1 ps at T = 443 K with the *x*-*y* positions of Li fixed. The time step is 0.97 fs with the velocity-Verlet algorithm for the dynamics. The CL interatomic potentials used are explained below.

We perform two simulation runs at T = 443 K with different initial arrangements of seven Li ions. In a run, seven (N_{Li} = 7) Li ions are placed, one by one, at the centers of neighboring six-membered C rings in the *x*-*y* view; we call it the C_2Li run (see the panel of t = 0 ps in Fig. 5.11). In the other run, seven Li ions assume a coarser grid initially, to be called the C_6Li run (see the panel of t = 0 ps in Fig. 5.13).

Figure 5.10 The simulation system in the C_2Li run at t = 0 ps viewed from y direction. Cyan, magenta, and gray spheres denote Li ions, QM-C, and CL-C atoms, respectively. Reprinted from Ref. [51], Copyright (2015), with permission from Elsevier.

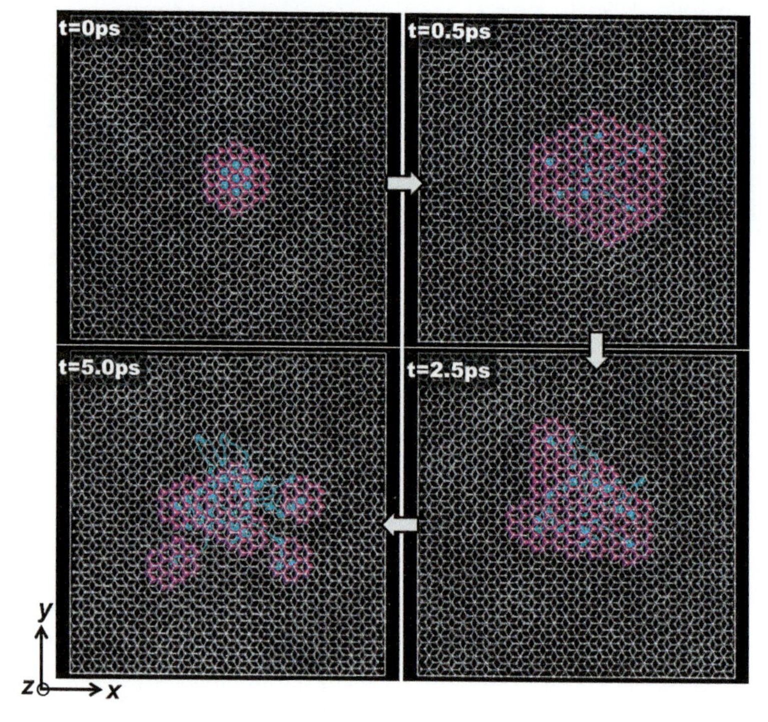

Figure 5.11 The snapshots in the C_2Li run. Cyan and magenta spheres are Li ions and QM-C atoms, respectively. Gray spheres are CL-C atoms. Cyan curves are the trajectories of the Li ions. Reprinted from Ref. [51], Copyright (2015), with permission from Elsevier.

Figure 5.11 depicts the time evolution of the total system in the C_2Li run. In Fig. 5.11, magenta and cyan correspond, respectively, to C and Li in the QM regions. At t = 5 ps, four QM regions are set. Since a Li ion has a positive charge of +e after the electron transfer to neighboring C atoms [46, 50], Li ions scatter each other isotropically during t = 0–0.5 ps. After about t = 1 ps, Li ions appear to diffuse thermally. As the trajectory curves of Li ions indicate, Li ions have a tendency to diffuse within a limited *x*-*y* area.

It is interesting to analyze possible deformation of the C layers sandwiching Li ions. Figure 5.12 depicts the *x*-*y* distribution of the distance d between the upper and lower C layers of Li ions at t = 2.5 ps in the C_2Li run at T = 443 K. Seven triangles in Fig. 5.13 are Li ions at t = 2.5 ps. For reference, the equilibrium values of d at T = 0 K and 443 K in our hybrid simulation method are 3.44 Å and 3.6 Å, respectively, for the AA-stacking of graphite without Li ions. In the C_2Li run, we find that the values of d where Li ions reside are larger than 3.6 Å at t > 0.5 ps; the average of d over the *x*-*y* area is 4 Å, forming a cage of the C layers to enclose all the Li ions.

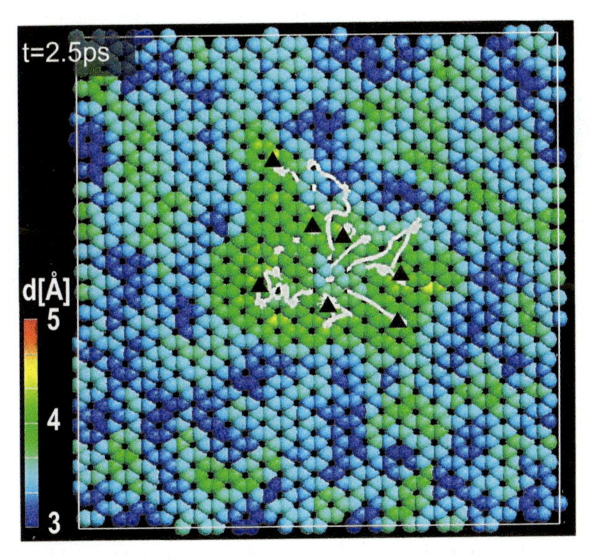

Figure 5.12 The *x*-*y* distribution of the interlayer distance (d) between the higher and lower layers sandwiching Li ions in the C_2Li run at t = 2.5 ps. Triangles denote Li ions, with their trajectories shown by gray curves. Reprinted from Ref. [51], Copyright (2015), with permission from Elsevier.

The diffusion processes of Li ions in the C_6Li run are shown in Fig. 5.13. In the run, Li ions thermalize faster than in the C_2Li run and diffuse thermally without significant repulsive scattering after t = 0.5 ps. It is because the initial Li-Li distance is 4.2 Å and therefore the positive charge of a Li ion is screened well at the distance. In a similar way to the C_2Li run, Li ions have a tendency to diffuse in a limited area except for a single Li ion, as seen in the panel of t = 5 ps. Our hybrid simulation method has the advantage of being able to analyze the electronic structure. We confirm that each Li atom is singly ionized. The degree of screening of the Li-ion charge by the surrounding C atoms is more significant at higher concentration of Li.

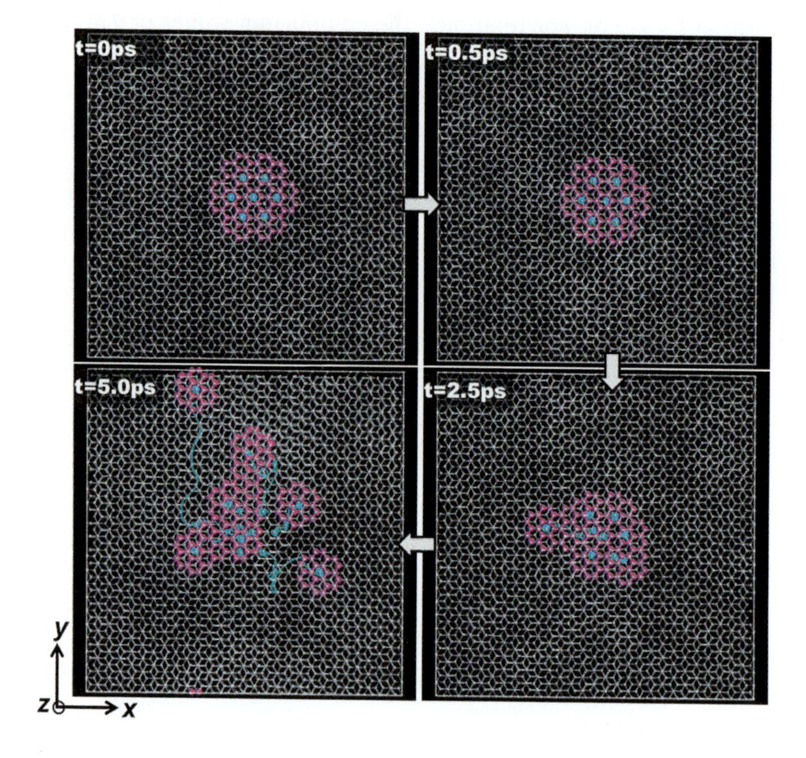

Figure 5.13 The same as Fig. 5.11 but in the C_6Li run. Reprinted from Ref. [51], Copyright (2015), with permission from Elsevier.

To further confirm the formation of the C cage, we define the 2D expansion radius of Li ions at time t as

$$R_{\exp}(t) = \frac{1}{N_{\text{Li}}} \sum_{i}^{N_{\text{Li}}} \sqrt{\left|\vec{r}_i(t) - \vec{r}_{\text{COM}}(t)\right|^2} \tag{5.23}$$

with the position of the center of mass (COM) of Li ions

$$\vec{r}_{\text{COM}}(t) = \frac{1}{N_{\text{Li}}} \sum_{i}^{N_{\text{Li}}} \vec{r}_i(t) \tag{5.24}$$

and analyze the time evolution of $R_{\exp}$ in both C_2Li and C_6Li runs. Figure 5.14 shows time evolutions of $R_{\exp}$ in both runs. If Li ions diffuse independently, $R_{\exp}$ should be proportional to $\sqrt{t}$. However, as is obvious in Fig. 5.14, $R_{\exp}$ saturates faster in both runs. In the C_2Li run, $R_{\exp}$ begins glowing from 0 to 0.5 ps, reflecting isotropic scattering via mutual Coulomb repulsion. At about t = 5 ps, a stable cage appears to form. In the C_6Li run, a similar cage is formed at about t = 3 ps. In both runs, $R_{\exp}$ gets saturated to 13.5 Å (t > 11 ps). We thereby find that $R_{\exp}$ gets saturated to about 14 Å for the larger-system run also after 10 ps [51]. The saturated expansion radius $R_{\exp}$ ≈ 14 Å for seven Li ions in the present and large-system runs that correspond to the interaction range (~10 Å) plus one or two times the diameter of the six-membered C ring.

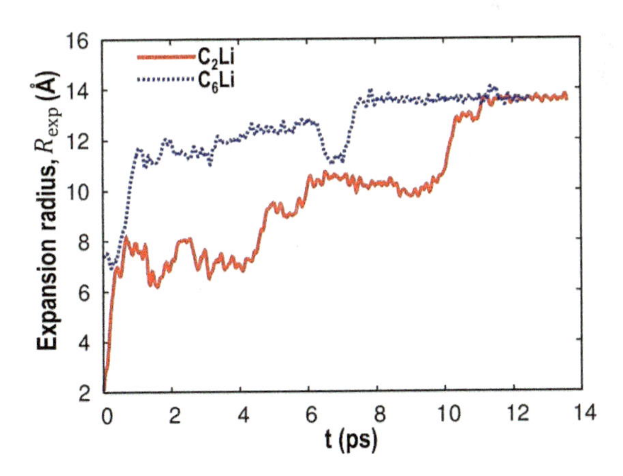

Figure 5.14 The time evolutions of the expansion radius ($R_{\exp}$) of Li ions in both C_2Li and C_6Li runs. Red (solid) and blue (dotted) curves are for the C_2Li and C_6Li runs, respectively. Reprinted from Ref. [51], Copyright (2015), with permission from Elsevier.

We now consider the mean squared displacement of Li ions

$$\sigma_{\mathrm{Li}}^2(t_\mathrm{e}) = \left\langle \frac{1}{N_{\mathrm{Li}}} \sum_i^{N_{\mathrm{Li}}} \left| \vec{r}_i(t_\mathrm{e}+t_0) - \vec{r}_i(t_0) \right|^2 \right\rangle_{t_0} \quad (5.25)$$

and the squared displacement of the COM of Li ions

$$\sigma_{\mathrm{COM}}^2(t_\mathrm{e}) = \left\langle \left| \vec{r}_{\mathrm{COM}}(t_\mathrm{e}+t_0) - \vec{r}_{\mathrm{COM}}(t_0) \right|^2 \right\rangle_{t_0} \quad (5.26)$$

at elapsed time t_e. The t_0 for the average $\langle \cdots \rangle$ are set as every 0.01 ps after the formation of a stable cage (after t = 5 ps and t = 3 ps for the C_2Li and C_6Li runs, respectively). The time (t_e) evolutions of σ_{Li}^2 and σ_{COM}^2 in both runs are shown in Fig. 5.15 for a numerically reliable range of t_e = 0–5 ps. For a reference, the squared displacement of a Li ion in the simulation run of a single Li ion inserted in graphite (AA-stacking) is drawn in Fig. 5.15 using the dashed curve. We find in Fig. 5.15 that the diffusivity of the COM (σ_{COM}^2) appears to be similar to that of the single-Li-ion run. The Li diffusivity in the C_2Li run agrees well with that in the C_6Li run, indicating high accuracy of the present results. It is remarkable that a Li ion diffuses much faster within the C cage in both C_2Li and C_6Li runs than in the single-Li-ion run.

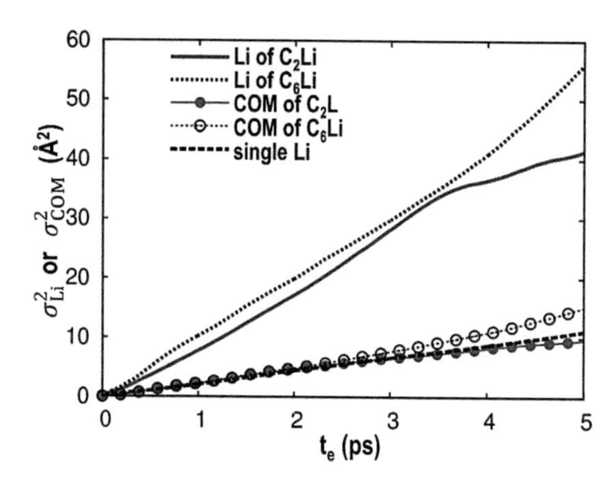

Figure 5.15 The time (t_e) evolution of the mean squared displacement (σ^2) after the formation of the stable cage. Solid and dotted curves show σ_{Li}^2 in the C_2Li and C_6Li runs, respectively. The dashed curve shows σ_{Li}^2 of an isolated Li ion. The mean squared displacements of the COMs of Li ions are shown for the C_2Li and C_6Li runs, respectively, by solid and dotted curves with symbols. Reprinted from Ref. [51], Copyright (2015), with permission from Elsevier.

We give the reasons for such fast Li diffusion within the C cage in both C_2Li and C_6Li runs. Remember the relation of Li diffusivity and interlayer distance (d) in graphite, which is obtained in the hybrid simulation runs in Refs. [46, 49]: (i) a Li ion diffuses faster for expanded d, which is demonstrated up to about 10% expansion, and (ii) when d is expand further, a Li ion shows a tendency to be trapped by either the upper or the lower layer of graphite, resulting in lower diffusivity. In both C_2Li and C_6Li runs, d within the C cage of radius R_{exp} =13.5 Å is about 4 Å, which corresponds to 9% expansion of the equilibrium value at T = 443 K. Therefore, each Li ion in the present runs diffuses fast within the vertically expanded C cage. Because of the confinement of Li ions in the C cage, σ_{Li}^2 should get saturated at much longer t_e (>30 ps). However, it is too long for us to simulate.

Figure 5.16 shows time (t_e) evolution of the ratio $\sigma_{\mathrm{Li}}^2/\sigma_{\mathrm{COM}}^2$ in both C_2Li and C_6Li runs. If each Li ion diffuses isotropically in a random manner, that is, without inter-Li correlation, the $\sigma_{\mathrm{Li}}^2/\sigma_{\mathrm{COM}}^2$ should be equal to N_{Li} = 7. As Fig. 5.16 shows, the ratio is 3–4 after reaching the thermal equilibration in the present runs. It means that the COM of Li ions diffuses faster in the C_2Li and C_6Li runs than in the random situation, which indicates that there is a correlation between Li ions. In other words, Li ions altogether have a tendency to diffuse cooperatively.

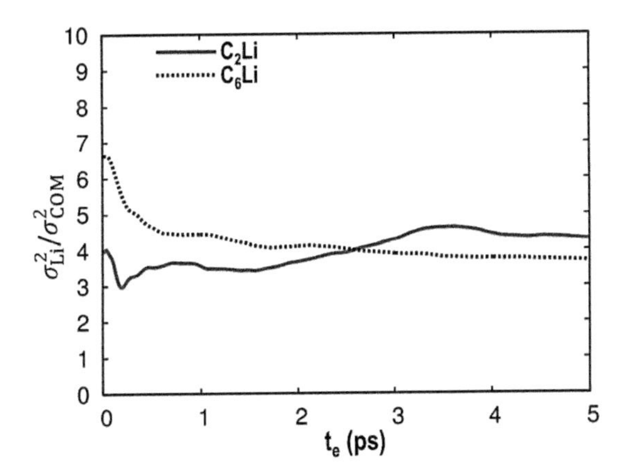

Figure 5.16 The time (t_e) evolution of the ratio $\sigma_{\mathrm{Li}}^2/\sigma_{\mathrm{COM}}^2$. Solid and dotted curves are for the C_2Li and C_6Li runs, respectively. Reprinted from Ref. [51], Copyright (2015), with permission from Elsevier.

Considering the behavior of σ_{Li}^2 in Fig. 5.15 and the C cage radius (R_{exp}), we expect that the diffusion constant for Li ions at longer times can be obtained in either C_2Li or C_6Li case if the run is continued to t = 30 ps at least. The diffusion constant of the COM, that is, $\sigma_{\mathrm{COM}}^2(t_{\mathrm{e}})/4t_{\mathrm{e}}$, at a short time of t_{e} = 5 ps is an overestimation of the long-time diffusion constant because the intracage diffusion of Li ions is included in σ_{COM}^2. In the following we evaluate the long-time diffusion constant of the case, which is equivalent to the long-time diffusion constant of Li ions. Introducing the central position of cage $\vec{r}_{\mathrm{cage}}$, the position of the ith Li ion is written as

$$\vec{r}_i = \vec{r}_{\mathrm{cage}} + d\vec{r}_i \tag{5.27}$$

where $d\vec{r}_i$ is the position of the ith Li ion measured from $\vec{r}_{\mathrm{cage}}$. Since $\vec{r}_{\mathrm{COM}} = \vec{r}_{\mathrm{cage}} + (1/N_{\mathrm{Li}})\sum_i^{N_{\mathrm{Li}}} d\vec{r}_i$,

$$\sigma_{\mathrm{COM}}^2 = \sigma_{\mathrm{cage}}^2 + \frac{1}{N_{\mathrm{Li}}}\sigma_{\mathrm{d}}^2 \tag{5.28}$$

if the correlation between $d\vec{r}_i$'s is ignored, which is appropriate at short times. Here σ_{dLi}^2 is the mean squared displacement with respect to { $d\vec{r}_i$ }. Since $\sigma_{\mathrm{cage}}^2 \ll \sigma_{\mathrm{dLi}}^2$, we may approximate $\sigma_{\mathrm{dLi}}^2 \approx \sigma_{\mathrm{Li}}^2$. We, therefore, obtain

$$\sigma_{\mathrm{cage}}^2 \approx \sigma_{\mathrm{COM}}^2 - \frac{1}{N_{\mathrm{Li}}}\sigma_{\mathrm{Li}}^2 \tag{5.29}$$

With N_{Li} = 7. Figure 5.17 shows σ_{cage}^2 in both C_2Li and C_6Li runs. The long-time diffusion constant of the cage is of the same order as or somewhat smaller than that of a Li ion in the single-Li-ion run. Since it takes longer to reach a thermal equilibrium in the C_2Li run in comparison with the C_6Li one, the reliable range of elapsed time for the diffusion analyses becomes shorter in the C_2Li run. Relating to this, the slopes of the red (solid) and blue (dotted) curves in Fig. 5.17 are nearly the same for up to 2 ps, indicating reliability. The long-time diffusion constant of Li ions is thereby predicted as (3 ± 1) × 10^{-5} cm^2/s, which is in good agreement with the experimental value of order 10^{-5} cm^2/s in the dilute Li density phase [20].

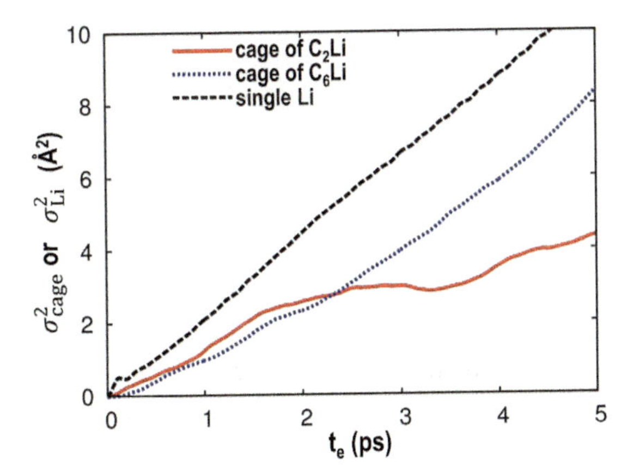

Figure 5.17 The time (t_e) evolution of σ^2_{cage}. Solid and dotted curves are for the C_2Li and C_6Li runs, respectively. For a reference, the dashed curve is σ^2_{Li} of an isolated Li ion. Reprinted from Ref. [51], Copyright (2015), with permission from Elsevier.

As explained above, the long-time diffusion constant of the C cage in the multiple-Li-ion run is of the same order as that of an isolated Li ion. The reasons for this are as follows. In the single-Li-ion run at $T = 443$ K, the C cage, defined as the region with $d > 3.67$ Å around Li ions, has a radius of about 4 Å. We regard a C–C pair between the higher and lower C layers sandwiching Li ions as bonded temporarily if $d < 3.67$ Å. Comparing the present seven-Li-ion and single-Li-ion runs, we find that the numbers of bonding pairs change by nearly the same extent with a cycle of 4 times per 1 ps, which corresponds to the vertical B_{2g}-optical vibration frequency of C layers at 3.81 THz, obtained via the neutron scattering experiment [17, 37]. This indicates that a local edge of the C cage fluctuates thermally in the *x-y* position in both runs. In addition, the edge shape of the C cage is not circular, as seen in Fig. 5.12. We, therefore, consider that the elementary process of the horizontal motion of the C cage is the shift of a local edge of the C cage following the movement of a Li ion inside the C cage. Other Li ions in the C cage should have a tendency to move with the Li ion in a cooperative manner. Therefore, the long-time diffusion constant of the C cage in the multiple-Li-ion case is of the same order and also substantially smaller than that in the single-Li-ion case.

5.3.2.2 Interplane correlation between Li ions

In this section, we perform the hybrid simulation run with 14 Li ions inserted, 7 by 7, into two interlayer spaces of graphite (10 C layers). The total system at the starting point of the simulation is shown in Figs. 5.18a and 5.18b as the top and side views, respectively. The stacking structures of the C layers that sandwich Li ions are AA-stacking, as depicted in Fig. 5.18b. The temperature of the total system is 443 K. Figure 5.18c shows the snapshot at t = 3.4 ps, where four QM regions are set. The largest QM region contains 625 C atoms and 11 Li ions; it takes about 9 minutes per time step using 180 CPUs of K-computer at AICS of RIKEN with the domain decomposition 2×2×1, the spatial decomposition of each domain 5×5×3, the band parallelization 2, and the OpenMP parallelization 4 set in the divide-and-conquer-type O(N) RGDFT (named DC-RGDFT) code [48]. Because of its large size, we can continue the simulation run up to 3.4 ps.

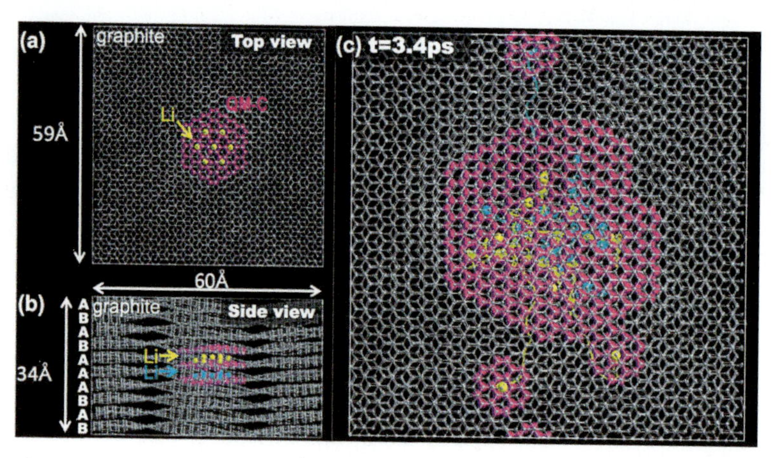

Figure 5.18 The snapshots in the simulation run with 14 Li ions inserted in graphite: (a) top view at t = 0 ps, (b) side view at t = 0 ps, and (c) top view at t = 3.4 ps. Yellow, cyan, magenta, and gray spheres denote Li ions in the upper layer, Li ions in the lower layer, QM-C atoms, and CL-C atoms, respectively. Yellow and cyan curves are the trajectories of Li ions in the upper and lower layers, respectively. Reprinted from Ref. [51], Copyright (2015), with permission from Elsevier.

The trajectories of Li ions up to t = 3.4 ps are depicted in Fig. 5.17c; yellow (cyan) curves are for the Li ions at higher (lower)-z.

The Li ions concentrated initially in the higher and lower spaces of C layer form domains first. The domains repel each other due to their repulsive interplane interaction. Then the Li ions in the higher-z continue to diffuse in the opposite direction to that in the lower-z for long time periods. Since the positive charge of Li ions is screened well at a distance of a few angstroms by the surrounding negatively charged C atoms, such an underlying repulsive interaction at a long distance should originate from deformation of the C layers that sandwich Li ions. This finding supports the hypothetical idea that the stage structure of Li ions (i.e., alternating Li-rich and Li-poor planes appear) forms to release the mechanical stress of the C layers due to the Li ions in the Li-GIC.

5.4 Conclusion and Future Scope

Electronic-state and MD calculations were performed to investigate the reaction mechanisms and dynamics of Li ions in the LIBs. To directly consider the electrode structural change, interface phenomenon, and potential influence due to Li-ion migration, it is necessary to develop a large-scale calculation method that links the electronic-level microscopic processes with the macroscopic dynamics processes. Therefore, we described two types of original atomistic calculation methods; (i) the order-N DC-RGDFT method and (ii) the hybrid QM-CL simulation method. Both methods appropriately treat boundary features of the divided system. These methods can then realize a universality and robustness with a high degree of precision, as shown in two applications in Section 5.3.1 (the simulation for microscopic mechanisms of Li-ion transfer between the organic SEI formed on the graphite anode and the liquid electrolyte) and Section 5.3.2 (an application of the hybrid QM-CL simulation to elucidate the thermal diffusion process of Li ions in the graphite anode). We also note our recent results of the hybrid quantum classic simulation in the investigation of the insertion process of Li ions at the graphite/EC electrolyte interface, which successfully disclosed the reductive decomposition reaction of EC catalyzed by Li-ion proximity leading to the SEI formation [52].

To deal with the migration of atoms and the chemical reaction simultaneously, FPMD calculations, which can account for the electronic structure of systems such as chemical bonding, are needed. Although the DC-RGDFT method introduced in this chapter can be applied to thousands of atomic scales using supercomputers, the simulation time is about 10 ps (time step of 10,000) at its best. In the MD simulation, the time step is set to about 1 fs, as it follows motion associated with the thermal vibration of atoms (ions). In the hybrid quantum-classical simulation method, which limits the spatial region described by the time-consuming DFT calculation, the spatial and time scale of the target system is greatly expanded compared to the FPMD calculations without degrading calculation accuracy. However, to do so, it is necessary to ensure the accuracy of the calculation of the classical region represented by empirical interaction potential. For example, for specific elements such as silicon and carbon, a force field (empirical interaction potential) develops, which can accurately describe various bonding environments corresponding to various crystal structures [8, 9, 65, 71]. These force fields express interatomic potentials in a relatively simple functional form and optimize the parameters to reproduce the results of the first-principles calculation. However, in order to create a highly reliable force field, enormous trial-and-error calculations are required for optimization of the parameters. Furthermore, its algorithm becomes complicated because the functional form of the potential is selectively used according to the combination of chemical bonds. Therefore, a method of applying a machine learning scheme such as a neural network or Gaussian approximation was recently developed to create an interatomic potential so as to match the result of the first-principles calculation [16, 26, 27, 29, 34, 36]. In machine learning potential, using atomic position as input and learning the result of the first-principles calculations on a relatively small scale as supervised data, it is possible to obtain high generalization performance corresponding to various coupling situations. An example of accurately evaluating chemical reactions on nanoparticles using machine learning potential has also been reported [26, 27], and it has become a new trend in large-scale calculations.

On the other hand, to track the time evolution of chemical reactions, it is also essential to expand the time scale as well as the spatial scale. For example, in order to investigate the details of the formation process of SEI films [77, 78] in an LIB, it is necessary to follow the chemical reaction for a long time. Using the hyper dynamics method [74], a bias is added to the interatomic potential to raise the potential of the system and urge the transition from the local stable state to the next state. The time scale of the transition is accelerated by $e^{\Delta V / k_B T}$ times due to the bias potential of ΔV (k_B is Boltzmann constant and T shows temperature). As a setting for the bias potential, a bond boost method [57] exists, which expands the strain from the equilibrium of the bonding length and cuts the atomic bond to accelerate the diffusion phenomenon. To effectively accelerate a desired phenomenon, it is necessary to appropriately set the bias potential, and the computational scheme is further developed.

Finally, the development of large-scale calculation methods has matured with the growth of computers. Along with the emergence of new architectures, such as those of quantum computers, it is expected that the algorithms of large-scale calculation methods will evolve as well.

Acknowledgments

This research was supported by MEXT Strategic Programs for Innovative Research (SPIRE), Computational Materials Science Initiative (CMSI), High Performance Computing Infrastructure (HPCI) of RIST (hp120123, hp13022, hp140096, hp140214, hp150041), and Grant-in-Aid for Scientific Research (Kakenhi: 23310074) of Japan. The computations were performed using Fujitsu FX10 at Information Technology Center of the University of Tokyo, Hitachi SR16000 at Institute of Material Research of Tohoku University, Fujitsu PRIMERGY at Research Center for Computational Science (Okazaki), Fujitsu FX10 at Institute for Solid State Physics of the University of Tokyo, Fujitsu FX1 and FX10 at Information Technology Center of Nagoya University, and K-computer at RIKEN.

References

1. Armand, M. and Tarascon, J.-M. (2008). Building better batteries, *Nature*, **451**, pp. 652.
2. Balbuena, P. B. and Wang, Y. (2004). *Lithium-Ion Batteries: Solid-Electrolyte Interphase* (Imperial College Press).

3. Bauschlicher Jr, C. W., Haskins, J. B., Bucholz, E. W., Lawson, J. W., and Borodin, O. (2014). Structure and energetics of $Li^+-(BF4^-)_n$, $Li^+-(FSI^-)_n$, and $Li^+-(TFSI^-)_n$: ab initio and polarizable force field approaches, *J. Phys. Chem. B*, **118**, pp. 10785–10794.
4. Borodin, O. and Smith, G. D. (2006). Mechanism of ion transport in amorphous poly (ethylene oxide)/LiTFSI from molecular dynamics simulations, *Macromolecules*, **39**, pp. 1620–1629.
5. Borodin, O. (2009). Polarizable force field development and molecular dynamics simulations of ionic liquids, *J. Phys. Chem. B*, **113**, pp. 11463–11478.
6. Borodin, O. and Smith, G. D. (2009). Quantum chemistry and molecular dynamics simulation study of dimethyl carbonate: ethylene carbonate electrolytes doped with $LiPF_6$, *J. Phys. Chem. B*, **113**, pp. 1763–1776.
7. Borodin, O., Zhuang, G. V., Ross, P. N., and Xu, K. (2013). Molecular dynamics simulations and experimental study of lithium ion transport in dilithium ethylene dicarbonate, *J. Phys. Chem. C*, **117**, pp. 7433–7444.
8. Brenner, D. W. (1990). Empirical potential for hydrocarbons for use in simulating the chemical vapor deposition of diamond films, *Phys. Rev. B*, **42**, pp. 9458.
9. Brenner, D. W. (1992). Err^{a}tum: Empirical potential for hydrocarbons for use in simulating the chemical vapor deposition of diamond films, *Phys. Rev. B*, **46**, pp. 1948.
10. Briggs, E., Sullivan, D., and Bernholc, J. (1995). Large-scale electronic-structure calculations with multigrid acceleration, *Phys. Rev. B*, **52**, pp. R5471.
11. Briggs, E., Sullivan, D., and Bernholc, J. (1996). Real-space multigrid-based approach to large-scale electronic structure calculations, *Phys. Rev. B*, **54**, pp. 14362.
12. Bylander, D., Kleinman, L., and Lee, S. (1990). Self-consistent calculations of the energy bands and bonding properties of B 12 C 3, *Phys. Rev. B*, **42**, pp. 1394.
13. Chelikowsky, J. R., Troullier, N., and Saad, Y. (1994). Finite-difference-pseudopotential method: electronic structure calculations without a basis, *Phys. Rev. Lett.*, **72**, pp. 1240.
14. Chelikowsky, J. R., Troullier, N., Wu, K., and Saad, Y. (1994). Higher-order finite-difference pseudopotential method: an application to diatomic molecules, *Phys. Rev. B*, **50**, pp. 11355–11364.
15. Dedryvère, R., Leroy, S., Martinez, H., Blanchard, F., Lemordant, D., and Gonbeau, D. (2006). XPS valence characterization of lithium salts as a

tool to study electrode/electrolyte interfaces of Li-ion batteries, *J. Phys. Chem. B*, **110**, pp. 12986–12992.

16. Deringer, V. L. and Csányi, G. (2017). Machine learning based interatomic potential for amorphous carbon, *Phys. Rev. B*, **95**, pp. 094203.
17. Eklund, P., Holden, J., and Jishi, R. (1995). Vibrational modes of carbon nanotubes; spectroscopy and theory, *Carbon*, **33**, pp. 959–972.
18. Fattebert, J.-L. and Gygi, F. (2004). Linear scaling first-principles molecular dynamics with controlled accuracy, *Comput. Phys. Commun.*, **162**, pp. 24–36.
19. Fattebert, J. (2008). Adaptive localization regions for O (N) density functional theory calculations, *J. Phys.: Condens. Matter*, **20**, pp. 294210.
20. Funabiki, A., Inaba, M., Ogumi, Z., Yuasa, S. i., Otsuji, J., and Tasaka, A. (1998). Impedance study on the electrochemical lithium intercalation into natural graphite powder, *J. Electrochem. Soc.*, **145**, pp. 172–178.
21. Ganesh, P., Kent, P., and Jiang, D.-E. (2012). Solid–electrolyte interphase formation and electrolyte reduction at Li-ion battery graphite anodes: insights from first-principles molecular dynamics, *J. Phys. Chem. C*, **116**, pp. 24476–24481.
22. Govind, N., Wang, Y. A., and Carter, E. A. (1999). Electronic-structure calculations by first-principles density-based embedding of explicitly correlated systems, *J. Chem. Phys.*, **110**, pp. 7677–7688.
23. Greengard, L. and Rokhlin, V. (1987). A fast algorithm for particle simulations, *J. Comput. Phys.*, **73**, pp. 325–348.
24. Gropp, W. D., Gropp, W., Lusk, E., and Skjellum, A. (1999). *Using MPI: Portable Parallel Programming with the Message-Passing Interface*, Vol. 1 (MIT Press).
25. Iwata, J.-I., Takahashi, D., Oshiyama, A., Boku, T., Shiraishi, K., Okada, S., and Yabana, K. (2010). A massively-parallel electronic-structure calculations based on real-space density functional theory, *J. Comput. Phys.*, **229**, pp. 2339–2363.
26. Jinnouchi, R. and Asahi, R. (2017). Predicting catalytic activity of nanoparticles by a DFT-aided machine-learning algorithm, *J. Phys. Chem. Lett.*, **8**, pp. 4279–4283.
27. Jinnouchi, R., Hirata, H., and Asahi, R. (2017). Extrapolating energetics on clusters and single-crystal surfaces to nanoparticles by machine-learning scheme, *J. Phys. Chem. C*, **121**, pp. 26397–26405.

28. Kleinman, L. and Bylander, D. (1982). Efficacious form for model pseudopotentials, *Phys. Rev. Lett.*, **48**, pp. 1425.
29. Kobayashi, R., Giofré, D., Junge, T., Ceriotti, M., and Curtin, W. A. (2017). Neural network potential for Al-Mg-Si alloys, *Phys. Rev. Mater.*, **1**, pp. 053604.
30. Kresse, G. and Furthmüller, J. (1996). Efficient iterative schemes for ab initio total-energy calculations using a plane-wave basis set, *Phys. Rev. B*, **54**, pp. 11169.
31. Lee, I.-H., Kim, Y.-H., and Martin, R. M. (2000). One-way multigrid method in electronic-structure calculations, *Phys. Rev. B*, **61**, pp. 4397.
32. Leung, K. and Budzien, J. L. (2010). Ab initio molecular dynamics simulations of the initial stages of solid–electrolyte interphase formation on lithium ion battery graphitic anodes, *Phys. Chem. Chem. Phys.*, **12**, pp. 6583–6586.
33. Leung, K. (2012). Electronic structure modeling of electrochemical reactions at electrode/electrolyte interfaces in lithium ion batteries, *J. Phys. Chem. C*, **117**, pp. 1539–1547.
34. Li, W., Ando, Y., and Watanabe, S. (2017). Cu diffusion in amorphous Ta_2O_5 studied with a simplified neural network potential, *J. Phys. Soc. Jpn.*, **86**, pp. 104004.
35. Louie, S. G., Froyen, S., and Cohen, M. L. (1982). Nonlinear ionic pseudopotentials in spin-density-functional calculations, *Phys. Rev. B*, **26**, pp. 1738.
36. Miwa, K. and Ohno, H. (2017). Interatomic potential construction with self-learning and adaptive database, *Phys. Rev. Mater.*, **1**, pp. 053801.
37. Mohr, M., Maultzsch, J., Dobardžić, E., Reich, S., Milošević, I., Damnjanović, M., Bosak, A., Krisch, M., and Thomsen, C. (2007). Phonon dispersion of graphite by inelastic x-ray scattering, *Phys. Rev. B*, **76**, pp. 035439.
38. Muralidharan, A., Chaudhari, M., Rempe, S., and Pratt, L. R. (2017). Molecular dynamics simulations of lithium ion transport through a model solid electrolyte interphase (SEI) layer, *ECS Trans.*, **77**, pp. 1155–1162.
39. Niehoff, P., Passerini, S., and Winter, M. (2013). Interface investigations of a commercial lithium ion battery graphite anode material by sputter depth profile X-ray photoelectron spectroscopy, *Langmuir*, **29**, pp. 5806–5816.
40. Ogata, S., Lidorikis, E., Shimojo, F., Nakano, A., Vashishta, P., and Kalia, R. K. (2001). Hybrid finite-element/molecular-dynamics/electronic-density-functional approach to materials simulations on parallel computers, *Comput. Phys. Commun.*, **138**, pp. 143–154.

41. Ogata, S., Shimojo, F., Kalia, R. K., Nakano, A., and Vashishta, P. (2002). Hybrid quantum mechanical/molecular dynamics simulation on parallel computers: density functional theory on real-space multigrids, *Comput. Phys. Commun.*, **149**, pp. 30–38.

42. Ogata, S., Campbell, T. J., Kalia, R. K., Nakano, A., Vashishta, P., and Vemparala, S. (2003). Scalable and portable implementation of the fast multipole method on parallel computers, *Comput. Phys. Commun.*, **153**, pp. 445–461.

43. Ogata, S. (2005). Buffered-cluster method for hybridization of density-functional theory and classical molecular dynamics: application to stress-dependent reaction of H_2O on nanostructured Si, *Phys. Rev. B*, **72**, pp. 045348.

44. Ogata, S., Abe, Y., Ohba, N., and Kobayashi, R. (2010). Stress-induced nano-oxidation of silicon by diamond-tip in moisture environment: a hybrid quantum-classical simulation study, *J. Appl. Phys.*, **108**, pp. 064313.

45. Ogata, S., Ohba, N., and Kouno, T. (2013). Multi-thousand-atom DFT simulation of Li-ion transfer through the boundary between the solid–electrolyte interface and liquid electrolyte in a Li-ion battery, *J. Phys. Chem. C*, **117**, pp. 17960–17968.

46. Ohba, N., Ogata, S., Tamura, T., Yamakawa, S., and Asahi, R. (2011). A hybrid quantum-classical simulation study on stress-dependence of Li diffusivity in graphite, *Comput. Model. Eng. Sci.*, **75**, pp. 247.

47. Ohba, N., Ogata, S., Kouno, T., and Asahi, R. (2012). A hybrid quantum-classical simulation study on the Li diffusion in Li-graphite intercalation compounds, *TSUBAME ESJ*, **8**, pp. 9–16.

48. Ohba, N., Ogata, S., Kouno, T., Tamura, T., and Kobayashi, R. (2012). Linear scaling algorithm of real-space density functional theory of electrons with correlated overlapping domains, *Comput. Phys. Commun.*, **183**, pp. 1664–1673.

49. Ohba, N., Ogata, S., Tamura, T., Kobayashi, R., Yamakawa, S., and Asahi, R. (2012). Enhanced thermal diffusion of Li in graphite by alternating vertical electric field: a hybrid quantum-classical simulation study, *J. Phys. Soc. Jpn.*, **81**, pp. 023601.

50. Ohba, N. and Ogata, S. (2014). Hybrid quantum-classical simulation of intercalation compounds, *R&D Rev. Toyota CRDL*, **45**, pp. 51–53.

51. Ohba, N., Ogata, S., Kouno, T., and Asahi, R. (2015). Thermal diffusion of correlated Li-ions in graphite: a hybrid quantum–classical simulation study, *Comput. Mater. Sci.*, **108**, pp. 250–257.

52. Ohba, N., Ogata, S., and Asahi, R. (2019). Hybrid quantum-classical simulation of Li ion dynamics and the decomposition reaction of electrolyte liquid at a negative-electrode/electrolyte interface, *J. Phys. Chem. C*, **123**, pp. 9673–9679.

53. Ono, T. and Hirose, K. (1999). Timesaving double-grid method for real-space electronic-structure calculations, *Phys. Rev. Lett.*, **82**, pp. 5016.

54. Payne, M. C., Teter, M. P., Allan, D. C., Arias, T., and Joannopoulos, A. J. (1992). Iterative minimization techniques for ab initio total-energy calculations: molecular dynamics and conjugate gradients, *Rev. Mod. Phys.*, **64**, pp. 1045.

55. Peled, E., Golodnitsky, D., Menachem, C., and Bar-Tow, D. (1998). An advanced tool for the selection of electrolyte components for rechargeable lithium batteries, *J. Electrochem. Soc.*, **145**, pp. 3482–3486.

56. Perdew, J. P. (1986). Density-functional approximation for the correlation energy of the inhomogeneous electron gas, *Phys. Rev. B*, **33**, pp. 8822.

57. Perez, D. and Voter, A. F. (2008). Accelerating atomistic simulations through self-learning bond-boost hyperdynamics, *J. Chem. Phys.*, Submitted for publication.

58. Press, W. H., Teukolsky, S. A., Vetterling, W. T., and Flannery, B. P. (1992). *Numerical Recipes in FORTRAN* (Cambridge Univ. Press).

59. Pulay, P. (1980). Convergence acceleration of iterative sequences: the case of SCF iteration, *Chem. Phys. Lett.*, **73**, pp. 393–398.

60. Pulay, P. (1982). Improved SCF convergence acceleration, *J. Comput. Chem.*, **3**, pp. 556–560.

61. Ramos-Sanchez, G., Soto, F., de la Hoz, J. M., Liu, Z., Mukherjee, P., El-Mellouhi, F., Seminario, J., and Balbuena, P. (2016). Computational studies of interfacial reactions at anode materials: initial stages of the solid-electrolyte-interphase layer formation, *J. Electrochem. Energy Convers. Storage*, **13**, pp. 031002.

62. Shimojo, F., Campbell, T. J., Kalia, R. K., Nakano, A., Vashishta, P., Ogata, S., and Tsuruta, K. (2000). A scalable molecular-dynamics algorithm suite for materials simulations: design-space diagram on 1024 Cray T3E processors, *Future Gener. Comput. Syst.*, **17**, pp. 279–291.

63. Shimojo, F., Kalia, R. K., Nakano, A., and Vashishta, P. (2005). Embedded divide-and-conquer algorithm on hierarchical real-space grids: parallel molecular dynamics simulation based on linear-scaling density functional theory, *Comput. Phys. Commun.*, **167**, pp. 151–164.

64. Shimojo, F., Kalia, R. K., Nakano, A., and Vashishta, P. (2008). Divide-and-conquer density functional theory on hierarchical real-space grids: parallel implementation and applications, *Phys. Rev. B*, **77**, pp. 085103.

65. Stillinger, F. H. and Weber, T. A. (1985). Computer simulation of local order in condensed phases of silicon, *Phys. Rev. B*, **31**, pp. 5262.

66. Sun, B., Mindemark, J., Morozov, E. V., Costa, L. T., Bergman, M., Johansson, P., Fang, Y., Furó, I., and Brandell, D. (2016). Ion transport in polycarbonate based solid polymer electrolytes: experimental and computational investigations, *Phys. Chem. Chem. Phys.*, **18**, pp. 9504–9513.

67. Takenaka, N., Suzuki, Y., Sakai, H., and Nagaoka, M. (2014). On electrolyte-dependent formation of solid electrolyte interphase film in lithium-ion batteries: strong sensitivity to small structural difference of electrolyte molecules, *J. Phys. Chem. C*, **118**, pp. 10874–10882.

68. Takenaka, N., Sakai, H., Suzuki, Y., Uppula, P., and Nagaoka, M. (2015). A computational chemical insight into microscopic additive effect on solid electrolyte interphase film formation in sodium-ion batteries: suppression of unstable film growth by intact fluoroethylene carbonate, *J. Phys. Chem. C*, **119**, pp. 18046–18055.

69. Takenaka, N., Fujie, T., Bouibes, A., Yamada, Y., Yamada, A., and Nagaoka, M. (2018). Microscopic formation mechanism of solid electrolyte interphase film in lithium-ion batteries with highly concentrated electrolyte, *J. Phys. Chem. C*, **122**, pp. 2564–2571.

70. Tang, M., Miyazaki, K., Abe, T., and Newman, J. (2012). Effect of graphite orientation and lithium salt on electronic passivation of highly oriented pyrolytic graphite, *J. Electrochem. Soc.*, **159**, pp. A634–A641.

71. Tersoff, J. (1988). New empirical approach for the structure and energy of covalent systems, *Phys. Rev. B*, **37**, pp. 6991.

72. Troullier, N. and Martins, J. L. (1991). Efficient pseudopotentials for plane-wave calculations, *Phys. Rev. B*, **43**, pp. 1993.

73. Van Der Spoel, D., Lindahl, E., Hess, B., Groenhof, G., Mark, A. E., and Berendsen, H. J. C. (2005). GROMACS: fast, flexible, and free, *J. Comput. Chem.*, **26**, pp. 1701–1718.

74. Voter, A. F. (1997). Hyperdynamics: accelerated molecular dynamics of infrequent events, *Phys. Rev. Lett.*, **78**, pp. 3908.

75. Waghmare, U., Kim, H., Park, I., Modine, N., Maragakis, P., and Kaxiras, E. (2001). HARES: an efficient method for first-principles electronic structure calculations of complex systems, *Comput. Phys. Commun.*, **137**, pp. 341–360.

76. Winter, M. (2009). The solid electrolyte interphase–the most important and the least understood solid electrolyte in rechargeable Li batteries, *Z. Phys. Chem.*, **223**, pp. 1395–1406.

77. Xu, K. (2004). Nonaqueous liquid electrolytes for lithium-based rechargeable batteries, *Chem. Rev.*, **104**, pp. 4303–4418.

78. Xu, K. and von Cresce, A. (2011). Interfacing electrolytes with electrodes in Li ion batteries, *J. Mater. Chem.*, **21**, pp. 9849–9864.

79. Xu, K. (2014). Electrolytes and interphases in Li-ion batteries and beyond, *Chem. Rev.*, **114**, pp. 11503–11618.

80. Yang, W. (1991). Direct calculation of electron density in density-functional theory, *Phys. Rev. Lett.*, **66**, pp. 1438.

81. Yang, W. and Lee, T. S. (1995). A density-matrix divide-and-conquer approach for electronic structure calculations of large molecules, *J. Chem. Phys.*, **103**, pp. 5674–5678.

82. Zhou, Z. (1993). Ab initio construction of Yang's divide-and-conquer method, *Chem. Phys. Lett.*, **203**, pp. 396–398.

83. Zhu, T., Pan, W., and Yang, W. (1996). Structure of solid-state systems from embedded-cluster calculations: a divide-and-conquer approach, *Phys. Rev. B*, **53**, pp. 12713.

Chapter 6

Large-Scale Simulation II: Atomistic and Coarse-Grained Simulations of Polyelectrolyte Membranes

Tomoyuki Kinjo[a] and Satoru Yamamoto[a,b]
[a]*Toyota Central R&D Laboratories, Inc., Nagakute, Aichi 480-1192, Japan*
[b]*Kyushu University, Fukuoka 819-0395, Japan*
e1308@mosk.tytlabs.co.jp

6.1 Introduction

As mentioned in the previous chapter, in first-principles calculation, forces acting on the atoms are obtained from the electronic structure under the given configuration of the atomic nuclei. The calculation of the electronic structure accounts for most of the calculation cost in the first-principles molecular dynamics simulation. On the other hand, in classical molecular dynamics simulations, interactions between two atoms are represented as a function of the interatomic distance. By preparing the functional form of interatomic interaction and the parameters contained in it in advance, we can greatly reduce

Multiscale Simulations for Electrochemical Devices
Edited by Ryoji Asahi

ISBN 978-981-4800-71-6 (Hardcover), 978-0-429-29545-4 (eBook)
www.jennystanford.com

the calculation cost. The functional form and parameters of these interatomic interactions are called molecular force fields [2, 14, 15, 23, 29, 30]. In conventional classical molecular dynamics simulation, all the atoms are represented explicitly as a set of mass points each of which has a coordinate and a momentum. This is often called the all-atom model to distinguish it from the coarse-grained (CG) model mentioned below. For some systems, such as colloidal liquids and polymer solutions, the all-atom model is too detailed and costly. In such a case, the degrees of freedom (DOFs) of the system are reduced in order to lower the calculation cost. This reduction in the DOFs is called coarse graining [6]. In this chapter we describe all-atom molecular dynamics and CG simulations and applications of polyelectrolyte membranes.

This chapter is organized as follows: Section 6.2 describes the all-atom model and the CG model. Section 6.3 describes the background of the calculation of polyelectrolyte membranes. Section 6.4 is devoted to the methodology of dissipative particle dynamics (DPD). In Section 6.5 a CG model of a polyelectrolyte membrane is discussed. In Section 6.6 the result of the CG simulation is shown. Section 6.7 describes the all-atom model briefly. In Section 6.8 the diffusion of oxygen in bulk Nafion is shown. Section 6.9 summarizes this chapter.

6.2 From Atomistic Description to Coarse-Grained Description

This section is devoted to brief descriptions of the all-atom model and the CG model. We start with a fine description of the system of atoms each of which is represented as a mass point. In classical mechanics the dynamics of the system are given by Hamilton's equations of motion [10].

$$\frac{\mathrm{d}\mathbf{r}_i}{\mathrm{d}t} = -\frac{\partial H}{\partial \mathbf{p}_i} \tag{6.1}$$

and

$$\frac{\mathrm{d}\mathbf{p}_i}{\mathrm{d}t} = \frac{\partial H}{\partial \mathbf{r}_i}, \tag{6.2}$$

where $\mathbf{r}_i$ and $\mathbf{p}_i$ are the coordinate and momentum of *i*th particle, respectively, and H is the Hamiltonian of the system, which is usually expressed as the sum of the kinetic energy and the potential energy U as follows:

$$H = \sum_i \frac{\mathbf{p}_i^2}{2m_i} + U(\mathbf{r}_1, \cdots, \mathbf{r}_N), \tag{6.3}$$

where m_i is the mass of the *i*th particle. The time evolution of the system is determined completely by Hamilton's equations of motion (6.1) and (6.2). These equations are equivalent to Newton's equation of motion

$$m_i \frac{\mathrm{d}^2 \mathbf{r}_i}{\mathrm{d}t^2} = \mathbf{F}_i, \tag{6.4}$$

where $\mathbf{F}_i = -\nabla_i U$ is the force acting on the *i*th particle. In classical molecular dynamics simulation, Eq. 6.4 is solved numerically [1, 8]. The solution of Eq. 6.4 under a specific initial condition, $\{\mathbf{r}_i(0), \mathbf{p}_i(0)\}$, is uniquely determined within numerical error. This solution $\{\mathbf{r}_i(t), \mathbf{p}_i(t)\}$ is called the trajectory and includes all the information of the classical system. In the all-atom model, the potential energy U consists of the interaction energy of the connection and of the nonbonded pair, which is discussed in Section 6.7.

However, in some cases, such as polymer solutions, the above description is too fine and costly. In such a case, a reduction in the DOFs by coarse graining is useful. Brownian dynamics (BD) [1] and DPD [11] are the representative models of coarse graining. For example, in BD simulation, the stochastic differential equation

$$\frac{\mathrm{d}\mathbf{p}_i}{\mathrm{d}t} = \mathbf{F}_i - \frac{\gamma}{m_i}\mathbf{p}_i + \mathbf{R}_i \tag{6.5}$$

is solved numerically, where γ is the friction constant and $\mathbf{R}_i$ is the random force acting on the *i*th coarse-grained particle. The second term and the third term represent dissipative force and random force, respectively [1, 3, 22]. These terms arise from eliminated DOFs. Equation 6.5 has an equivalent partial differential equation for the correspondent distribution function $\rho(\{\mathbf{r}_i, \mathbf{p}_i\})$,

$$\frac{\partial \rho}{\partial t} = \left(\sum_i -\frac{\mathbf{p}_i}{m_i} \cdot \frac{\partial}{\partial \mathbf{r}_i} - \mathbf{F}_i \cdot \frac{\partial}{\partial \mathbf{p}_i} + \frac{\gamma}{m_i} \frac{\partial}{\partial \mathbf{p}_i} \cdot \mathbf{p}_i + \frac{\sigma^2}{2} \frac{\partial^2}{\partial p_i^2} \right) \rho. \tag{6.6}$$

Equation 6.6 is Fokker–Plank equation for the Brownian particles [22]. The random force satisfies $\langle R_{i\mu}(t)R_{j\nu}(0)\rangle = \sigma^2\delta_{ij}\delta_{\mu\nu}\delta(t)$ and $\langle R_{i\mu}(t)\rangle = 0$, where δ_{ij} represents Kronecker delta, which is 1 for $i = j$ and 0 for $i \neq j$. The subscripts of Greek letters represent components of Cartesian coordinates. The parameter σ corresponds to the strength of the random force. The relation between the parameters γ and σ is given by the fluctuation-dissipation theorem,

$$\gamma = \frac{\sigma^2}{k_B T}, \tag{6.7}$$

where k_B is the Boltzmann constant. These relations yield the canonical distribution of the equilibrium state as a solution of Eq. 6.6 under the steady-state condition ($\partial\rho/\partial t = 0$).

$$\rho(\{\mathbf{r}_i, \mathbf{p}_i\}) = \frac{1}{Z}\exp\left\{-\frac{1}{k_B T}\left(\sum_i \frac{\mathbf{p}_i^2}{2m_i} + U\right)\right\}, \tag{6.8}$$

where Z is the partition function. As will be discussed in the following, the equation of motion for the DPD is similar in form to Eq. 6.5. However, in BD and DPD, the eliminated DOFs are different [6, 18]. The BD method is typically applied to a colloidal system in which the DOFs of the solvent molecules are eliminated. On the other hand, in DPD the DOFs of the internal motion of the CG particles are eliminated. It has been shown that the stochastic equation of motion for the CG method can be derived by the projection operator method from a fine atomistic description [24, 26]. Kinjo and Hyodo obtained the general equation of motion for CG particles by the projection operator as follows [18, 19]:

$$\frac{\mathrm{d}\mathbf{p}_i}{\mathrm{d}t} = \mathbf{F}_i^{\mathrm{mean}} - \beta\sum_j \int_0^t \mathrm{d}s \left\langle \delta\mathbf{F}_i(t-s)\delta\mathbf{F}_j(0)^T \right\rangle \frac{\mathbf{p}_j(s)}{m_j} + \delta\mathbf{F}_i(t), \tag{6.9}$$

where $\mathbf{F}_i^{\mathrm{mean}}$ is the mean force acting on the ith particle, $\delta\mathbf{F}_i$ is the fluctuating force that stems from internal DOF of the CG particles. Equation 6.9 encompasses BD and DPD, and it becomes BD or DPD depending on the fluctuating force term [19]. For BD, the α component of the fluctuating force is modeled as

$$\delta F_{i\alpha}(t) = \sigma\theta_{i\alpha}(t), \tag{6.10}$$

where $\theta_{i\alpha}$ is a random number that satisfies

$$\langle\theta_{i\alpha}(t)\theta_{j\beta}(0)\rangle = \delta_{ij}\delta_{\alpha\beta}\delta(t) \tag{6.11}$$

and

$$\langle \theta_{i\alpha}(t) \rangle = 0. \tag{6.12}$$

By substitution of Eq. 6.10 into Eq. 6.9, we can obtain the stochastic differential equation for Brownian particles (Eq. 6.5).

For DPD, the fluctuating force is assumed to be pairwise additive,

$$\delta \mathbf{F}_i(t) = \sum_{i \neq j} \delta \mathbf{f}_{ij}(t). \tag{6.13}$$

Each element is

$$\delta \mathbf{f}_{ij}(t) = \sigma w(r_{ij}) \theta_{ij}(t) \mathbf{n}_{ij}, \tag{6.14}$$

where

$$\langle \theta_{ij}(t) \theta_{kl}(0) \rangle = (\delta_{ik}\delta_{jl} + \delta_{il}\delta_{jk})\delta(t) \tag{6.15}$$

and

$$\langle \theta_{ij}(t) \rangle = 0. \tag{6.16}$$

By substitution of Eq. 6.13 into Eq. 6.9, we can obtain the stochastic differential equation for DPD,

$$\frac{\mathrm{d}\mathbf{p}_i}{\mathrm{d}t} = \mathbf{F}_i^{\mathrm{mean}} - \gamma \sum_{j \neq i} w^2(r_{ij})(\mathbf{n}_{ij} \cdot \mathbf{v}_{ij})\mathbf{n}_{ij} + \sum_{j \neq i} \delta \mathbf{f}_{ij} \tag{6.17}$$

Derivations of CG models described above are well defined. However, conventional DPD simulation relies on the heuristic method that uses the similarity between the soft potential model and the Flory–Huggins lattice model of polymer solutions [11]. The methodology will be discussed in the following section.

6.3 Polyelectrolyte Membranes

The following sections describe CG simulation and application of all-atom simulation to an electrolyte membrane.

Polymer electrolyte fuel cells (PEFCs) for automotive applications have rapidly progressed over the past decades [4, 17, 31]. Improvement of the proton conductivity of the electrolyte membrane and the activity of the platinum catalyst are major factors for improving the performance of the PEFC. In the mesoscale structure of hydrated Nafion, which is used as the electrolyte membrane in PEFCs, it is presumed from the monomer molecular structure, small-angle X-ray scattering experiments, etc., that the water clusters surrounded by the sulfonic acid groups are linked in a sponge-like manner; further, the material's proton conductivity is understood to be expressed as proton movement through the water channel. However, although several models have been proposed to describe the detailed structure, these are still being debated; with respect to the mechanism of proton conductivity as well, although hopping conductivity via sulfonic acid groups is the most likely explanation, the details are not yet understood [9, 13]. Here, the mesoscale structure of the hydrated Nafion membrane is calculated using DPD, and comparisons with other experiments reported to date are made [32]. In addition, anisotropy in the swelling behavior of some hydrocarbon-based electrolyte membranes has been reported recently, and the possibility of an anisotropic mesostructure has been identified [16, 25]. A characteristic of these hydrocarbon-based electrolyte membranes is that their main chain skeletons are rigid; hence, their molecular chains become oriented easily. So the formation of an anisotropic structure can be expected. Therefore, we investigated the mechanism of expression of anisotropy in hydrocarbon-based electrolyte membranes by incorporating molecular rigidity into the CG model. In Section 6.6 we describe the mesostructure of the electrolyte membrane by the CG model.

6.4 Dissipative Particle Dynamics

The original DPD model was introduced by Koelman and Hoogerbrugge on the basis of a precedent method of the lattice gas model [20]. After that, the DPD model was formulated to satisfy the fluctuation-dissipation theorem by Espanol and Warren [5]. DPD is a method proposed to study the hydrodynamic behavior of complex fluids. The method is based on the dynamics of soft particles interacting via conservative, dissipative, and random forces [11]. By introducing bonds connecting each particle, the method is extended to polymer systems [11, 12]. In this section we give an outline of the model and the time evolution algorithm of the DPD.

Now, we consider that the particles are subject to conservative, dissipative, random forces, and spring forces for connecting spheres. The time evolution of the system is obtained by solving the stochastic equation of motion,

$$\frac{d\mathbf{r}_i}{dt} = \mathbf{v}_i \tag{6.18}$$

and

$$m_i \frac{d\mathbf{v}_i}{dt} = \mathbf{F}, \tag{6.19}$$

where $\mathbf{r}_i$, $\mathbf{v}_i$, and m_i are the position, velocity, and mass of the ith particle, respectively. For simplicity, the masses and diameters of the particles are scaled, relative to 1, in the following. The force $\mathbf{F}_i$ contains three parts of the original DPD formula and an additional spring force for a polymer system. The interaction between two particles can be written as the sum of these forces

$$\mathbf{F}_i = \sum_{i \neq j} \left(\mathbf{F}_{ij}^{\mathrm{C}} + \boldsymbol{F}_{ij}^{\mathrm{D}} + \boldsymbol{F}_{ij}^{\mathrm{R}} + \boldsymbol{F}_{ij}^{\mathrm{S}} \right). \tag{6.20}$$

The first three forces of the original DPD are considered within a certain cutoff radius r_c, as follows:

The conservative force $\mathbf{F}_{ij}{}^{\mathrm{C}}$ is a soft repulsive force acting along the line of the centers and is given by

$$\mathbf{F}_{ij}^{\mathrm{C}} = \begin{cases} -a_{ij}\left(r_{\mathrm{c}} - r_{ij}\right)\mathbf{n}_{ij} & r_{ij} < r_{\mathrm{c}} \\ 0 & r_{ij} \geq r_{\mathrm{c}} \end{cases}, \tag{6.21}$$

where a_{ij} is the maximum repulsion force between particles i and j, $\mathbf{r}_{ij} = \mathbf{r}_j - \mathbf{r}_i$, $r_{ij} = |\mathbf{r}_{ij}|$, and $\mathbf{n}_{ij} = \mathbf{r}_{ij}/|\mathbf{r}_{ij}|$. Groot and Warren found that the soft potential described above yields the quadratic form of pressure as a function of density and, consequently, free energy similar to the lattice model of the polymer solution (i.e., the Flory–Huggins model [28]). In this study, following Ref. [11], the repulsion parameter between water particles is set to 25 kT for a density ρ = 3 to match the compressibility of liquid water at room temperature. Other repulsion parameters between particles of the same type are chosen as the same value as the water particles. The repulsion parameters between particles of different types correspond to the mutual solubility, expressed as the Flory–Huggins χ parameter. We determined these parameters to reproduce the hydrophobic interaction between Nafion and water. The details will be described below.

The dissipative force $\mathbf{F}_{ij}{}^{\mathrm{D}}$ is a hydrodynamic drag and is given by

$$\mathbf{F}_{ij}^{\mathrm{D}} = \begin{cases} -\gamma\omega^{\mathrm{D}}(r_{ij})(\mathbf{n}_{ij} \cdot \mathbf{v}_{ij})\mathbf{n}_{ij} & r_{ij} < r_{\mathrm{c}} \\ 0 & r_{ij} \geq r_{\mathrm{c}} \end{cases}, \tag{6.22}$$

where γ is the friction parameter, $\omega^{\mathrm{D}}(r_{ij})$ is the weighting function, and $\mathbf{v}_{ij} = \mathbf{v}_i - \mathbf{v}_j$. The friction parameter is related to the viscosity of the system and acts in such a manner as to slow the relative velocity of two particles and to remove the kinetic energy.

The random force $\mathbf{F}_{ij}^{\mathrm{R}}$ corresponds to the thermal noise and is governed by the parameter σ and a different weighting function $\omega^{\mathrm{R}}(r_{ij})$ as follows:

$$\mathbf{F}_{ij}^{\mathrm{R}} = \begin{cases} \sigma\omega^{\mathrm{R}}(r_{ij})\zeta_{ij}\Delta t^{-1/2}\mathbf{n}_{ij} & r_{ij} < r_{\mathrm{c}} \\ 0 & r_{ij} \geq r_{\mathrm{c}} \end{cases}. \tag{6.23}$$

The randomness is contained in ζ_{ij}, which is a randomly fluctuating variable with Gaussian statics,

$$\langle \zeta_{ij}(t) \rangle = 0, \tag{6.24}$$

and

$$\langle \zeta_{ij}(t)\zeta_{kl}(t') \rangle = (\delta_{ik}\delta_{jl} + \delta_{il}\delta_{jk})\delta(t - t'). \tag{6.25}$$

They are assumed to be uncorrelated for different pairs of particles and time. There is a relation between the two weighting functions and two parameters,

$$\omega^{\mathrm{D}}(r_{ij}) = \left[\omega^{\mathrm{R}}(r_{ij})\right]^2 \tag{6.26}$$

and

$$\sigma^2 = 2\gamma kT \tag{6.27}$$

to guarantee the canonical distribution of the equilibrium state [5, 11]. In our simulation, according to custom, we choose the weighting functions as follows:

$$\omega^{\mathrm{D}}(r_{ij}) = \left[\omega^{\mathrm{R}}(r_{ij})\right]^2 = \begin{cases} (r_{\mathrm{c}} - r_{ij})^2 & r_{ij} < r_{\mathrm{c}} \\ 0 & r_{ij} \geq r_{\mathrm{c}} \end{cases}, \tag{6.28}$$

and the parameters σ and γ are chosen to be 3 and 4.5, respectively. Therefore, $kT = 1$. The appearance of $\Delta t^{-1/2}$ in Eq. 6.14 is due to ensuring the consistent diffusion of particles independent of the step size of the numerical integration as discussed by Groot and Warren [11]. The spring force $\mathbf{F}_{ij}^{\mathrm{S}}$ for a polymer is considered as harmonic springs for an equilibrium bond distance r_s,

$$\mathbf{F}_{ij}^{S} = -C(r_s - r_{ij})\mathbf{n}_{ij}, \tag{6.29}$$

and the spring constant C is set at 100 to obtain a very stiff chain for stretching.

In this study, we set both the cutoff radius r_c and the bond distance r_s at 1, which equals the diameter of the particle. The time evolution of the system is calculated by the Verlet algorithm with time steps of $\Delta t = 0.05$. We implemented an original DPD program that is used for the simulations discussed in the following sections.

6.5 Coarse-Grained Model of Polyelectrolyte Membrane

Nafion, whose chemical structure is shown below, is a fluorinated polyethylene main chain with a fluorinated vinyl ether side chain that has a sulfonic acid group attached to its end.

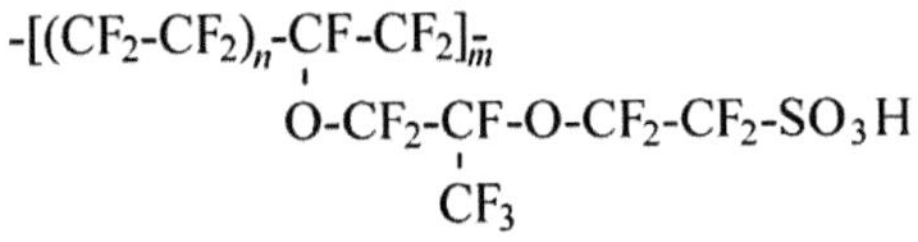

$$\text{-[(CF}_2\text{-CF}_2)_n\text{-CF-CF}_2]_{\bar{m}}$$
$$\text{O-CF}_2\text{-CF-O-CF}_2\text{-CF}_2\text{-SO}_3\text{H}$$
$$\text{CF}_3$$

Here, n and m generally fall within the range of 5 to 14 and 200 to 1000, respectively, and vary on the basis of the length of the side chains and the spacing at which they are bonded to the main chain. The sulfonic acid group at the end of the side chain easily dissociates in the presence of water to form SO_3^-. Nafion is thermally, chemically, and mechanically stable. Using the molecular structure of the Nafion monomer as a reference, DPD was used to construct a CG molecular model as shown in Fig. 6.1. In DPD, a single group of atoms is expressed as a sphere and, in this case, sphere A largely corresponds to $-CF_2CF_2CF_2CF_2-$, sphere B to $-OCF_2C(CF_3)FO-$, and sphere C to $-CF_2CF_2SO_3H$. The size of each sphere is approximately 0.5 nm. Nafion's actual degree of polymerization is approximately 230, but due to limitations on the model size used for calculation, a model with a degree of polymerization of 5 was used. If the length of the main chain is too short, there is a concern that its polymer properties may not be expressed. However, in the phase diagram of the block copolymer obtained by mean field theory, the conditions under which phase separation occurs are such that when the number of phase segments N reaches a certain size, further increase in N causes no change in phase structure [11]. Hence, if a stable structure can be obtained by the model used, it may be assumed that the relevant polymer properties have been adequately considered. In fact, since the same structure formation was confirmed in the dimer model used for preliminary calculations (refer to the results described later), it may be concluded that the 5-mer model shown in the diagram is fully adequate. For water, sphere W was used, as it is the same size as the spheres constituting the Nafion model.

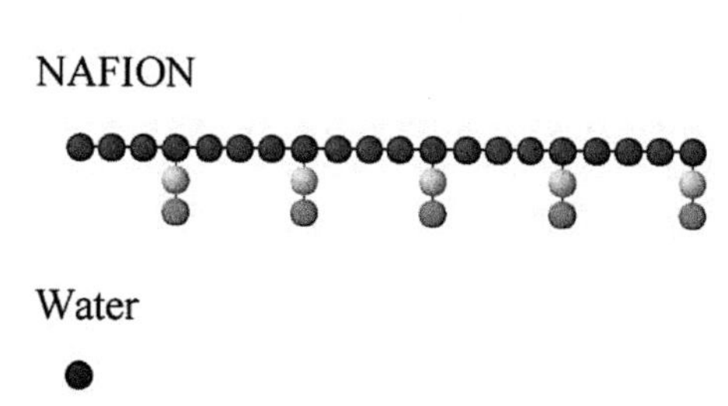

Figure 6.1 DPD models of Nafion and water.

When DPD calculation is performed, the coefficients for the interaction between each sphere must be decided; these are determined as shown below. The DPD conservative force a_{ij} is related to the parameter χ_{ij} for the interaction between chemical species as follows (in the case of a particle density of 3) [11, 12]:

$$a_{ij} = a_{ii} + 3.27\chi_{ij} \tag{6.30}$$

Furthermore, χ_{ij} is related to the solubility parameter δ obtained from the cohesive energy for each chemical species [33].

$$\chi_{ij} = \frac{V_{\text{seg}}}{RT}(\delta_i - \delta_j)^2 \tag{6.31}$$

$$\delta = \left(\frac{E_{\text{coh}}}{V}\right)^{0.5} \tag{6.32}$$

Here, V_{seg} is the volume of each segment making up the polymer. Cohesive energy can be calculated by subtracting the energy of the chemical species without the cell from the energy when the chemical species is in the cell with the periodic boundary conditions [7].

Table 6.1 shows the results of calculation of the solubility parameter from the cohesive energy of the three types of spheres constituting the Nafion. Materials Studio (Dassault Systemes Biovia) was used for this calculation. Following this, Table 6.2 shows the results of determination—from the results in Table 6.1—of the interaction coefficients. Strongly hydrophobic A has poor affinity with strongly hydrophilic C, and so it is expected that a strong separation of A and C will occur. In addition, since the sulfonic acid group dissociates and dissolves in water, W (water) and C were considered to be similar particles, and this was used to determine the interaction coefficient with respect to water.

Table 6.1 Solubility parameters of spheres constituting the Nafion DPD model

	Corresponding segment	δ (J/cm^3)$^{0.5}$
A	$-CF_2CF_2CF_2CF_2-$	12.49
B	$-OCF_2C(CF_3)FO-$	19.80
C	$-CF_2CF_2SO_3H$	33.36

Table 6.2 Interaction parameters, a_{ij}

	a_{ij}
A-B	27
A-C	41
B-C	31.8

6.6 Morphology of a Polyelectrolyte Membrane

The results of calculation of the mesostructure under a condition of 20 vol% moisture content are shown in Fig. 6.2a. In the figure, water is depicted as spheres, but the A, B, and C spheres that constitute the Nafion are represented only by lines indicating the bonds connecting them. Small water clusters form in the initial stage; these gradually become larger as time progresses, reaching a steady state at a time of approximately 200, at which point the water clusters have formed a dispersed structure. The formation of water clusters surrounded by sulfonic acid groups can be observed. From the results of detailed observation of the internal structure, it can be seen that areas with water have a sponge-like structure containing connected channels. It is clear that the mesostructured is isotropic with the help of radial distribution $g(r)$ of water as shown in Fig. 6.3a.

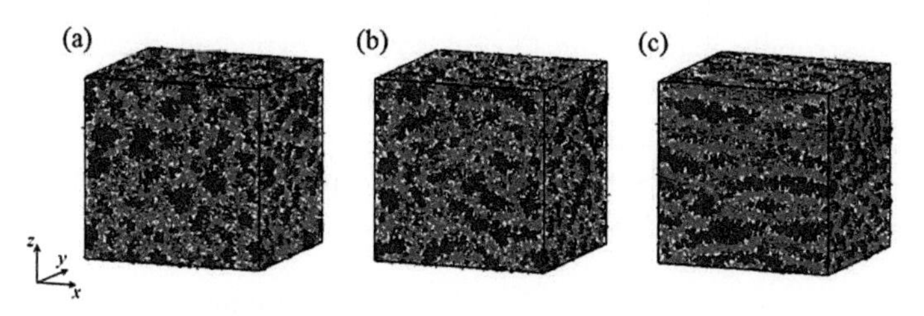

Figure 6.2 Mesostructure of electrolyte membrane: (a) with no rigidity or shear, (b) without rigidity and with shear, and (c) with rigidity and shear (moisture content: 20 vol%).

In a manner similar to the case of 20 vol% moisture content, results were also calculated under conditions of 10 vol% and 30 vol%; doing so revealed that water cluster size increases as moisture content increases. It was already clear from past small-angle X-ray scattering experiments, etc., that changing the moisture content causes the size and spacing of water clusters to change, and it has been reported that changing the moisture content from 5 wt% to 20 wt% causes the cluster spacing to increase from 3 nm to 5 nm. Thus, the radial distribution $g(r)$ of water was calculated for each model, and the dependency of water cluster size and spacing on moisture content was investigated.

For water clusters, the cluster size (diameter) is defined such that the radius is given by the point at which $g(r)$ becomes smaller than 1 near the first peak. The results of investigating its relationship to moisture content indicate that when moisture content changes from 10 vol% (5.3 wt%) to 30 vol% (17.6 wt%), cluster size changes from 5.9 to 9.5. Considering the sphere size in the DPD model, this corresponds approximately to a change from 3 to 4.7 nm.

In addition, the results of the investigation of the relationship between cluster spacing and moisture content indicate that depending on moisture content, cluster spacing changes from 7 to 11, corresponding approximately to a change from 3.5 nm to 5.5 nm.

These calculation results agree with experimental reports on cluster spacing and moisture content dependency.

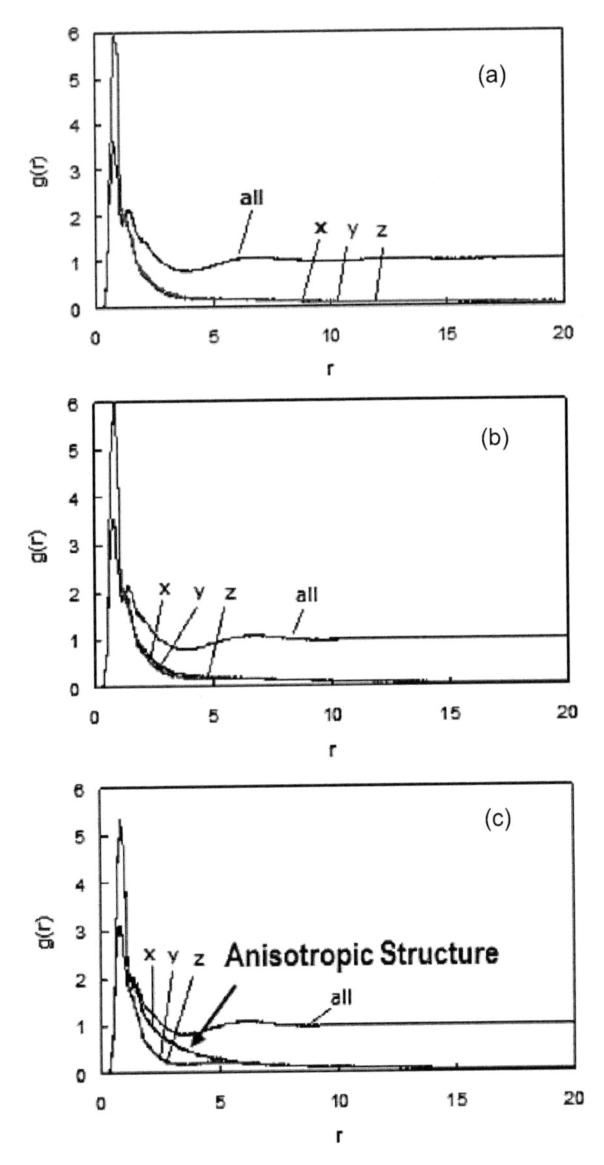

Figure 6.3 Radial distribution $g(r)$ of water: (a) without rigidity and no shear, (b) without rigidity and with shear, and (c) with rigidity and shear.

Next, the mesostructure was calculated for the case in which the rigidity is equivalent to that of the main chain poly(*p*-phenylene) (PPP) skeleton in a hydrocarbon-based electrolyte membrane. Here, a comparison was made presuming rigidity is conferred using the same chemical structure as that of Nafion. Rigidity is represented by constraining the equilibrium value of the main chain A–A–A bond angle and side chain A–B–C bond angle to 180° and the equilibrium value of the side chain bond angle with respect to the main chain (A–A–B bond angle) to 90°. The results of analyzing the obtained mesostructure indicated that rigidity alone is insufficient to cause anisotropy to appear. Hence, it is possible that, in the actual membrane production process, shearing that occurs due to the flow of solvent may trigger the appearance of anisotropy in the mesostructure; therefore, after imposing a shear flow in the x direction and calculating the mesostructure, the shear was removed and calculation was performed until equilibrium was reached. The results are shown in Figs. 6.2b and 6.2c. In case (b), in which molecular rigidity was not considered, the mesostructure was isotropic even though a shear was imposed, but in case (c), which considered molecular rigidity, the polymer and water regions were extended in the x direction, as clearly confirmed by examination of the water particle radial distribution function as shown in Figs. 6.3b and 6.3c. The reason for this can be assumed to be that if the molecules are rigid, the molecular chains have a strong tendency to become oriented in the shear direction, which forms a mesostructure extended in the x direction. A mesostructure with this kind of anisotropy is expected to be strong with respect to deformation in the in-plane direction so that, in wet/dry cycling, the dimensional changes in the direction of thickness will be larger than those in the in-plane direction, a result that supports the swelling anisotropy [16, 25].

6.7 Force Field in Classical Molecular Dynamics Simulation

To analyze behaviors at the atomistic scale, such as diffusion of gas molecules in polymers, the all-atom model is more suitable than the CG model. In classical molecular dynamics simulation, Newton's equation of motion is numerically solved on the basis of the force acting on each atom [1, 8].

$$m_i \frac{d^2\mathbf{r}_i}{dt^2} = \mathbf{F}_i \tag{6.33}$$

The force acting on each atom is expressed by the sum of functions corresponding to each term of nonbonded pair, bonded pair, angle, and dihedral angle as follows:

$$\mathbf{F}_i = -\nabla_i U \tag{6.34}$$

and

$$U = \sum_{ij} u_{\text{nonbond}}(r_{ij}) + \sum_{ij} u_{\text{bond}}(r_{ij}) + \sum_{ijk} u_{\text{angle}}(\theta_{ijk}) + \sum_{ijkl} u_{\text{dihedral}}(\phi_{ijkl}), \tag{6.35}$$

where r_{ij}, θ_{ijk}, and ϕ_{ijkl} are the distances between atoms, the bending angle of three sequencing atoms, and the dihedral angle consisting four atoms, respectively (see Fig. 6.4). Contrary to DPD, the force does not include random and dissipative force. Each term is represented by a function containing several parameters. The set of these functions and parameters is called force field [2, 14, 15, 23, 29, 30]. Many force fields employ the Lennard–Jones potential for the nonbonded term,

$$u_{\text{nonbond}}(r) = 4\epsilon \left[\left(\frac{\sigma}{r} \right)^{12} - \left(\frac{\sigma}{r} \right)^{6} \right], \tag{6.36}$$

and the harmonic potential for the bond and angle terms,

$$u_{\text{bond}}(r) = K_{\text{b}}(r - r_0)^2 \tag{6.37}$$

and

$$u_{\text{angle}}(\theta) = K_{\text{a}}(\theta - \theta_0)^2. \tag{6.38}$$

In a Class II force field [14, 23], unharmonic terms and crossing terms were introduced to enhance versatility, such as

$$u_{\text{bond}}(r) = k_1(r - r_0)^2 + k_2(r - r_0)^3 + k_3(r - r_0)^4 \tag{6.39}$$

and

$$u_{\text{bond-angle}}(r, \theta) = k_{\text{ba}}(r - r_0)(\theta - \theta_0). \tag{6.40}$$

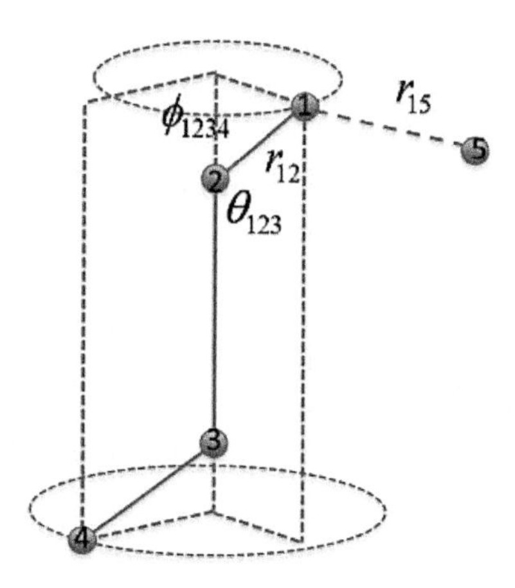

Figure 6.4 The inter- and intravariables for each force field term.

In this study we used the COMPASS force field [30], which is categorized as a Class II force field. A Class II force field contains more parameters than a Class I force field, such as AMBER [27] or OPLS [15]. The details of each force field are reviewed in Refs. [14, 23, 30].

6.8 Self-Diffusion Coefficient of Oxygen and Water in Bulk Nafion

In this section we discuss the self-diffusion coefficient of oxygen and water molecules in bulk Nafion. The molecular dynamics program Discover and the COMPASS [30] force field in Materials Studio (Dassault Systemes Biovia) are used for molecular dynamics simulation. A simulation cell of a Nafion membrane containing water and oxygen is prepared by using the Amorphous cell module in Materials Studio and equilibrated before simulation is performed. Simulations are performed for two kinds of Nafion with different equivalent weights (EW = 680 and 980) with the water content λ = 3, 6, and 10.

The self-diffusion coefficient is determined from the slope of the mean square displacement of oxygen and water molecules. Diffusion coefficients are calculated by

$$D = \lim_{t \to \infty} \frac{\left\langle (\mathbf{r}(t) - \mathbf{r}(0))^2 \right\rangle}{t}. \tag{6.41}$$

Figure 6.5 shows the results of different water contents λ for EW = 980 and 680. The self-diffusion coefficient of water increases with increasing water content. This is thought to be due to the increased connectivity of the water clusters formed in the electrolyte membrane. On the other hand, the self-diffusion coefficient of oxygen molecules does not differ as much as the self-diffusion coefficient of water molecules. The diffusion coefficient of oxygen in the electrolyte membrane is measured using the potential step method by Kudo et al. [21]. The correlation between the calculation result and the measured diffusion coefficient is shown in Fig. 6.6. The result shows qualitatively good agreement.

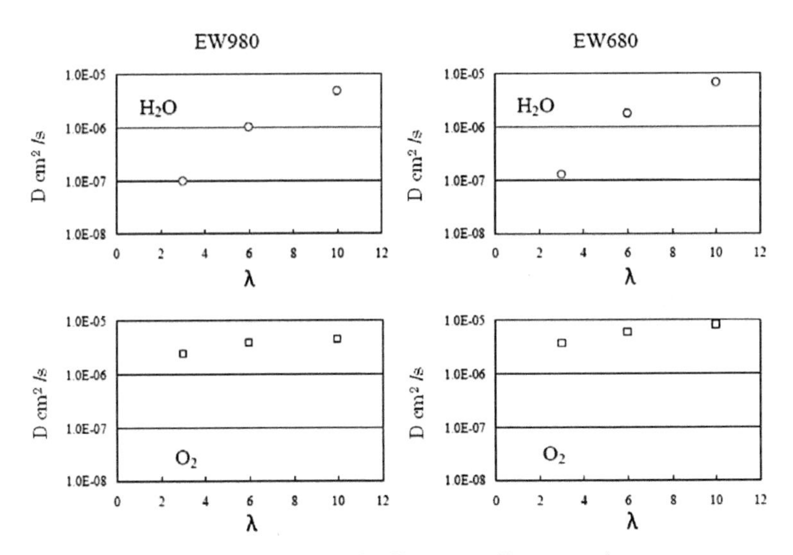

Figure 6.5 Dependence of the self-diffusion coefficient on the water content.

To investigate the path oxygen molecules traverse as they diffuse in the electrolyte membrane, the number of atoms within a certain distance of each oxygen molecule is counted as follows:

$$n(t) = \int_0^{r_c} \sum_j \delta\left(r_c - \left|\mathbf{r}_i(t) - r_0\right|\right) \mathrm{d}r, \tag{6.42}$$

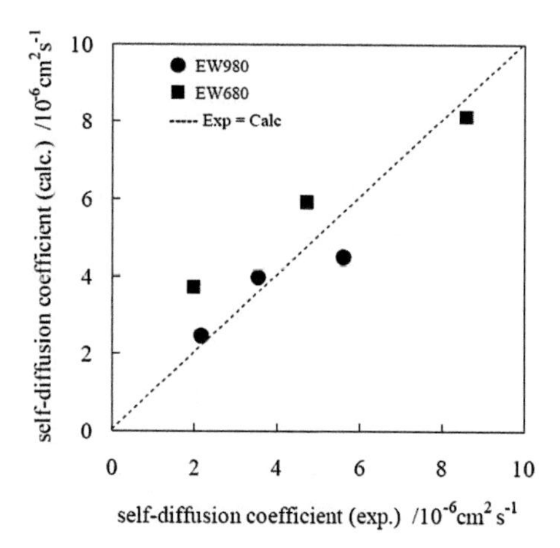

Figure 6.6 Comparison of experimental [21] and calculated self-diffusion coefficient values.

where $\mathbf{r}_0$ is the coordinate of an oxygen molecule of interest. The cutoff length used for counting is 4 Å. Figure 6.7 shows changes in the number of atoms around each oxygen molecule with time. The darkly shaded part corresponds to the number of fluorine atoms of the polymer, and the brightly shaded part corresponds to the number of hydrogen atoms of the water near the oxygen molecules. Although each simulation cell contains 50 oxygen molecules, the typical results of 4 molecules are shown here. The numbers shown in this figure are sequential numbers assigned to molecules. It can be seen from the figure that there are many dark areas overall. This indicates that oxygen molecules tend to be present in the vicinity of the polymer. Since there are few parts that are light gray, oxygen molecules are rarely in the center of the water cluster. Figure 6.8 shows the correlation between the square of the displacement of oxygen molecules after a certain time period and the fraction of fluorine atoms around that oxygen molecule at that time. As seen in Figs. 6.7 and 6.8, oxygen molecules are present in the vicinity of the polymer in a higher fraction, but in that case the amount of movement of the molecules is small. Furthermore, as mentioned above, oxygen molecules are rarely located in the central part of the water cluster, so it is presumed that oxygen molecules are mainly moving around the interface between water and polymer.

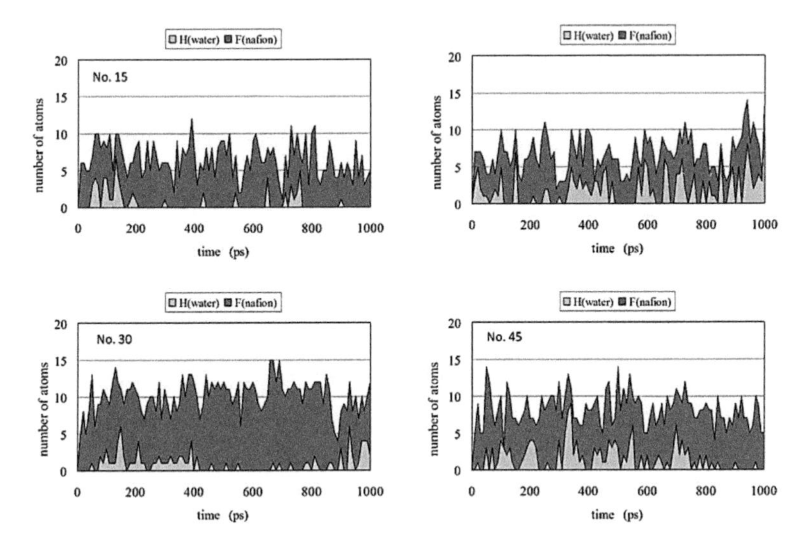

Figure 6.7 Time variation of the number of atoms around an oxygen molecule. Of the 50 molecules, 4 are shown. Dark gray is the number of fluorine atoms of the polymer skeleton, and light gray is the number of hydrogen atoms contained in the water molecule.

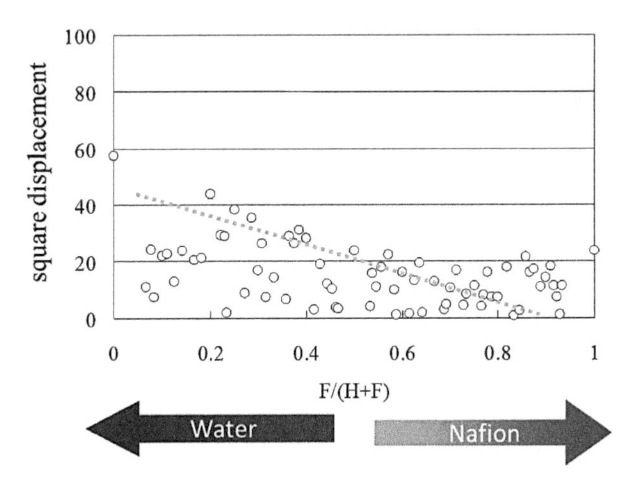

Figure 6.8 Correlation between the proportion of macromolecules around oxygen molecules and the amount of migration.

6.9 Conclusions

In this chapter we described all-atom molecular dynamics and CG simulations and applications of polyelectrolyte membranes. The structure of the polymer electrolyte membrane Nafion was calculated using DPD, which is a mesoscale calculation method. A DPD model was created on the basis of the molecular structure of the Nafion monomer, with the interaction coefficients between the DPD particles determined by calculating the cohesive energy of the segments corresponding to each particle type. The results of calculating the structure of Nafion starting from random initial conditions showed the appearance of a sponge-like structure similar to what has been postulated from experiments. The water cluster size and distribution spacing are approximately equal to those from experiments, and the results of experiments involving changes in structure depending on moisture content were also reproduced. From the above, it can be considered that a model that possibly represents the actual structure of Nafion has been successfully demonstrated through calculation. Results from the current model also suggest that the anisotropy observed in hydrocarbon-based electrolyte membranes may be explained by taking into consideration molecular rigidity.

To investigate the path oxygen molecules traverse as they diffuse in the electrolyte membrane, molecular dynamics simulation was executed. It is shown that oxygen molecules tend to be present in the vicinity of the polymer and that oxygen molecules are mainly moving around the interface between water and polymer.

References

1. Allen, M. P. and Tildesley, D. J. (1987). *Computer Simulation of Liquids* (Clarendon Press; Oxford University Press), xix, 385 p.
2. Burkert, U. and Allinger, N. L. (1982). *Molecular Mechanics*, ACS Monograph Series (American Chemical Society Washington, DC).
3. Chandrasekhar, S. (1943). Stochastic problems in physics and astronomy, *Rev. Mod. Phys.*, **15**, pp. 1–89.
4. Debe, M. K. (2012). Electrocatalyst approaches and challenges for automotive fuel cells, *Nature*, **486**, pp. 43–51.

5. Espanol, P. and Warren, P. (1995). Statistical-mechanics of dissipative particle dynamics, *Europhys. Lett.*, **30**, pp. 191–196.
6. Espanol, P. and Warren, P. B. (2017). Perspective: dissipative particle dynamics, *J. Chem. Phys.*, **146**, pp. 16.
7. Fan, C. F., Olafson, B. D., Blanco, M., and Hsu, S. L. (1992). Application of molecular simulation to derive phase-diagrams of binary-mixtures, *Macromolecules*, **25**, pp. 3667–3676.
8. Frenkel, D. and Smit, B. (2002). *Understanding Molecular Simulation: From Algorithms to Applications*, 2nd Ed. Computational Science: From Theory to Applications/series editors, Frenkel, D., et al. Vol. 1 (Academic Press, Harcourt), xxii, 638 p.
9. Gierke, T. D., Munn, G. E., and Wilson, F. C. (1981). The morphology in Nafion perfluorinated membrane products, as determined by wide-angle and small-angle x-ray studies, *J. Polym. Sci., Part B: Polym. Phys.*, **19**, pp. 1687–1704.
10. Goldstain, H. (1980). *Classical Mechanics*, 2nd Ed. (Addison-Wesley, Reading).
11. Groot, R. D. and Warren, P. B. (1997). Dissipative particle dynamics: bridging the gap between atomistic and mesoscopic simulation, *J. Chem. Phys.*, **107**, pp. 4423–4435.
12. Groot, R. D. and Madden, T. J. (1998). Dynamic simulation of diblock copolymer microphase separation, *J. Chem. Phys.*, **108**, pp. 8713–8724.
13. Haubold, H. G., Vad, T., Jungbluth, H., and Hiller, P. (2001). Nano structure of NAFION: a SAXS study, *Electrochim. Acta*, **46**, pp. 1559–1563.
14. Hwang, M. J., Stockfisch, T. P., and Hagler, A. T. (1994). Derivation of class-II force-fields. 2. Derivation and characterization of a class-II force-field, CFF93, for the alkyl functional-group and alkane molecules, *J. Am. Chem. Soc.*, **116**, pp. 2515–2525.
15. Jorgensen, W. L., Maxwell, D. S., and TiradoRives, J. (1996). Development and testing of the OPLS all-atom force field on conformational energetics and properties of organic liquids, *J. Am. Chem. Soc.*, **118**, pp. 11225–11236.
16. Kaneko, K., Takeoka, Y., Rikukawa, M., and Sanui, K. (2005). Syntheses and fuel cell performance of hydrocarbon polymer electrolytes: durability of polymer electrolytes, *Polym. Prepr. Jpn.*, **54**, pp. 4515.
17. Kawatsu, S. (1998). Advanced PEFC development for fuel cell powered vehicles, *J. Power Sources*, **71**, pp. 150–155.

18. Kinjo, T. and Hyodo, S. (2007). Linkage between atomistic and mesoscale coarse-grained simulation, *Mol. Simul.*, **33**, pp. 417–420.

19. Kinjo, T. and Hyodo, S. A. (2007). Equation of motion for coarse-grained simulation based on microscopic description, *Phys. Rev. E*, **75**, pp. 051109.

20. Koelman, J. and Hoogerbrugge, P. J. (1993). Dynamic simulations of hard-sphere suspensions under steady shear, *Europhys. Lett.*, **21**, pp. 363–368.

21. Kudo, K., Suzuki, T., and Morimoto, Y. (2010). Analysis of oxygen dissolution rate from gas phase into Nafion surface and development of an agglomerate model, in *Polymer Electrolyte Fuel Cells 10, Pts 1 and 2*, Gasteiger, H. A., et al., eds. (Electrochemical Soc Inc, Pennington), pp. 1495–1502.

22. Kubo, R., Toda, M., and Hashitsume, N. (1995). *Statistical Physics II* (Springer-Verlag, Heidelberg).

23. Maple, J. R., Hwang, M. J., Stockfisch, T. P., Dinur, U., Waldman, M., Ewig, C. S., and Hagler, A. T. (1994). Derivation of class-II force-fields. 1. Methodology and quantum force-field for the alkyl functional-group and alkane molecules, *J. Comput. Chem.*, **15**, pp. 162–182.

24. Mori, H. (1965). Transport, collective motion, and Brownian motion, *Prog. Theor. Phys.*, **33**, pp. 423–455.

25. Mumby, S. J., Sher, P., and Eichinger, B. E. (1993). Phase-diagrams of quasi-binary polymer-solutions and blends, *Polymer*, **34**, pp. 2540–2545.

26. Nordholm, S. and Zwanzig, R. (1975). A systematic derivation of exact generalized Brownian motion theory, *J. Stat. Phys.*, **13**, pp. 437–371.

27. Ponder, J. W. and Case, D. A. (2003). Force fields for protein simulations, *Protein Simul.*, **66**, pp. 27–+.

28. Rubinstein, M. and Colby, R. H. (2003). *Polymer Physics* (Oxford University Press), xi, 440 p.

29. Sato, F., Hojo, S., and Sun, H. (2003). On the transferability of force field parameters: with an ab initio force field developed for sulfonamides, *J. Phys. Chem. A*, **107**, pp. 248–257.

30. Sun, H. (1998). COMPASS: an ab initio force-field optimized for condensed-phase applications - Overview with details on alkane and benzene compounds, *J. Phys. Chem. B*, **102**, pp. 7338–7364.

31. Wagner, F. T., Lakshmanan, B., and Mathias, M. F. (2010). Electrochemistry and the future of the automobile, *J. Phys. Chem. Lett.*, **1**, pp. 2204–2219.

32. Yamamoto, S. and Hyodo, S. A. (2003). A computer simulation study of the mesoscopic structure of the polyelectrolyte membrane Nafion, *Polym. J.*, **35**, pp. 519–527.

33. Zeng, W., Du, Y., Xue, Y., and Frish, H. L. (2007). Solubility parameters, in *Physcal Properties of Polymers Handbook*, Mark, J. E., ed. (Springer, New York).

Chapter 7

Phase-Field Models for the Microstructural Characterization of Electrode Materials

Shunsuke Yamakawa
Toyota Central R&D Labs., Inc., Nagakute, Aichi 480-1192, Japan
e1044@mosk.tytlabs.co.jp

7.1 Introduction

Many concepts for new types of advanced high-performance batteries and fuel-cell systems have recently been proposed. The performance of a battery or a fuel cell, as characterized by its capacity, power density, and operating conditions (including temperature), is dependent on the properties and structure of the electrode materials employed, in the nanometer-to-micrometer range [41]. Importantly, such parameters are not homogeneous on this so-called mesoscale, and this heterogeneity is an important factor in terms of improving the initial capabilities and lifetimes of high-performance cells.

Many simulation methods have been developed to address numerous macroscopic problems involving structural mechanics,

Multiscale Simulations for Electrochemical Devices
Edited by Ryoji Asahi

ISBN 978-981-4800-71-6 (Hardcover), 978-0-429-29545-4 (eBook)
www.jennystanford.com

fluid dynamics, heat transport, and so on. At present, simulations based on continuum models are commonly used for the analysis of such topics, allowing system design based on the resulting information. At the other extreme, commercial software that allows quantum mechanical and/or atomic simulations is also available and has practical applications.

Despite this, there are no mesoscopic simulation techniques that cover wide spatial and temporal ranges. This is unfortunate because such methods could be employed (i) for postimplementation assessments of the extent to which specific structures lead to the desired properties, (ii) for examination of design trade-offs (i.e., between initial performance and durability), and (iii) as process engineering tools to predict the microstructures of newly developed materials.

The lack of mesoscopic simulation methods has led to significant interest in the development of such methods. As such, several new and useful techniques for mesoscopic material simulation have recently been reported [31, 46]. These new methods can be applied not only to the simulation of specific mesoscopic structures but also to a variety of scenarios that bridge macroscopic and microscopic technologies, as shown in Fig. 7.1. However, further advances in this technology are essential.

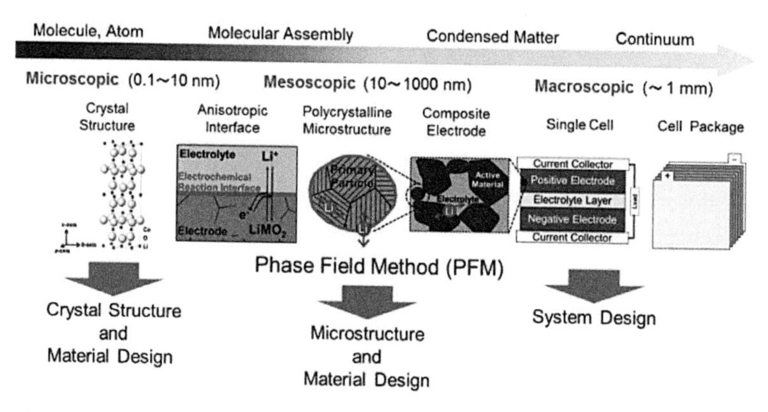

Figure 7.1 Simulation techniques proposed for multiscale problems.

Approaches to simulating mesoscale material structures can be roughly divided into two classes. The first category models the time evolution of microstructures by directly simulating atomic motions.

In the second category, microstructural evolution is expressed as the transformation of a set of field variables. Monte Carlo methods and molecular dynamics based on first-principles or classical potentials typically belong to the first category.

The phase-field method (PFM) [13], which belongs to the second group and is a continuum model assuming localized thermodynamic equilibrium, has been employed since the late 1980s to study processes such as dendritic growth and spinodal decomposition, as well as the resulting microstructures and grain growth. The PFM has also been used to predict microstructures and to elucidate driving forces producing such structures. To improve the performance of polymer electrolyte fuel cells (PEFCs) and lithium (Li)-ion secondary batteries (LIBs), the PFM has been applied to predict the microstructures of electrode materials on length scales ranging from nanometers to micrometers. Moreover, current research seeks to determine the correlation between microstructures and functional characteristics on the basis of various structures calculated by the PFM.

The main characteristics of the PFM are explained in Section 7.2. In this process, the time evolution of the microstructure is calculated on the basis of the free energy function, taking into account various factors that affect microstructure formation. The effective formulation of the free energy function is the key to the successful calculation of microstructures on the basis of the actual material parameters. Means of determining the free energy function are also described in this section.

In Section 7.3, a PFM approach to the morphological characterization of PEFCs is demonstrated. As shown in Fig. 7.2, platinum (Pt) supported on a carbon carrier is widely used as a catalyst for PEFCs. The catalytic activity of this material is significantly affected by the size distribution and morphologies of the Pt particles [56]. Therefore, the PFM has been employed to describe the deposition of Pt particles onto the substrate. It is also necessary to employ a PFM approach that effectively describes the compositional variation and phase transformation within a single Pt-based alloy catalyst nanoparticle in a PEFC and that provides information regarding crucial factors that affect the internal structures of the catalyst particles [16, 22].

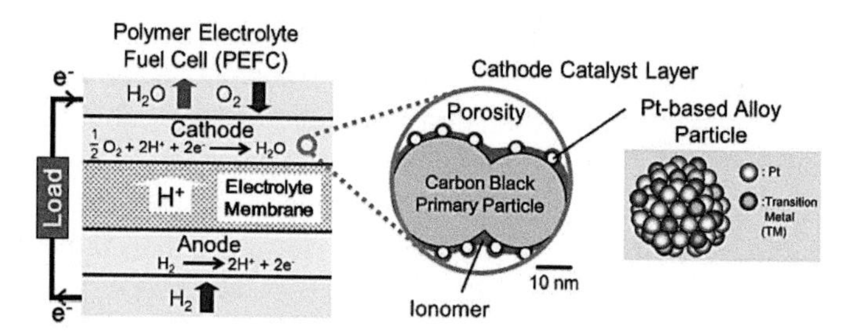

Figure 7.2 Schematic illustrations of a PEFC system and an electrocatalyst supported on a carbon substrate.

In Section 7.4, the PFM is applied to the numerical simulation of Li diffusion in 2D polycrystalline $LiCoO_2$ in LIBs. The information obtained regarding the essential properties of the associated anisotropic materials is then used to quantitatively evaluate the Li transport properties [44, 54]. An integrated computational approach that simultaneously simulates the microstructure and evaluates the transport properties is helpful in obtaining a quantitative understanding of each basic step in the Li transport process. Li diffusivity can be quantitatively predicted using a model that incorporates the kinetic parameters for the Li diffusion and electrochemical Li insertion processes, as well as the crystallographic misfit obtained from theoretical calculations or from empirical data. The PFM technique is also applied to the quantitative assessment of the effect of the microstructure on the stress generated during the charging of a polycrystalline secondary Li_xCoO_2 particle, taking into account the crystallographic anisotropies of the Li self-diffusion coefficient and elastic properties. The scope of future research is discussed in Section 7.5.

7.2 Simulation Methodology

This section explains the methodologies regarding the numerical simulations of nano- and microstructural characteristics. Microstructural evolution was simulated using the PFM on the basis of the free energy function. In this process, the conservative and nonconservative variables, which define the atomic concentrations and the structural heterogeneities, respectively, were initially set in the simulation area and the total free energy was defined using these variables. The time evolutions of the conservative variables were then evaluated using the Cahn–Hilliard equation [12]

$$\frac{\partial c_i}{\partial t} = \nabla \cdot \left[\sum_{j=1}^{n-1} M_{ij} \nabla \left(\frac{\delta G_{\text{sys}}}{\delta c_j} \right) \right]. \tag{7.1}$$

Here, c is the atomic concentration, G_{sys} is the total free energy of the targeted system, and M_{ij} is the mobility of the component, which is related to the self-diffusion coefficient, D.

The time evolutions of the nonconservative variables were calculated using the Allen–Cahn equation [6]

$$\frac{\partial \phi_i}{\partial t} = -\sum_j L_{ij} \frac{\delta G_{\text{sys}}}{\delta \phi_j}, \tag{7.2}$$

where ϕ is the phase-field parameter describing the phase interface or grain boundary (GB) and L_{ij} refers to the mobility at the interface or boundary. The time evolution of the system was solved using a practical and simple approach. Firstly, conservative and nonconservative variables (which were based on the atomic concentration and the phase state, respectively) were introduced for each discretized grid point. Subsequently, G_{sys} was defined at each local point in relation to these variables, as shown in Fig. 7.3. Subsequently, the Cahn–Hilliard and Allen–Cahn partial differential equations were solved numerically, using finite volume, finite difference, and finite element algorithms. The variables were evolved sequentially and discretely, on the basis of the limitation imposed by the stability criterion at each time step. Finally, as a result of the time evolution, the stationary state of the microstructure giving the minimum free energy was determined. The expression for G_{sys} varied depending on the type of interactions in the targeted system, and G_{sys} can be written as the volume integral of the local free energy over the entire volume as

$$G_{\text{sys}} = \frac{1}{V_{\text{m}}} \int_{\mathbf{r}} \left(G_{\text{chem}} + E_{\text{grad}} + E_{\text{str}} \right) d\mathbf{r}. \tag{7.3}$$

Here, $\mathbf{r}$ is the position vector and V_{m} is the molar volume. Typically, G_{sys} contains G_{chem}, E_{grad}, and E_{str}, which represent the bulk chemical free energy, the gradient energy, and the elastic strain energy, respectively. The details of these terms are discussed in the following sections.

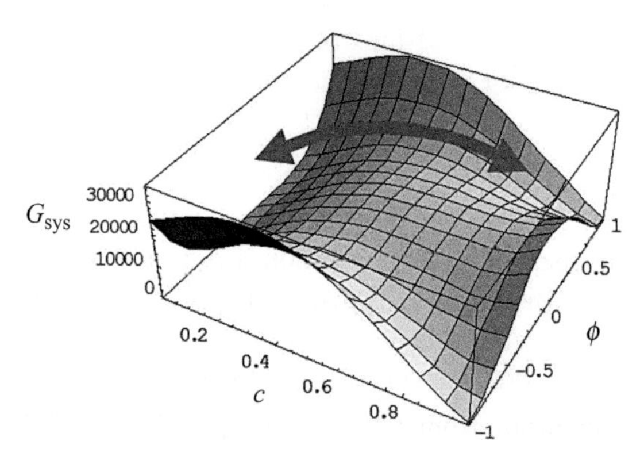

Figure 7.3 Total free energy, G_{sys}, plotted against the conservative variable, c, and the nonconservative variable, ϕ.

7.2.1 Bulk Free Energy

The bulk chemical free energy, G_{chem}, is equivalent to the molar Gibbs free energy and is determined assuming the spatial uniformity of the thermodynamic fields. The CALculation of PHAse Diagrams (CALPHAD) method is a useful means of evaluating the atomic interactions on the basis of thermodynamic models [37], and this technique has recently been applied to various alloy systems. In this approach, certain parameters are estimated to obtain an appropriate representation of the thermodynamic properties and the theoretically estimated or experimentally observed phase boundaries. As a representative example, the molar free energy was defined as a binary substitutional solution

$$G_{\text{chem}} = \sum_{i=\text{AorB}} c_i\, {}^{\circ}G_i + RT \sum_{i=\text{AorB}} c_i \ln c_i + c_A c_B L_{AB}. \tag{7.4}$$

Here, c_A and c_B are the mole fractions of components A and B, respectively, ${}^{\circ}G_i$ is the Gibbs formation energy of component i, and T and R are the absolute temperature and gas constant, respectively. The second and last terms represent the ideal entropy of mixing and the excess Gibbs free energy, respectively, while L_{AB} indicates the extent of the binary interaction. The Redlich–Kister (RK) power series [24] gives the composition dependence of L_{AB}, written as

$$L_{AB} = \sum_{\upsilon=0}^{k} (c_A - c_B)^{\upsilon}\, {}^{\upsilon}L_{AB}. \tag{7.5}$$

Here, ${}^{\upsilon}L_{\mathrm{AB}}$ is a series of parameters in the RK series. The ordered phase was described using two or more sublattices. A binary alloy using four sublattices can be expressed as

$$(\mathrm{A},\mathrm{B})_{0.25}(\mathrm{A},\mathrm{B})_{0.25}(\mathrm{A},\mathrm{B})_{0.25}(\mathrm{A},\mathrm{B})_{0.25}. \tag{7.6}$$

The Gibbs free energy for this ordered system was divided into two parts that express the contributions from disordering, $G_{\mathrm{chem}}^{\mathrm{dis}}(c_i)$, and long-range ordering (LRO), $\Delta G_{\mathrm{chem}}^{\mathrm{ord}}(y_i^{(s)})$. This was written as

$$G_{\mathrm{chem}} = G_{\mathrm{chem}}^{\mathrm{dis}}(c_i) + \Delta G_{\mathrm{chem}}^{\mathrm{ord}}(y_i^{(s)}). \tag{7.7}$$

The contribution from $G_{\mathrm{chem}}^{\mathrm{dis}}(c_i)$ is defined by Eq. 7.4. Here, the variable $y_i^{(s)}$ refers to the site fraction of constituent i in the sublattice s, which was determined using the equation

$$c_i = 0.25\sum_{s=1}^{4} y_i^{(s)}. \tag{7.8}$$

To ensure that $\Delta G_{\mathrm{chem}}^{\mathrm{ord}}(y_i^{(s)})$ was equal to zero in the disordered state, the LRO term for a four sublattice model was expressed by

$$\Delta G_{\mathrm{chem}}^{\mathrm{ord}}\left(y_i^{(s)}\right) = G_{\mathrm{chem}}^{\mathrm{ord}}\left(y_i^{(s)}\right) - G_{\mathrm{chem}}^{\mathrm{ord}}(c_i). \tag{7.9}$$

This same term for an arbitrary number of components can be expressed as

$$\begin{aligned} G_{\mathrm{chem}}^{\mathrm{ord}} &= \sum_i\sum_j\sum_k\sum_l y_i^{(1)} y_j^{(2)} y_k^{(3)} y_l^{(4)}\,{}^{\circ}G_{i:j:k:l} \\ &+0.25RT\sum_s\sum_i y_i^{(s)}\ln y_i^{(s)} + {}^{E}G_{\mathrm{chem}}. \end{aligned} \tag{7.10}$$

Here, ${}^{\circ}G_{i:j:k:l}$ is the Gibbs free energy with components i, j, k, and l. The excess free energy can be defined as

$$\begin{aligned} {}^{E}G_{\mathrm{chem}} &= \sum_{i_1}\sum_{i_2>i_1}\sum_j\sum_k\sum_l y_{i_1}^{(r)} y_{i_2}^{(r)} y_j^{(s)} y_k^{(t)} y_l^{(u)} L_{i_1,i_2:j:k:l} + \cdots + \\ &\sum_{i_1}\sum_{i_2>i_1}\sum_{j_1}\sum_{j_2>j_1}\sum_k\sum_l y_{i_1}^{(r)} y_{i_2}^{(r)} y_{j_1}^{(s)} y_{j_2}^{(s)} y_k^{(t)} y_l^{(u)} L_{i_1,i_2:j_1,j_2:k:l} + \cdots + \end{aligned} \tag{7.11}$$

Similar to a binary alloy model, the interface between the condensed matter and the vapor phase can be modeled by introducing the variables c_{M} and c_{Va}, which represent the mole fractions of metal

(M) and vacancies (Va), with the restriction that $c_M + c_{Va} = 1$. Here, a region having a finite thickness over which the primary occupation species changes from M to Va is equivalent to the interface between the metal condensed phase and the gas phase. At each lattice point, the value of G_{chem} was calculated on the basis of the regular solution model for a M-Va binary alloy [71] as

$$G_{chem} = c_M \, {}^{\circ}G_M + RT(c_M \ln c_M + c_{Va} \ln c_{Va}) + c_M c_{Va} L_{M\text{-}Va}. \quad (7.12)$$

Here, $^{\circ}G_M$ is the standard formation energy of M (which varies according to whether M is in a solid phase or a liquid one) and $L_{M\text{-}Va}$ is the M-Va interaction energy.

7.2.2 Gradient Energy

If the thermodynamic field is spatially non-uniform, G_{chem} must be accompanied by a gradient energy term, E_{grad}, to take into account the energetic interactions between the system and the surroundings. By introducing E_{grad}, the variables change smoothly from one domain or phase to another [33] at the boundary, which is likely to diffuse in actual microstructures. By introducing this term into the PFM method, the numerical difficulty of tracking a moving boundary can be mitigated. To derive this term, the Taylor series about G_{chem} with respect to the derivatives of the variable, c, was used, written as [25]

$$\begin{aligned} G_{chem}(c, \nabla c, \nabla^2 c) \approx G_{chem}(c,0,0) + \mathbf{K}_0(c)\cdot(\nabla c) + K_1(c)\cdot(\nabla^2 c) \\ + K_2(c)\cdot(\nabla c)^2. \end{aligned} \quad (7.13)$$

The first term on the right is equivalent to the homogeneous energy. To ensure that the energy value was invariant in the case of a coordinate transformation, $\mathbf{K}_0$ was given a value of zero. Therefore, E_{grad} was expressed as

$$\begin{aligned} E_{grad} &= K_1(c)\cdot(\nabla^2 c) + K_2(c)\cdot(\nabla c)^2 \\ &= \left[-\frac{\partial K_1}{\partial c} + K_2(c) \right](\nabla c)^2 = \frac{1}{2}\kappa(c)(\nabla c)^2. \end{aligned} \quad (7.14)$$

Here, the term κ corresponds to

$$\kappa(c) = 2\left[-\frac{\partial K_1}{\partial c} + K_2(c) \right]. \quad (7.15)$$

However, in practical applications, κ is treated as a coefficient having a constant value and defined by the relationship between the surface and the excess energy in the interface region [11, 50]. Assuming a surface with a gradient only in one direction, the excess energy across the interface, $\Delta G_{\text{interface}}$, is defined as

$$\Delta G_{\text{interface}} = \frac{1}{V_{\text{m}}} \int_{\mathbf{r}} \left[\Delta f(c) + \frac{1}{2} \kappa(c)(\nabla c)^2 \right] d\mathbf{r} = S\gamma, \tag{7.16}$$

where

$$\gamma = \frac{1}{V_{\text{m}}} \int_{-\infty}^{\infty} \left[\Delta f(c) + \frac{1}{2} \kappa(c)(dc/dx)^2 \right] dx. \tag{7.17}$$

Here, S is the surface area in m^2 and γ is the surface energy in $\text{J}\cdot\text{m}^{-2}$. The function, Δf, refers to the difference between the linear combinations of the free energies of two homogeneous phases and the chemical free energy, G_{chem}, at the interface. In addition, the coefficient κ is represented in $\text{J}\cdot\text{m}^2\cdot\text{mol}^{-1}$ to use units consistent with other energy terms. It can be explained that the more frequently used representation in $\text{J}\cdot\text{m}^{-1}$ is multiplied by the V_{m} in $\text{m}^3\cdot\text{mol}^{-1}$ [71]. At equilibrium, if c is regarded as stationary points of γ, it is possible to use a variational method to derive the Euler–Lagrange equation

$$I - \left(\frac{dc}{dx} \right) \left[\frac{\partial I}{\partial (dc/dx)} \right] = 0, \tag{7.18}$$

where

$$I = \Delta f(c) + \frac{\kappa}{2}(dc/dx)^2. \tag{7.19}$$

Equation 7.18 can be used to obtain the relationship

$$\Delta f(c) - \frac{\kappa}{2}(dc/dx)^2 = \text{const.} \tag{7.20}$$

Since both Δf and dc/dx are zero at $x = \pm\infty$, the term “const” in Eq. 7.20 is zero and so we can write

$$\Delta f(c) = \frac{\kappa}{2}(dc/dx)^2. \tag{7.21}$$

Introducing Eq. 7.21 into Eq. 7.17 yields

$$\gamma = \frac{2}{V_{\text{m}}} \int_{-\infty}^{\infty} \Delta f(c)\, dx, \tag{7.22}$$

and modifying Eq. 7.21 gives

$$dx = \sqrt{\frac{\kappa}{2\Delta f(c)}}dc. \tag{7.23}$$

Therefore, Eq. 7.22 can be expressed as

$$\gamma = \frac{1}{V_{\mathrm{m}}}\int_{c_{\mathrm{B}}}^{c_{\mathrm{A}}} \sqrt{2\kappa\Delta f(c)}dc, \tag{7.24}$$

where the variables c_{A} and c_{B} refer to the equilibrium concentrations of the two coexisting phases and κ is determined using the excess energy, Δf, and γ. The width of the interface area, Δd, can be approximately obtained by maximizing the concentration gradient at which Δf is maximized:

$$\Delta d = (c_{\mathrm{A}} - c_{\mathrm{B}})/(dc/dx) = (c_{\mathrm{A}} - c_{\mathrm{B}})\left[\kappa/(2\Delta f_{\max})\right]^{1/2} \tag{7.25}$$

7.2.3 Elastic Strain Energy

In some cases of microstructural formation, the elastic strain energy, E_{str}, caused by the lattice mismatch affects the total free energy. E_{str} can be expressed using the Einstein summation convention [34] as

$$E_{\mathrm{str}} = \frac{1}{2}C_{ijkl}(\mathbf{r})\left[\varepsilon_{ij}^{\mathrm{c}}(\mathbf{r},t) - \varepsilon_{ij}^{0}(\mathbf{r},t)\right]\left[\varepsilon_{kl}^{\mathrm{c}}(\mathbf{r},t) - \varepsilon_{kl}^{0}(\mathbf{r},t)\right]. \tag{7.26}$$

Here, C_{ijkl} is the elastic stiffness constant and $\varepsilon_{ij}^{\mathrm{c}}(\mathbf{r},t)$ and $\varepsilon_{ij}^{0}(\mathbf{r},t)$ are the constrained strain and eigenstrain, respectively. E_{str} can be estimated using the $\varepsilon_{ij}^{\mathrm{c}}$ value for a given C_{ijkl} and ε_{ij}^{0}. $\varepsilon_{ij}^{\mathrm{c}}$, in turn, is obtained from the stress equilibrium equation. In the above, $C_{ijkl}(\mathbf{r})$ is the component representing the stiffness tensor and $\varepsilon_{ij}^{\mathrm{c}}(\mathbf{r},t)$ and $\varepsilon_{ij}^{0}(\mathbf{r},t)$ are expressed as

$$\varepsilon_{ij}^{\mathrm{c}}(\mathbf{r},t) = \bar{\varepsilon}_{ij}^{0}(t) + \delta\varepsilon_{ij}^{\mathrm{c}}(\mathbf{r},t) \tag{7.27}$$

and

$$\varepsilon_{ij}^{0}(\mathbf{r},t) = A_{im}(\mathbf{r})A_{jn}(\mathbf{r})\eta_{mn}(\mathbf{r},t), \tag{7.28}$$

where $A_{ij}(\mathbf{r})$ is the rotation matrix and $\eta_{mn}(\mathbf{r},t)$ and $\bar{\varepsilon}_{ij}^{0}(t)$ are the lattice misfit and the mean strain, respectively. The second term in Eq. 7.27, $\delta\varepsilon_{ij}^{c}(\mathbf{r},t)$, is the deviation of the local strain value from $\bar{\varepsilon}_{ij}^{0}(t)$ and is obtained by the iterative scheme described below [26] for inhomogeneous elasticity. The equilibrium equations are satisfied by using the equation

$$\sigma^{\text{el}}_{ij,j}(\mathbf{r},t)=\frac{\partial\sigma^{\text{el}}_{ij}(\mathbf{r},t)}{\partial x_j}=C_{ijkl}\left[\frac{\partial\delta\varepsilon^{\text{c}}_{kl}(\mathbf{r},t)}{\partial x_j}-\frac{\partial\varepsilon^{0}_{kl}(\mathbf{r},t)}{\partial x_j}\right]$$
$$+\frac{\partial\left\{\Delta C_{ijkl}(\mathbf{r})\left[\bar{\varepsilon}^{\text{c}}_{kl}+\delta\varepsilon^{\text{c}}_{kl}(\mathbf{r},t)-\varepsilon^{0}_{kl}(\mathbf{r},t)\right]\right\}}{\partial x_j}=0, \tag{7.29}$$

and modifying Eq. 7.29 gives

$$C_{ijkl}\frac{\partial\delta\varepsilon^{\text{c}}_{kl}(\mathbf{r},t)}{\partial x_j}=C_{ijkl}\frac{\partial\varepsilon^{0}_{kl}(\mathbf{r},t)}{\partial x_j}$$
$$-\frac{\partial\left\{\Delta C_{ijkl}(\mathbf{r})\left[\bar{\varepsilon}^{\text{c}}_{kl}+\delta\varepsilon^{\text{c}}_{kl}(\mathbf{r},t)-\varepsilon^{0}_{kl}(\mathbf{r},t)\right]\right\}}{\partial x_j}. \tag{7.30}$$

Next, the relationship between $\delta\varepsilon^{\text{c}}_{ij}(\mathbf{r},t)$ and $u_i(\mathbf{r},t)$ is defined as

$$\delta\varepsilon^{\text{c}}_{kl}(\mathbf{r},t)=\frac{1}{2}\left[\frac{\partial u_k(\mathbf{r},t)}{\partial x_l}+\frac{\partial u_l(\mathbf{r},t)}{\partial x_k}\right]. \tag{7.31}$$

Here, $u_i(\mathbf{r},t)$ is displacement. Substituting $\delta\varepsilon^{\text{c}}_{ij}(\mathbf{r},t)$ in Eq. 7.30 with Eq. 7.31 followed by executing a Fourier transform gives

$$C_{ijkl}k_jk_lu_k(\mathbf{k},t)$$
$$=-ik_j\left\{C_{ijkl}\varepsilon^{0}_{kl}(\mathbf{k},t)-\left[\Delta C_{ijkl}(\mathbf{r})\left(\bar{\varepsilon}^{\text{c}}_{kl}+\delta\varepsilon^{\text{c}}_{kl}(\mathbf{r},t)-\varepsilon^{0}_{kl}(\mathbf{r},t)\right)\right]_{\text{k}}\right\}. \tag{7.32}$$

Here, **k** is the reciprocal space vector and $[]_{\text{k}}$ indicates that the Fourier transforms within the parentheses are executed after the calculation in real space. Equation 7.32 can be modified using $G_{ik}(\mathbf{k}) = (C_{ijkl}k_jk_l)^{-1}$ to obtain

$$u_k(\mathbf{k},t)$$
$$=-iG_{ik}k_j\left\{C_{ijkl}\varepsilon^{0}_{kl}(\mathbf{k},t)-\left[\Delta C_{ijkl}(\mathbf{r})\left(\bar{\varepsilon}^{\text{c}}_{kl}+\delta\varepsilon^{\text{c}}_{kl}(\mathbf{r},t)-\varepsilon^{0}_{kl}(\mathbf{r},t)\right)\right]_{\text{k}}\right\}. \tag{7.33}$$

Substituting Eq. 7.33 into the Fourier transform of Eq. 7.31 and executing the subsequent inverse Fourier transform yields

$$\delta\varepsilon^{\text{c(n)}}_{kl}(\mathbf{r},t)=\frac{1}{2}\left[\frac{\partial u_k(\mathbf{r},t)}{\partial x_l}+\frac{\partial u_l(\mathbf{r},t)}{\partial x_k}\right]$$
$$=\frac{1}{2}\int_k i\left\{k_lu_k(\mathbf{k},t)+k_ku_l(\mathbf{k},t)\right\}\exp(i\mathbf{k}\cdot\mathbf{r})\frac{d\mathbf{k}}{(2\pi)^3}$$

$$=\frac{1}{2}\int_{k}\{n_l\Omega_{ik}(\mathbf{n})+n_k\Omega_{il}(\mathbf{n})\}n_j$$

$$\left\{C_{ijkl}\varepsilon^{0}_{kl}(\mathbf{k},t)-\left[\Delta C_{ijkl}(\mathbf{r})\left(\bar{\varepsilon}^{c}_{kl}+\delta\varepsilon^{c(n-1)}_{kl}(\mathbf{r},t)-\varepsilon^{0}_{kl}(\mathbf{r},t)\right)\right]_{\mathbf{k}}\right\}$$

$$\exp(i\mathbf{k}\cdot\mathbf{r})\frac{d\mathbf{k}}{(2\pi)^3}. \tag{7.34}$$

Here, $\Omega_{ij}(\mathbf{n}) = k^2G_{ij}(\mathbf{k})$ and $\mathbf{n} = \mathbf{k}/k$ is the unit vector in the direction of the reciprocal space vector. Since the right-hand side of Eq. 7.34 includes $\delta\varepsilon^{c}_{ij}(\mathbf{r},t)$, and an iterative calculation is required to obtain a self-consistent value of $\delta\varepsilon^{c}_{ij}(\mathbf{r},t)$, the variable n in $\delta\varepsilon^{c(n)}_{ij}(\mathbf{r},t)$ represents the iteration number.

7.3 Morphological Characterization of Electrocatalysts in a PEFC

7.3.1 Phase-Field Model for the Deposition Process of Pt Nanoparticles on a Carbon Substrate

In the catalyst layer of a PEFC, Pt in the form of particles several nanometers in size is supported on a carbon black surface. Carriers with low defect density, such as highly crystalline carbon, provide excellent corrosion resistance, but adequate dispersion of Pt particles remains a challenge. It is important to elucidate the correlation between the surface structures of carriers and the precipitation mechanism to enable the design of carbon carrier surfaces with suitable initial performance and durability [35, 78]. The effect of surface structure heterogeneity on the precipitated morphology of Pt particles, such as the shape, size, and dispersion of the particles, was examined using the PFM. The influence of the means by which the Pt atoms are supplied on the competition between nucleation and nucleus growth was also investigated.

7.3.1.1 Computational details

As shown in Fig. 7.4, the surface on one end of a calculation region was treated as the carbon black surface. The concentrations of Pt and vacancies (Va) at each grid point were represented by the variables c_{Pt} and c_{Va}, with the constraint of $c_{\mathrm{Pt}} + c_{\mathrm{Va}} = 1$. To explicitly consider a phase state, the nonconservative variable θ, having values of 0 and 1 in the liquid and solid phases, respectively, was introduced. In this manner, the total free energy, G_{sys}, of the Pt–Va system was represented as the integral of the local free energy and the gradient energy throughout the system [70], as in the series of equations

$$G_{\mathrm{sys}} = \frac{1}{V_{\mathrm{m}}} \int_r \left[h(\theta) G_{\mathrm{chem}}^{(\mathrm{S})} + (1 - h(\theta)) G_{\mathrm{chem}}^{(\mathrm{L})} + E_{\mathrm{int}}^{\mathrm{S-L}} + E_{\mathrm{grad}} + E_{\mathrm{Pt\text{-}C}} \right] d\mathbf{r}, \tag{7.35}$$

$$E_{\mathrm{int}}^{\mathrm{S-L}} = g(c)\Omega\theta^2(1-\theta)^2, \tag{7.36}$$

$$E_{\mathrm{grad}} = \frac{1}{2}\left[h(\theta)\kappa_{c_{\mathrm{S}}} + (1 - h(\theta))\kappa_{c_{\mathrm{L}}} \right](\nabla c)^2 + g(c)\frac{\kappa_\theta}{2}(\nabla\theta)^2, \tag{7.37}$$

$$h(\theta) = \theta^3(6\theta^2 - 15\theta + 10), \tag{7.38}$$

and

$$g(c) = c^3(6c^2 - 15c + 10). \tag{7.39}$$

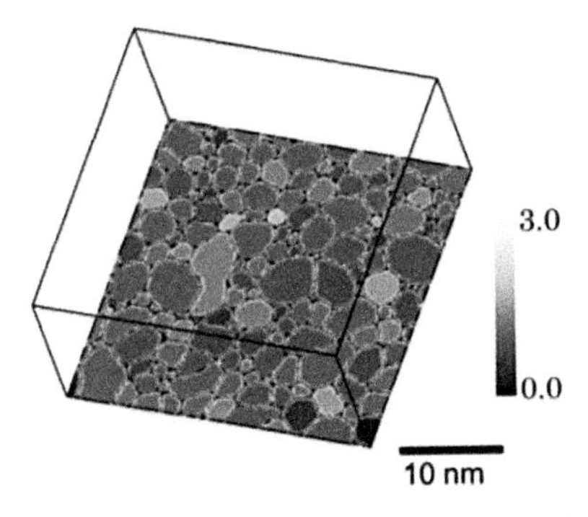

Figure 7.4 A simulation model including a carbon black surface at the base. The varying range of Pt–substrate interactions (normalized by the Pt–graphene interaction) is shown by the gray tone.

Here, $G_{\mathrm{chem}}^{(\mathrm{S})}$ and $G_{\mathrm{chem}}^{(\mathrm{L})}$ represent the chemical free energies in the solid–vapor and liquid–vapor states, respectively, and follow the basic form in Eq. 7.12:

$$G_{\mathrm{chem}}^{(\mathrm{S})} = (H_f - T\Delta S_f)c(1-c) + RT[c\ln c + (1-c)\ln(1-c)], \tag{7.40}$$

and

$$G_{\mathrm{chem}}^{(\mathrm{L})} = \Delta G_m c + \varepsilon(H_f - T\Delta S_f)c(1-c) + RT[c\ln c + (1-c)\ln(1-c)]. \tag{7.41}$$

Here, H_f is the mono vacancy formation energy (1.51 eV or 145,693 J·mol^{-1} [7]) and ΔS_f (=1.32R [7]) is the variation in the entropy term due to the relaxation of lattice vibration due to vacancy formation. Here, R and T are the gas constant and absolute temperature, respectively. The formation energy of a pure substance is normalized as zero. The coefficient ε obtained from the latent heat of solidification, ΔH_m, is 19,700 J·mol^{-1} [1] (melting point: 2042 K) and the latent heat of vaporization, ΔH_v, is 510,600 J·mol^{-1} [1] (vaporization point: 4100 K). The equation for this term is

$$\varepsilon = 1 - \frac{\Delta H_m}{\Delta H_m + \Delta H_v}. \tag{7.42}$$

Here, the molar volume, V_m, is approximately 9.09 × 10^{-6} m^3·mol^{-1} [1]. In Eq. 7.35, the contributions of $G^{(S)}_{chem}$ and $G^{(L)}_{chem}$ are determined by the function, h, which is correlated with θ and represents the fraction of solid–vapor and liquid–vapor equilibrium. Equations 7.38 and 7.39 are continuous functions for which the first derivative in the case that θ = 0 or 1 (or c = 0 or 1) is 0. The value of the free energy change related to the latent heat, ΔG_m, has been reported in the literature [18]. The coefficients κ_{c_S} and κ_{c_L} were determined using the excess energy, Δf, and the surface free energy, γ, in Eq. 7.24. In the case of a solid–liquid interface, there are two unknown parameters, κ_θ and Ω. To determine these parameters, $d_{SL} \approx d_{LV}$ was assumed. The expression for the surface energy of the liquid–vapor interface of Pt, γ_{LV}, has already been determined experimentally to be γ_{LV} = 1.8 – 0.17 × (T – T_m)/1000 J·m^{-2} [1]. In addition, the surface energy of the solid–vapor phase, γ_{SV}, has been shown to be approximately 20% larger than that of the liquid–vapor interface [59]. The remaining term, γ_{SL}, is determined by subtracting γ_{LV} from γ_{SV}.

The variable $E_{Pt\text{-}C}$ is the interaction energy between Pt and the carbon carrier surface, and its value varies depending on the defect site. When the Pt coordination state is unsaturated, the Pt atoms are expected to be attracted to the substrate surface. Therefore, the interaction potential energy per Pt atom, $V_{Pt\text{-}C}$, will vary with the coordination number. To estimate the interaction energy, first-principles calculations were performed using a model including a graphene sheet and a Pt cluster [42], and the energy value was found to decrease as the coordination number increased. The data also showed no significant interaction between the graphene sheet and Pt. Using a Pt concentration normalized by the bulk Pt concentration, $V_{\text{Pt-C}}$ was represented as

$$V_{\text{Pt-C}} = V_0\left(1 - c^{(1/2.4)}\right)^{2.4}, \tag{7.43}$$

where

$$V_0 = \frac{2}{3}V_{\text{Pt-C}}(\text{bridge}) + \frac{1}{3}V_{\text{Pt-C}}(\text{hollow}) \approx -1.85\,(\text{eV/atom}). \tag{7.44}$$

Here, V_0 is the monoatom adsorption energy. The interaction energy $E_{\text{Pt-C}}$, which is introduced into the free energy functions, was found by multiplying $V_{\text{Pt-C}}$ by the Pt concentration, as in

$$E_{\text{Pt-C}} = cV_{\text{Pt-C}}. \tag{7.45}$$

In addition, $E_{\text{Pt-C}}$ was incorporated into the free energy function only at the grid points within a vertical distance of 2 Å from the graphite plane. In this model, the effect of the dispersion forces between Pt and carriers is not considered. Figure 7.4 shows the calculation model. A temperature of 293.15 K was applied, and the dimensions of the simulation region were 40 × 40 × 20 nm, with a grid spacing of 0.2 nm. Supersaturated Pt vapor was introduced over the surface of the carbon substrate, and the carbon black surface was modeled so as to be as consistent as possible with experimental observations. The energetic fraction of the surface sites and the crystallite size were evaluated from gas adsorption isotherms [52] and X-ray diffraction patterns [60], respectively. Since the carbon atoms have dangling bonds at the grain boundaries, the interaction energy, $E_{\text{Pt-C}}$, at the grain boundaries was assumed to be roughly two or three times greater than that on the graphene sheet.

The equilibrium state of a multiphase system corresponding to the minimum energy was obtained by numerically solving the Cahn–Hilliard equation (Eq. 7.1) for c_{Pt} and the Allen–Cahn equation (Eq. 7.2) for θ, such that

$$\frac{\partial c}{\partial t} = \nabla\left(M\nabla\frac{\delta G_{\text{sys}}}{\delta c}\right) \tag{7.46}$$

and

$$\frac{\partial \theta}{\partial t} = -L_\theta\frac{\delta G_{\text{sys}}}{\delta \theta}. \tag{7.47}$$

The parameter L_θ is related to the mobility of the solid–liquid interface, and M is related to atomic diffusivity. To ensure that the mobility of the solid–liquid interface did not limit the rate of atomic diffusion, L_θ was given a value larger than the change in atomic concentration.

7.3.1.2 Results and discussion

Figure 7.5 shows the time evolution of the density profile of Pt over the carrier surface. Initial nuclei islands appeared at the grain boundaries due to fluctuations in the Pt concentration because of the effect of the heterogeneity of $E_{\text{Pt-C}}$. Subsequently, the adsorbed Pt formed 3D islands, indicating the Volmer–Weber growth mode. These Pt particles were almost hemispherical or spherical, with sizes of less than 10 nm. Electrochemically deposited Pt particles on highly oriented pyrolytic graphite have been observed by atomic force microscopy [35, 78]. The microstructural evolution in the present work was consistent with these prior experimental studies in that the Pt nanoparticles grew both at step edges and on terraces on the basis of the same growth mode. However, nucleation at the step edges was evidently preferred. The particle sizes and shapes in the current study also agree with experimental high-resolution images obtained from transmission electron microscopy [5]. Thus, the PFM can accurately reproduce microstructural evolution.

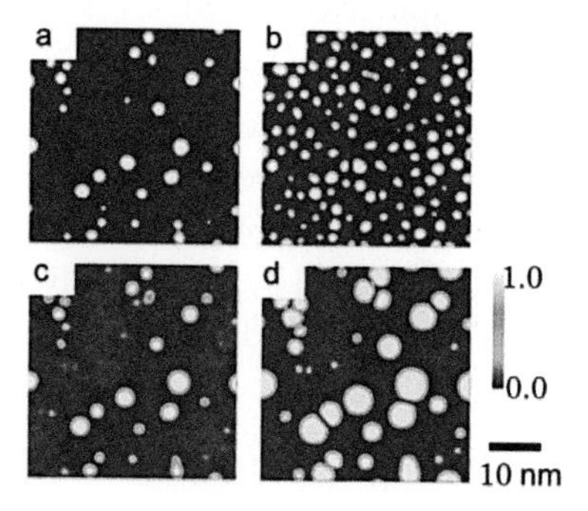

Figure 7.5 Cross-sectional images of Pt particles on carbon substrate. (a) and (c) correspond to the results of instantaneous and progressive nucleation, respectively, with a Pt loading of 1.9 mg·m^{-2}, while (b) and (d) correspond to a loading of 7.8 mg·m^{-2}. The varying range of the normalized Pt concentration is shown by the color tone. Reprinted from Ref. [70]. © IOP Publishing. Reproduced with permission. All rights reserved.

Figure 7.6 shows the dependency of the Pt particle size and the interparticle distance on the Pt supply conditions, using two different conditions. In one, the total amount of Pt atoms was supplied during the initial state while maintaining mass conservation, giving instantaneous nucleation. In the other, the Pt concentration at the opposite end from the substrate was fixed and Pt atoms were gradually supplied, leading to the progressive nucleation of Pt particles. The differences between the two conditions are clearly shown in Fig. 7.6. In the case of instantaneous nucleation, the Pt particle size was nearly constant and an increase in Pt loading resulted in an increase in the number of Pt particles. The interparticle distance gradually shrank to equal the average diameter of the graphitic crystallites. This occurred because the nucleation and growth processes proceeded separately while the growth duration of each particle was approximately the same. As a result, well-dispersed particle deposition was realized according to the surface heterogeneity of the carbon substrate. In contrast, the gradual supply condition increased both the mean diameter and the standard deviation of the Pt particles as the total loading increased. Because the nucleation and growth processes compete with one another, size distribution and coarsening are greatly increased. It is therefore confirmed that particle size distribution is affected by the heterogeneity of the carbon surface and the competitive particle formation process. Thus, numerical calculations can help to evaluate the influence of nucleation conditions and surface heterogeneity on Pt particle morphology.

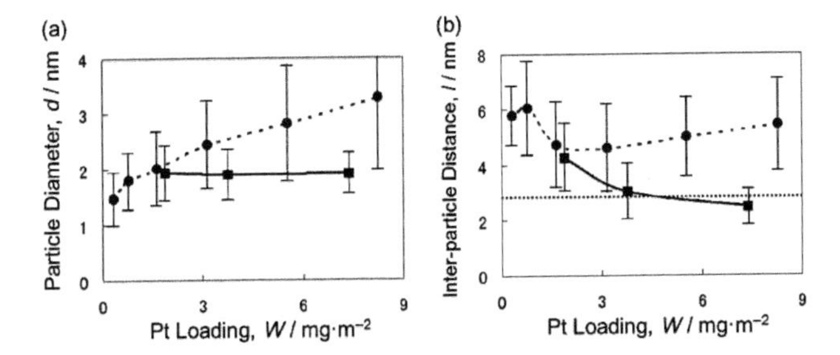

Figure 7.6 The effects of the Pt loading conditions on microstructural characteristics. (a) Pt particle size and (b) interparticle distance based on the nearest neighbor. Filled circles and filled squares show the results of progressive and instantaneous nucleation, respectively. The dotted line in (b) indicates the mean diameter of the graphitic crystallites. Reprinted from Ref. [70].

7.3.2 Surface Segregations in Pt-Based Alloy Nanoparticles

In the previous section, the PFM was proposed as a means of describing the formation of pure Pt nanoparticles on a carbon substrate. In this section, the PFM is employed to describe the phase transformations and compositional variations within a single nanoparticle of a Pt-based binary alloy. The results explain variations in the phase transformations and atomic segregation with the alloy composition, heat-treatment temperature, and particle size to give a designed alloy nanoparticle. To establish the validity of this model, the calculation results are compared with prior experimental and theoretical results for the effect of heat-treatment temperature on the solid–liquid and ordered–disordered transitions in an FePt particle. Subsequently, the radial distributions of the atomic compositions and the phase states of Pt-based-alloy (CrPt, FePt, CoPt, NiPt, CuPt, PdPt, IrPt, and AuPt) nanoparticles are examined.

7.3.2.1 Computational details

Three types of field variables relating to the atomic concentration, c; LRO, s; and solid–liquid phase transition, θ, were introduced [73]. The total free energy, G_{sys}, was determined as

$$G_{\text{sys}} = \frac{1}{V_{\text{m}}}\int [h(\theta)G_{\text{chem}}^{(\text{S})} + (1-h(\theta))G_{\text{chem}}^{(\text{L})} + W\theta(1-\theta) + \frac{1}{2}\sum_{i=1}^{n}\kappa_i(\nabla c_i)^2 + h(c)\frac{\kappa_s}{2}\sum_{i=1}^{3}(\nabla s_i)^2 + \frac{\kappa_\theta}{2}(\nabla\theta)^2]d\mathbf{r}, \tag{7.48}$$

where

$$c = \sum_{i=1}^{n-1} c_i, \tag{7.49}$$

$$h(x) = x^3(6x^2 - 15x + 10),\ x = \theta \text{ or } c, \tag{7.50}$$

$$W = \frac{1}{c}\sum_{i=1}^{n-1} c_i W_i, \tag{7.51}$$

and

$$\kappa_\theta = \frac{1}{c}\sum_{i=1}^{n-1} c_i \kappa_{\theta,i}. \tag{7.52}$$

Here, c_i is the atomic concentration of component i, normalized by the maximum concentration, and satisfies the condition

$$\sum_{i=1}^{n} c_i = 1 . \tag{7.53}$$

The binary TM-Pt alloy was characterized by three components (n = 3): TM (i = 1), Pt (i = 2), and vacancies, Va (i = 3). The variable $\mathbf{S}$ = (s_1, s_2, s_3) were expressed as three-component LRO parameters [10]. The parameters for complete $L1_0$ and $L1_1$ ordering were represented by the values (1, 0, 0), while $L1_2$ ordering was represented by (0.5, 0.5, 0.5). The functions $G_{\text{chem}}^{(\text{S})}$ and $G_{\text{chem}}^{(\text{L})}$ are the chemical free energy of the solid and liquid phases, respectively, and (S) and (L) refer to the solid and liquid phases, respectively. The values of $G_{\text{chem}}^{(\text{S})}$ and $G_{\text{chem}}^{(\text{L})}$ for the disordered state were determined assuming a regular solution model (as discussed in Section 7.2.1) as

$$G_{\text{chem}}^{\text{dis\_(k)}} = \sum_{i=1}^{n-1} c_i \,{}^{\circ}G_i^{(\text{k})} + \sum_{i=1}^{n-1} \sum_{j=2, j>i}^{n} L_{ij}^{(\text{k})} c_i c_j + RT \sum_{i=1}^{n} c_i \ln c_i \text{ , k = S or L,} \tag{7.54}$$

where ${}^{\circ}G_i^{(\text{k})}$ is the Gibbs formation energy of the element i in phase k and $L_{ij}^{(\text{k})}$ is the interaction between element i and j. The coefficient W is the interfacial energy between the solid and liquid phases and the coefficients κ_i, κ_s, and κ_θ are the gradient energy coefficients. The coefficients κ_1 and κ_2 are treated as the same variable referred to as κ_{12}. The coefficients $L_{in}^{(\text{k})}$ and κ_n, in relation to the solid–vapor interface, are obtained as

$$L_{in}^{(\text{k})} = \Delta H_{\text{f},i}^{(\text{k})} - T\Delta S_{\text{f}} \text{ , k = S or L,} \tag{7.55}$$

$$\kappa_n = \frac{1}{c} \sum_{i=1}^{n-1} c_i (\kappa_{ni} - \kappa_{12}), \tag{7.56}$$

and

$$\kappa_{ni} = h(\theta)\kappa_{ni}^{(\text{S})} + [1 - h(\theta)]\kappa_{ni}^{(\text{L})} . \tag{7.57}$$

In Eq. 7.55, $\Delta H_{f,i}^{(\text{k})}$ and ΔS_{f} are the enthalpy and entropy pertaining to mono-vacancy formation, respectively. The ΔS_{f} value is estimated using the simple approximate 1.32R, which has been employed for pure Pt [7].

The coefficients $L_{12}^{(S)}$ and $L_{12}^{(L)}$ in Eq. 7.54 and $\Delta G_{\text{chem}}^{\text{ord}}$ in Eq. 7.7 were represented as polynomial equations in relation to the atomic concentrations. These were based on thermodynamic assessments of binary alloys using the CALPHAD approach (Cr–Pt [45], Fe–Pt [20], Co–Pt [30], Ni–Pt [36], Cu–Pt [2], Pd–Pt [58], Ir–Pt [3], and Au–Pt [23]). Using Eqs. 7.24 and 7.25, the terms $\Delta H_{f,i}^{(S)}$ and $\kappa_{ni}^{(S)}$ were simultaneously evaluated on the basis of the (111) surface energy of the fcc metal [15, 67] and the solid–vapor interface width was assumed to be 3×10^{-10} m. The calculations used solid surface energy, $\gamma_i^{(S)}$; values for the (111) surfaces of Pt, Fe, and Co were obtained from first-principles calculations [15], while those for Ni, Cu, Pd, Ir, and Au were obtained from embedded atom calculations [67]. The Pt value in Ref. [67] was reconciled with the value in Ref. [15] by scaling the former. The surface energy of fcc-Cr was determined on the basis of the relative values of bcc-Cr and α-Fe in Ref. [21] and the relative values of γ-Fe and α-Fe in Ref. [15]. $\kappa_{ni}^{(L)}$ is estimated to reproduce $\Delta H_{f,i}^{(L)}$ and $\gamma_i^{(L)}$. The $\Delta H_{f,i}^{(L)}$ value is determined relevant to $\Delta H_{f,i}^{(S)}$ [71]. The liquid surface energy, $\gamma_i^{(L)}$, is determined using the relation [59] explained in Section 7.3.1.1. The values of the solid–liquid interfacial energy barrier, W_i, and $\kappa_{\theta,i}$ were estimated by adapting the solid–liquid interfacial energy and the interface width, as described by Eqs. 7.25 and 7.26 in Section 7.2.2. The resulting values are shown in Table 7.1, and the estimated values of $\Delta H_{f,i}^{(S)}$ agree well with the experimental data [38, 43, 51].

The coefficient κ_s represents the gradient energy coefficient in relation to the ordered phase–disordered phase interface. Using $\Delta G_{\text{chem}}^{\text{ord}}$ and assuming the antiphase boundary width to be 1×10^{-9} m, the values of κ_s for CrPt ($L1_0$), FePt ($L1_0$), CoPt ($L1_0$), and CuPt ($L1_1$) were found to be 0.303×10^{-15}, 1.63×10^{-15}, 0.160×10^{-15}, and 0.137×10^{-15} J·m^2·mol^{-1}, respectively. Using the interfacial energy heights obtained from the CALPHAD and an assumed interface width of 1×10^{-9} m, the values of κ_{12} for the Au–Pt and Ir–Pt binary alloys were estimated to be 1.67×10^{-15} and 1.66×10^{-15} J·m^2·mol^{-1}, respectively. The κ_{12} values for the other binary alloys were set at 5×10^{-17} J·m^2·mol^{-1}.

Table 7.1 Estimated parameters of the alloy elements at 973.15 K

	Cr	Fe	Co	Ni	Cu	Pd	Ir	Pt	Au
Surface energy, $\gamma_i^{(S)}$ / $J \cdot m^{-2}$	2.62	2.05	1.88	1.66	1.11	1.11	2.41	1.35	0.67
Molar volume,									
$V_{m,i}^{(S)} / 10^{-6} \mathrm{m}^3 \cdot \mathrm{mol}^{-1}$	6.60	7.28	6.76	6.65	7.20	8.94	8.62	9.15	10.3
$V_{m,i}^{(L)} / 10^{-6} \mathrm{m}^3 \cdot \mathrm{mol}^{-1}$	8.20	7.83	7.51	7.34	7.88	10.0	9.46	10.1	11.3
Monovacancy formation energy, $\Delta H_{f,i}^{(S)} / 96485 \mathrm{J} \cdot \mathrm{mol}^{-1}$	1.85	1.65	1.45	1.31	1.05	1.22	2.15	1.42	0.96
Gradient energy coefficient,									
$\kappa_{ni}^{(S)} / 10^{-15} \mathrm{J} \cdot \mathrm{m}^2 \cdot \mathrm{mol}^{-1}$	6.76	5.81	5.03	4.38	2.89	3.86	8.15	4.92	2.74
$\kappa_{ni}^{(L)} / 10^{-15} \mathrm{J} \cdot \mathrm{m}^2 \cdot \mathrm{mol}^{-1}$	7.87	5.08	4.72	4.08	2.65	3.72	7.45	4.57	2.50
$\kappa_{\theta,i} / 10^{-15} \mathrm{J} \cdot \mathrm{m}^2 \cdot \mathrm{mol}^{-1}$	1.13	0.947	0.797	0.681	0.439	0.636	1.33	0.804	0.426
Solid–liquid interface energy, $W_i / \mathrm{kJ} \cdot \mathrm{mol}^{-1}$	25.0	19.0	17.0	15.0	11.0	13.0	27.0	16.0	9.3

The conservative time evolutions of the atomic concentrations can be simulated by solving Eq. 7.1. The coefficient M_{ij} is the mobility of component *i* due to the gradient of the functional derivative of G_{sys} with regard to the concentration of component *j*. This is represented as

$$M_{ii} = c_i(1-c_i)D/RT \text{ and } M_{ij} = M_{ji} = -c_i c_j D/RT. \quad (7.58)$$

Here, D is the self-diffusion coefficient.

The distributions of the order-disorder phase were calculated by solving the equations

$$\frac{\partial s_i}{\partial t} = -L_s \frac{\delta G_{sys}}{\delta s_i}, \quad (7.59)$$

and

$$\frac{\partial \theta}{\partial t} = -L_\theta \frac{\delta G_{sys}}{\delta \theta}, \quad (7.60)$$

where L_s and L_θ denote the mobility of the ordered phase–disordered phase and solid phase–liquid phase interfaces, respectively. The numerical calculations of the radial coordinates of a nanoparticle were performed using the finite-volume method with an initial grid spacing of 0.005 nm. Depending on the local molar volume, the grid spacing was updated for each time step. The boundary conditions were set such that the spatial derivatives of the variables were zero, with $r = 0$ and $r = l$. The initial concentrations of Pt and the alloying element were equal, and the initial value of θ was set to the value c_{Pt+M}. In this study, it was important to determine the steady-state microstructure at the selected temperature. Therefore, to reduce the calculation time, variable time steps were employed on the basis of the concentration flux.

7.3.2.2 Results and discussion

The phase transformation results for this model, which affect the surface segregation, were compared with experimental results and other theoretical calculations. The radial distribution of the LRO parameter was initially simulated for binary alloys, indicating the order-disorder phase transition. Figure 7.7 summarizes the effect of particle size on the $L1_0$ ordering averaged over an FePt particle at 973.15 K. A reduction in the particle size evidently decreased the degree of $L1_0$ ordering. Since the curvature of the order–disorder interface increases as the particle size decreases, the gradient energy term multiplied by κ_S promotes this phase transformation in the direction that decreases the interface energy. When the antiphase boundary width was approximated as 1 nm, according to experimental results [65], the simulated gradation degree of $L1_0$ ordering was in agreement with experimental results [39, 48, 49]. This result indicates that the present PFM method allows sufficient accuracy when describing the order–disorder phase transition.

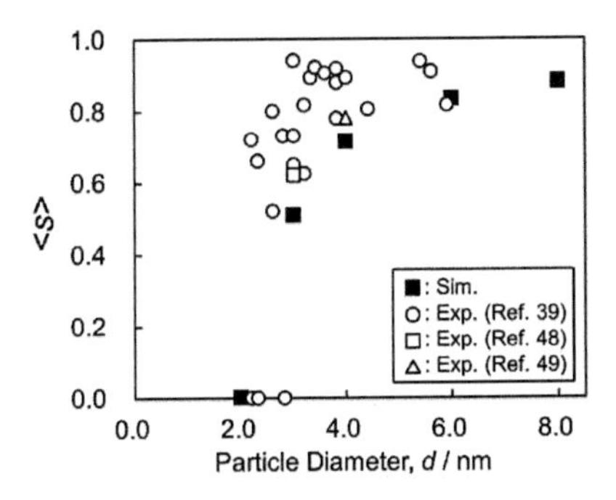

Figure 7.7 A comparison of calculated and experimental results for the effect of particle size on the $L1_0$ ordering of FePt particles. Reprinted from Ref. [71], with permission from The Japan Institute of Metals and Materials.

Another goal was to assess the size dependence of the solid–liquid phase transition temperature. Figure 7.8 shows the solid–liquid phase boundary of an FePt particle with regard to the particle size and the holding temperature. When the particle size is on the nanometer scale, the effect of the solid–liquid interface energy on the chemical potential through κ_θ is pronounced. Therefore, a liquid phase is more likely to occur at the particle surface even if the simulated temperature is lower than the bulk melting point. Furthermore, when the stability of the solid phase of the particle interior balances with the solid–liquid interface energy near the surface portion, a type of core-shell structure is formed because of particle surface premelting [64]. In Fig. 7.8, the open and filled circles indicate completely melted particles and partially melted particles, respectively, and the crossover from open to filled circles denotes the solid–liquid transition temperature. The decrease in the temperature for particle diameters less than 10 nm is more evident than that for larger particles. In addition, Fig. 7.8 shows the results obtained using a conventional theoretical model (commonly referred to as Pawlow's model) [9], determined as

$$T_{\mathrm{m}} = T_{\mathrm{m}}^{\mathrm{Bulk}} \left\{ 1 - \frac{2\sigma_{\mathrm{s}} v_{\mathrm{s}}}{\Delta H_{\mathrm{m}} r_{\mathrm{s}}} \left[1 - \frac{\sigma_{\mathrm{l}}}{\sigma_{\mathrm{s}}} \left(\frac{v_{\mathrm{l}}}{v_{\mathrm{s}}} \right)^{2/3} \right] \right\}. \tag{7.61}$$

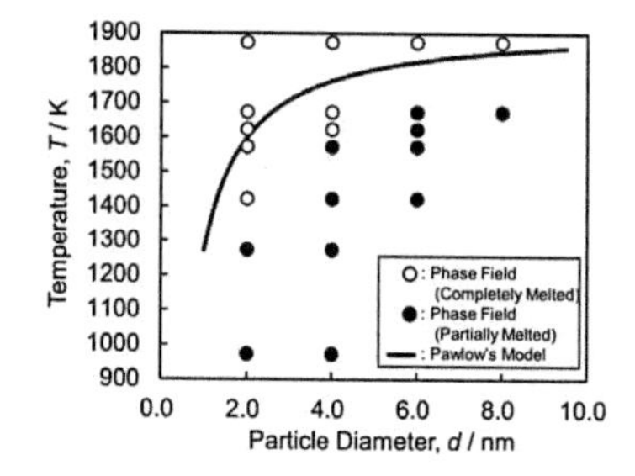

Figure 7.8 Phases of FePt particles calculated on the basis of particle size (circles), together with the phase boundaries estimated using Pawlow's model [9] (solid line). Reprinted from Ref. [71], with permission from The Japan Institute of Metals and Materials.

Here, the melting temperature, T_{m}, is determined by the thermodynamic balance between a solid particle and a liquid particle. The parameter r_{s} is the particle radius, ΔH_{m} is the molar heat of fusion, and $T_{\mathrm{m}}^{\mathrm{Bulk}}$ is the melting temperature of the bulk material. The parameters v and σ are the molar volume and surface tension, respectively, and the subscripts s and l indicate the solid state and the liquid state, respectively. Each parameter was set to the average value for Pt and Fe, and the simulated phase boundary was at a slightly lower temperature than that obtained using Pawlow's model. This discrepancy in the phase boundary temperature is attributed to particle surface premelting, which is not considered in Pawlow's model. The molecular dynamics reported for Ag nanoparticles [77] corroborate this conclusion.

Figure 7.9 shows the radial distributions of the c_{Pt}, c_{TM}, and $c_{\mathrm{Pt+TM}}$ values for the atomic concentration, $|\mathbf{S}| \times c_{\mathrm{Pt+TM}}$, associated with order-disorder and the θ term of the solid–liquid transition in nanoparticles with diameters of 2, 4 and 6 nm at 973.15 K. The composition ratio of the alloyed metal to Pt in these calculations was 1:1. The region at which the summation of the concentration of the component metal elements, c_{Pt+TM}, changes from 0 to 1 corresponds to the surface of the nanoparticle. The difference in the composition ratios between the particle center and the surface was obtained from the distributions of c_{Pt} and c_{TM} against the horizontal axis in this figure, and the FePt and CoPt particles showed no significant concentration decomposition, which is consistent with Monte Carlo predictions [40]. In contrast, the surfaces of the IrPt and AuPt particles were found to consist solely of one of the two component metals. As an example, the IrPt nanoparticles exhibited surface enrichment of Pt. These results are in agreement with values obtained from Monte Carlo simulations using the embedded-atom method potential [16], which showed segregation of Pt on flat (100) and (111) surfaces. In the cases of CrPt, NiPt, PdPt, and CuPt alloys, the concentration of one element was slightly lower than that of the other but both elements were present on the surface. Both Monte Carlo simulations using the modified embedded-atom method (MEAM) [53] and the energetic tight-binding Ising model (TBIM) [29] predicted Cu enrichment of the (110) and (100) surfaces of the CuPt due to the low surface energy of Cu. However, in the case of the (111) surface, the Monte Carlo model with MEAM predicted Pt segregation while the TBIM did not. Schurmans et al. [53] proposed that a decrease in elastic strain energy causes the Pt segregation. The PdPt was found to exhibit a slight surface enrichment of Pd, again in agreement with the Monte Carlo simulations [17]. Moreover, the difference between the Pt and Ni surface energies was determined to be very small. In the case of NiPt, slight Ni enrichment on the surface was predicted, in contrast to the results of Monte Carlo simulations [63]. Therefore, similar to the NiPt particles, it appears that if the surface energy difference is small, the actual surface composition may be determined by atomic size mismatch, which is not considered in the present model. Figure 7.9 also shows the phase states of the particles. In general, as the particle size is reduced, the stable phase changes from an ordered to a disordered state. The appearance of an ordered phase also changes depending on the alloy component. The $L1_0$ ordered phase is stable in the interiors of FePt, CoPt, and CrPt particles with diameters greater than 4 nm while in the case of the CrPt, the $L1_2$ ordered phase exists in particles with diameters below 4 nm. However, the $L1_1$ ordered phase is stable in CuPt particles with diameters above 4 nm. In the case of NiPt and PdPt, the ordered phase was less stable than the disordered phase.

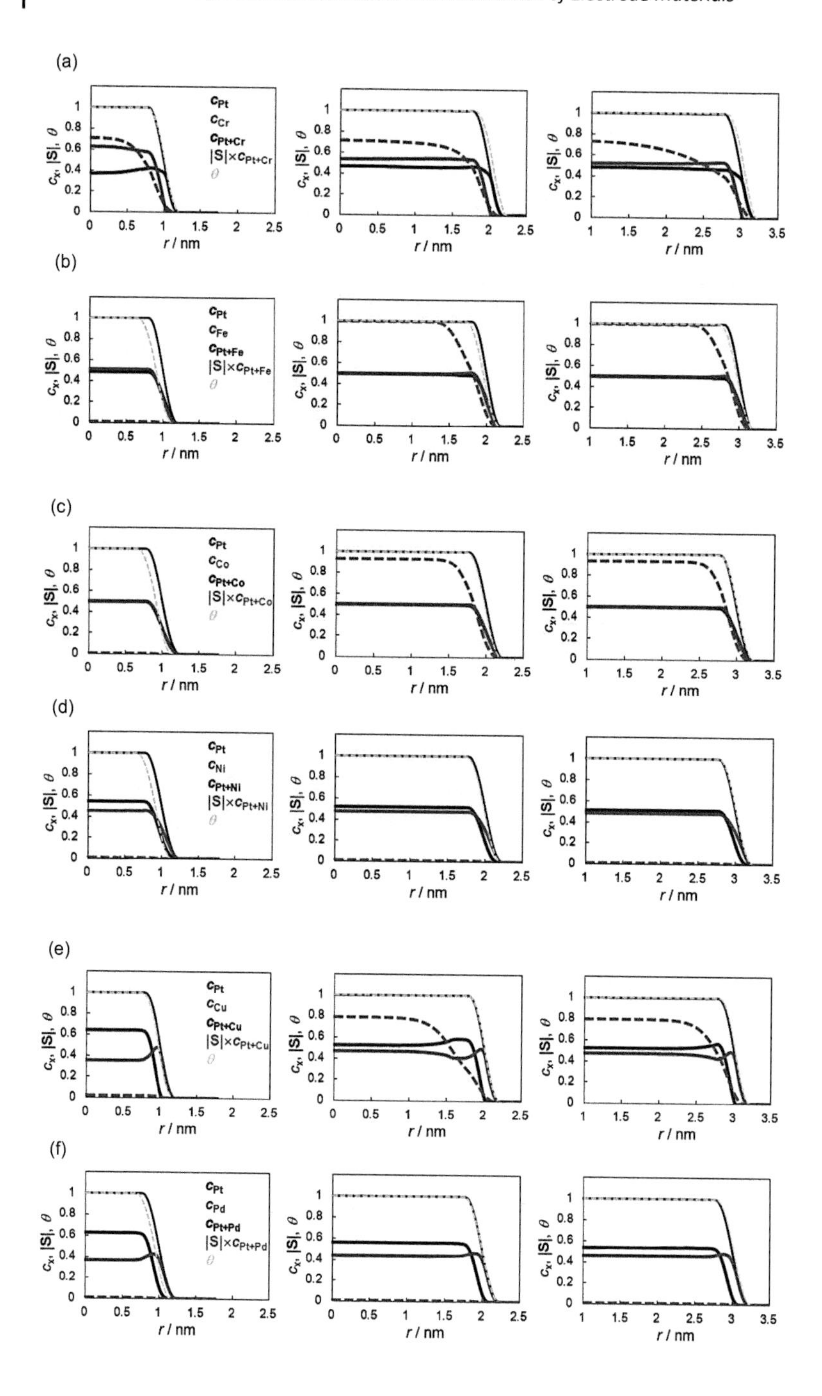

(a)
(b)
(c)
(d)
(e)
(f)
c_{Pt}
c_{Cr}
c_{Pt+Cr}
$|S| \times c_{Pt+Cr}$
c_{Fe}
c_{Pt+Fe}
$|S| \times c_{Pt+Fe}$
c_{Co}
c_{Pt+Co}
$|S| \times c_{Pt+Co}$
c_{Ni}
c_{Pt+Ni}
$|S| \times c_{Pt+Ni}$
c_{Cu}
c_{Pt+Cu}
$|S| \times c_{Pt+Cu}$
c_{Pd}
c_{Pt+Pd}
$|S| \times c_{Pt+Pd}$
θ
c_x, |S|, θ
r / nm

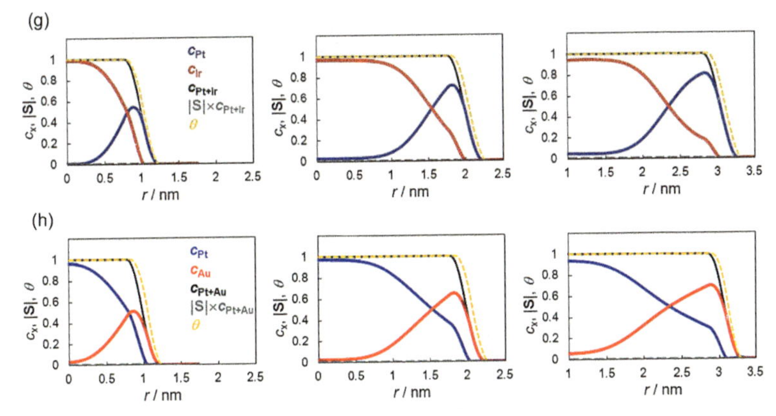

Figure 7.9 The radial distributions of the normalized atomic concentration, c_x; the LRO parameter, |**S**|; and the solid–liquid phase transition, θ, within (a) CrPt, (b) FePt, (c) CoPt, (d) NiPt, (e) CuPt, (f) PdPt, (g) IrPt, and (h) AuPt nanoparticles with diameters of 2 nm, 4 nm, and 6 nm at 973.15 K. The horizontal axis indicates the distance from the particle center, and an |**S**| value of 0 corresponds to a disordered phase. With the exception of the CrPt particles with a diameter of less than 4 nm, an |**S**| value of 1 corresponds to the $L1_0$ (or $L1_1$) ordered phase. For CrPt particles with a diameter of less than 4 nm, an |**S**| value of 0.866 corresponds to the $L1_2$ ordered phase.

Thus, in these binary alloy particles, the surface segregation did not depend solely on the difference in surface energies between the two metal elements, which acted as a driving force for surface segregation, but also on the particle volume. The degree of surface segregation was affected by the attractive interaction between the two elements (which acted to suppress segregation) as well as the repulsive interaction (which was a promoting factor).

7.4 Effect of Electrode Material Microstructure on LIB Performance

7.4.1 Effect of Microstructure on the Discharge Properties of Polycrystalline $LiCoO_2$

The PFM method was applied to investigate Li diffusivity in randomly oriented polycrystalline $LiCoO_2$, the active material in the positive electrodes in LIBs. In this process, each crystal grain in a secondary

$LiCoO_2$ particle was given a randomly assigned crystallographic orientation. The apparent Li diffusion coefficient was determined on the basis of the relative orientations of neighboring grains and the GB diffusivity. This apparent diffusivity is reflected in the electrode performance. In this section, the relationships between various microstructural characteristics and the constant-current discharge properties of $LiCoO_2$ are presented.

7.4.1.1 Computational details

The total free energy, G_{sys}, was represented as a combination of G_{chem} and E_{grad} as

$$G_{sys} = \frac{1}{V_m}\int_{\mathbf{r}}\left[G_{chem} + \frac{\kappa}{2}(\nabla c)^2\right]d\mathbf{r}. \tag{7.62}$$

Here, G_{chem} is the Gibbs free energy of the Li_xCoO_2 system and the second term in the square bracket is the gradient energy term. The variable c is the normalized concentration of Li (between 0 and 1). The value of gradient energy coefficient, κ, was set to 1×10^{-16} $J \cdot m^2 \cdot mol^{-1}$. This value was chosen so as to minimize the effect on the Li concentration profile while avoiding numerical instability. The Cahn–Hilliard equation for the time evolution of the Li concentration can be written as

$$\frac{\partial c}{\partial t} = \nabla \cdot \left[\frac{c(1-c)D_{self}}{RT}\nabla\mu\right]. \tag{7.63}$$

The diffusion potential, μ, can be expressed as the functional derivative of the total free energy, G_{sys}, with regard to the local concentration as

$$\mu = \frac{\delta G_{sys}}{\delta c}. \tag{7.64}$$

Here, $\delta G_{chem}/\delta c$ is defined using the electrochemical potential

$$\frac{\delta G_{chem}}{\delta c} = -n_{Li}F\left[U_{LiCoO_2}(c) - U_{ave}\right], \tag{7.65}$$

where n_{Li} is the number of electrons in the electrochemical reaction (i.e., $n_{Li} = 1$, $Li^+ + e^- = Li$) and F is Faraday's constant. The term $U_{LiCoO_2}(c)$ is the equilibrium potential of $LiCoO_2$ as determined using the equation [76]

$$c = \sum_{i=1}^{n} \frac{c_{\max,i}}{1+\exp\left[\frac{F}{\xi RT}\left(U_{\mathrm{LiCoO_2}} - U_i^0\right)\right]}. \tag{7.66}$$

The parameters $c_{\max,i}$, ξ, and U_i^0 were obtained by fitting the function to open-circuit potential data [76]. The average potential, U_{ave}, is given as

$$U_{\mathrm{ave}} = \left({}^{\circ}G_{\mathrm{LiCoO_2}} - {}^{\circ}G_{\mathrm{CoO_2}} - {}^{\circ}G_{\mathrm{Li}}\right) / \left(-n_{\mathrm{Li}}F\right). \tag{7.67}$$

As an alternative to experimental work, G_{chem} can be determined experimentally. The thermodynamic parameters for the individual phases in the $\mathrm{Li}_x\mathrm{CoO_2}$ system have already been reported [4]. The interaction parameters of the free energy functions were obtained so as to reproduce the formation enthalpy, ΔH, on the basis of first-principles calculations [61, 68]. Substituting Eq. 7.64 into Eq. 7.63, the Cahn–Hilliard equation can be rewritten as

$$\frac{\partial c}{\partial t} = \nabla \cdot \left[\frac{c(1-c)D_{\mathrm{self}}}{RT} \nabla \left(\frac{\partial G_{\mathrm{chem}}}{\partial c} - \kappa \nabla^2 c\right)\right]. \tag{7.68}$$

The Li flux, R_{ecr}, arising from the electrochemical reaction at the $\mathrm{LiCoO_2}$–electrolyte interface was estimated by solving the Butler–Volmer type equation [8]

$$R_{\mathrm{ecr}} = k\left\{\exp\left[\frac{\alpha F \eta_{\mathrm{Li}}}{RT}\right] - \exp\left[-\frac{\alpha F \eta_{\mathrm{Li}}}{RT}\right]\right\}, \tag{7.69}$$

where

$$\eta_{\mathrm{Li}} = \phi - U_{\mathrm{LiCoO_2}} - U_{\mathrm{str}}. \tag{7.70}$$

Here, α (=0.5) is the transfer coefficient, η_{Li} is the overpotential, and k is the kinetic rate constant, which is affected by the local Li concentrations in both the electrolyte and the $\mathrm{LiCoO_2}$. In Eq. 7.70, the electrode potential is indicated by ϕ and the contribution from the elastic strain energy is indicated by U_{str} $(=-(\partial E_{\mathrm{str}}/\partial x)/F)$.

The accuracy of the present PFM was verified by estimating the thermodynamic factor, ω, on the basis of the ratio of the chemical diffusion coefficient (D_{chem}) to a given D_{self} for homogeneous $\mathrm{LiCoO_2}$. The potentiostatic intermittent titration technique [66] was reproduced using a 1D model with an electrode thickness of 1 μm. The applied potential step was 2 mV, and the temperature was set at 300.15 K. D_{chem} was estimated from the time dependence of the current. Figure 7.10 compares the calculated ω values to the experimental results [28, 69]. There is a considerable change in ω near the composition of $\delta = 0.5$ ($Li_{0.5}CoO_2$), attributed to an increase in the degree of Li vacancy ordering [47]. The results shown as (a) in Fig. 7.10 were obtained from Eq. 7.66 using parameters adapted to match the experimental potential value. The results shown as (b) were simulated based on the theoretically assessed thermodynamic parameters in the Gibbs energy functions of the individual phases (O1-Disorder, H1-3, O3-Order, and O3-Disorder) in the Li_xCoO_2 system [4]. In general, the simulated results based on theoretically and experimentally determined values for G_{chem} were consistent with the experimental results.

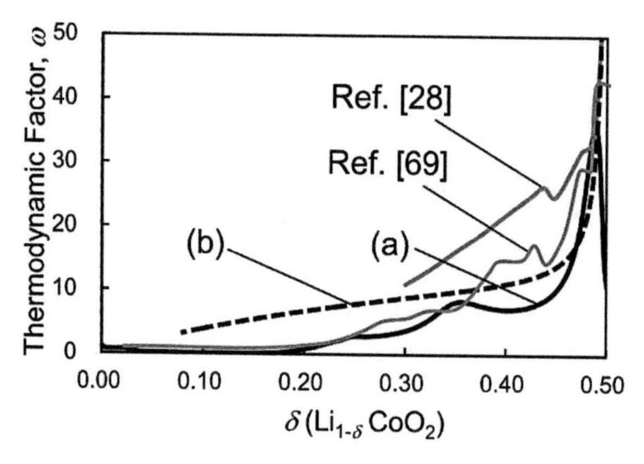

Figure 7.10 Variations in the thermodynamic factor for Li with Li composition. (a) The solid black line and (b) the dashed black line denote simulation results based on experimental and theoretical values for G_{chem}, respectively. The gray line denotes the experimental results. Reprinted from Ref. [72], Copyright (2013), with permission from Elsevier.

The effects of various microstructures on the discharge properties were evaluated by combining the 2D microstructure and the electrochemical-reaction–Li-diffusion model. The c axis of each crystal grain was involved in this 2D model. Figure 7.11 presents the features of the microstructure in this model on the basis of the crystal structure of $LiCoO_2$. Each GB was assumed to be a thin layer between adjacent grains. Fisher et al. [19] determined the activation energy for Li diffusion in a twin boundary core using first-principles calculations and obtained nearly twice the value for a single crystal.

In the present work, the GB width was set to 2 nm, which is roughly equal to the width of the region over which the activation energy deviates from the interior of the particle. The D_{self_basal} and $D_{self\_c\text{-}axis}$ values for the inner grains were calculated as 1×10^{-9} and 1×10^{-11} $cm^2 \cdot s^{-1}$, respectively [62, 69]. The Li diffusion coefficients across and along the GB thin layer, $D_{gb\perp}$ and $D_{gb//}$, were defined as $D_{gb\perp} = 0.01$ D_{self_basal} and $D_{gb//} = D_{self\_basal}$, respectively.

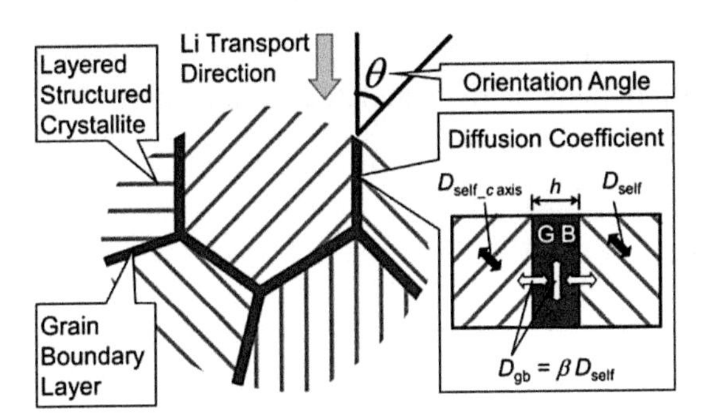

Figure 7.11 A schematic illustration of the microstructural characteristics considered in this simulation. Reprinted from Ref. [72], Copyright (2013), with permission from Elsevier.

Anisotropic Li diffusion was simulated using the diffusion equation

$$\frac{\partial c}{\partial t} = \nabla \cdot \left[\frac{c(1-c)}{RT} \mathbf{D}' \cdot \nabla \frac{\delta G_{sys}}{\delta c} \right]. \tag{7.71}$$

Equation 7.71 was numerically calculated by applying the finite volume method with periodic boundary conditions. The microstructural characteristics were expressed by the components of the diffusion tensor, $D'_{ij}(\mathbf{r})$. In a crystal grain, $D'_{ij}(\mathbf{r})$ is represented as

$$D'_{ij}(\mathbf{r}) = A_{im}(\mathbf{r}) A_{jn}(\mathbf{r}) \delta_{mn} D_{mn}. \tag{7.72}$$

The variables D_{11} ($=D_{22}$) and D_{33} are equivalent to D_{self_basal} and $D_{self\_c\text{-}axis}$, respectively, where D_{self_basal} and $D_{self\_c\text{-}axis}$ denote the self-diffusion coefficients in the direction along the basal plane of the hexagonal structure and in the direction perpendicular to the basal plane, respectively. The coefficient, δ_{mn}, is the Kronecker delta.

Figure 7.12 presents the 2D microstructures generated from multiphase-field modeling [27, 32], with the y axis parallel to the direction of Li transport. The effect of the connectivity of the conduction pathway between crystal grains on the Li diffusivity was evaluated by obtaining the area-averaged value, taking the relative orientation between adjacent grains into account [72], as

$$\theta_{\text{relative}} = 90 - \sum_{n_y=1}^{n_{y\_\max}} \left| \sum_{n_x=1}^{n_{x\_\max}} \left[90 - \theta(n_x, n_y) \right] \right| / (n_{x\_\max} \times n_{y\_\max}), \tag{7.73}$$

where $n_{x\_\max}$ and $n_{y\_\max}$ represent the number of grid points along the x and y axis of the simulation region, respectively. The fine and coarse grain microstructures were studied at the median, lower limit, and upper limit of θ_{relative}. The values in Table 7.2 confirm that the apparent Li diffusion coefficient decreases as θ_{relative} increases. The grid spacing was 0.02 μm, and the number of grid points along each axis was 256, although finer grid spacing was employed near the GBs to better model Li diffusion along the GB thin layer.

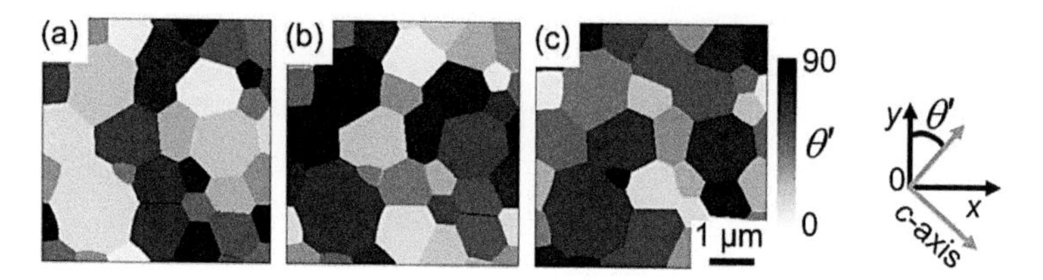

Figure 7.12 Two-dimensional models of a randomly oriented polycrystalline microstructure. The orientation angle, θ', of each grain is shown by its gray tone. The value of θ' is determined as $\theta' = 90 - |90 - \theta|$ ($\theta = 0$ to 180°). Reprinted from Ref. [74], Copyright (2014), with permission from Elsevier.

Table 7.2 Apparent diffusion coefficients of Li

		D_{app} (×10^{-9} $\text{cm}^2\cdot\text{s}^{-1}$)	
Model	θ_{relative} (°)	$D_{\text{gb//}} = D_{\text{self\_basal}}$	$D_{\text{gb//}} = 0.01\, D_{\text{self\_basal}}$
(a)	63.6	1.29	0.78
(b)	71.9	1.05	0.61
(c)	79.8	0.62	0.39

Note: The value of θ_{relative} is defined in Eq. 7.73.

The discharge capacity under practical discharge conditions was investigated by explicit processing of the electrochemical reaction. A Butler–Volmer-type equation (Eq. 7.69) was introduced into the model, and the Li flux across the electrochemical interface at $y = 0$ and $y = L$ was determined. Here, L denotes the length of the simulation region. The equilibrium potential of $LiCoO_2$ a function of Li concentration has been reported in Ref. [76]. The initial voltage, V, was approximately 4.026 V, equivalent to a Li concentration of 0.6. The current density was set at 3.45 $mA \cdot cm^{-2}$, corresponding to a discharge rate of approximately 20 C, and the temperature was set to 298.15 K. The objective of this study was to determine the extent to which the microstructure of $LiCoO_2$ affects its discharge characteristics.

7.4.1.2 Results and discussion

Figure 7.13 plots the changes in potential as a function of the Li concentration for a constant current discharge, on the basis of the microstructure shown in Fig. 7.12. The discharge properties under the conditions of $D_{gb} = D_{self\_basal}$ and $D_{gb} = 0.01\ D_{self\_basal}$ are indicated by the black and blue lines, respectively. The variation in the discharge capacity of each microstructure appears to be due to Li segregation at the electrochemical reaction surface. Figure 7.14 summarizes the Li concentration distribution for an average concentration of 0.65. In the case of the coarse grain structure, a pronounced variation in the Li concentration appears as a result of the initial assignment of the orientation angle distribution. If the GB acts as a diffusion barrier, the discharge capacity of the fine grain model at a high current density should be less than that of the coarse grain model. Thus, the discharge capacity of $LiCoO_2$ at a high discharge rate (e.g., 20 C) varies with the microstructure, specifically the crystal grain size, as well as the spatial distribution of the crystal orientations of each grain and the GB diffusivity. The simulated results imply that both the GB diffusivity and the angle mismatch between neighboring grains are important parameters when estimating the apparent discharge characteristics.

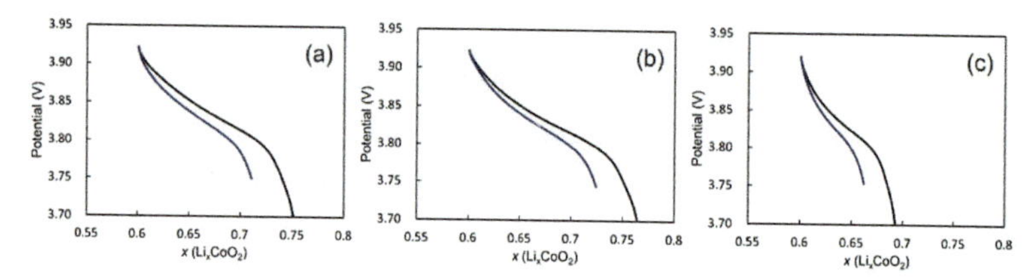

Figure 7.13 Constant-current discharge properties at 298.15 K and a current density of 3.45 mA·cm^{-2}. The simulation results under the conditions of $D_{gb} = D_{self\_basal}$ and $D_{gb} = 0.01\ D_{self\_basal}$ are represented by the black and blue lines, respectively. Reprinted from Ref. [74], Copyright (2014), with permission from Elsevier.

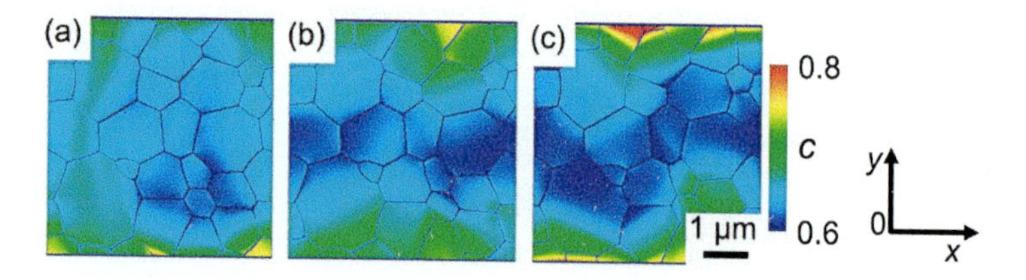

Figure 7.14 Spatial distributions of Li concentration during constant-current discharge. The y axis is parallel to the global Li transport direction. The average Li concentration is approximately 0.65. D_{gb} is assumed to be equivalent to D_{self_basal}. Reprinted from Ref. [74], Copyright (2014), with permission from Elsevier.

7.4.2 Phase-Field Modeling of Stress Generation in Polycrystalline $LiCoO_2$

The Li intercalation process in a layered transition metal oxide is considered to affect battery performance. Thus, the PFM was used to quantitatively evaluate the effect of the microstructure of a secondary polycrystalline Li_xCoO_2 particle on the stress generated during the charging process, taking into account anisotropic elastic properties.

7.4.2.1 Computational details

To explicitly treat the elastic strain energy, the total free energy G_{sys} [74, 75] was defined as

$$G_{sys} = \frac{1}{V_m}\int_{\mathbf{r}} (G_{chem} + E_{grad} + E_{str})d\mathbf{r}. \tag{7.74}$$

Here, G_{chem}, E_{grad}, and E_{str} refer to the chemical part of the free energy, gradient energy, and elastic strain energy, respectively. G_{chem} is expressed by

$$G_{\text{chem}} = c\,{}^{\circ}G_{\text{LiCoO}_2} + (1-c)\,{}^{\circ}G_{\text{CoO}_2} + \Delta g_{\text{chem}}, \tag{7.75}$$

where

$$\Delta g_{\text{chem}} = c(1-c)\sum_{m=1}^{3} L_m (2c-1)^{m-1} + RT\left[c\ln c + (1-c)\ln(1-c)\right]. \tag{7.76}$$

Here, c is the local concentration ratio of Li and is equivalent to x in Li_xCoO_2, with a value between 0 (CoO_2) and 1 (LiCoO_2). The terms ${}^{\circ}G_{\text{CoO}_2}$ and ${}^{\circ}G_{\text{LiCoO}_2}$ are the standard formation free energies of CoO_2 and LiCoO_2, respectively. Previously [57], the parameters L_m for the RK power series in Eq. 7.76 were determined so as to reproduce the two-phase coexisting region ($0.79 < c < 0.97$) such that $L_1 = -7.350 \times 10^4$ J·mol^{-1}, $L_2 = 4.696 \times 10^4$ J·mol^{-1}, and $L_3 = -7.058 \times 10^3$ J·mol^{-1}. The function E_{grad} was expressed as

$$E_{\text{grad}} = \frac{\kappa}{2}(\nabla c)^2. \tag{7.77}$$

The gradient energy coefficient, κ, is affected by the spatial resolution of the numerical model and the individual phase boundary [11], and the correct value of κ is still uncertain for the phase boundary between the Li-poor phase and Li-rich phase. In the present study, therefore, the value of κ was set to 1×10^{-13} J·m^2·mol^{-1}, such that the effect of gradient energy on the moving behavior was negligible. E_{str} can be written as

$$E_{\text{str}} = \frac{1}{2} C_{ijkl}(\mathbf{r})\left[\varepsilon_{ij}^{c}(\mathbf{r},t) - \varepsilon_{ij}^{0}(\mathbf{r},t)\right]\left[\varepsilon_{kl}^{c}(\mathbf{r},t) - \varepsilon_{kl}^{0}(\mathbf{r},t)\right]. \tag{7.78}$$

Here, $C_{ijkl}(\mathbf{r})$ is the component element of the fourth-order stiffness tensor. The variables $\varepsilon_{ij}^{c}(\mathbf{r},t)$ and $\varepsilon_{ij}^{0}(\mathbf{r},t)$ were introduced in Section 7.2.3. The rotation matrix $A_{ij}(\mathbf{r})$ in Eq. 7.28 is randomly allocated to each crystal grain, and the variable $\eta_{mn}(\mathbf{r},t)$ is the lattice misfit. The 2D plane strain was examined, with the crystallographic c axis of each crystal grain based on this 2D model. The values η_{11} ($=\eta_{22}$) and η_{33} represent the lattice mismatches of the a, b, and c axis, respectively. These variables were extracted from the lattice parameters using X-ray diffraction data [14].

The Li flux across the $LiCoO_2$–electrolyte interface was determined by the electrochemical reaction calculation described in Section 7.4.1.1. Each specific equilibrium state of charge was equivalent to the state in which the Faraday current density was close to zero. Anisotropic Li diffusion in $LiCoO_2$ was calculated by solving the diffusion equation introduced in Section 7.4.1.1.

The fracture surface ratio of the GB was estimated on the basis of the stress distribution in the microstructure. The criteria for crack initiation were set so that the maximum principal stress at the GB exceeded the value of the instability criterion, σ_f, according to the linear elastic fracture mechanics theory.

7.4.2.2 Results and discussion

The Li concentration and stress distribution in a 2D polycrystalline particle were examined on the basis of the explicit morphology created from multiphase-field simulations [75]. The orientation angle, θ, assigned to each grain is represented as

$$\theta = m\Delta\theta - 90. \tag{7.79}$$

Here, $\Delta\theta$ is defined as 180° divided by the total number of grains (N_{gb}) and the variable m, an integer between zero and N_{gb}, is randomly assigned to each grain. On the basis of this process of assigning orientation angles, it is expected that the total stress generated will be affected by the individual assignments. Figure 7.15a shows the variation in the mean stress values determined for 100 arrangement patterns. A single microstructure (Fig. 7.15b) equivalent to the mode of the stress variation in Fig. 7.15a was adopted for the calculation of concentration and stress distribution over the range from $x = 0.51$ to $x = 1$ for Li_xCoO_2. The $LiCoO_2$ was assumed to be in a stress-free state, and a plane strain state was also assumed. The grid interval and the total number of grid points in each direction were set to 0.01 nm and 512, respectively, while the temperature was set to 298.15 K. The electrolyte surrounding the secondary particle was regarded as having isotropic elasticity, and Young's modulus and Poisson's ratio were set to 70 GPa and 0.2, respectively, in this region. As explained in Section 7.4.2.1, a positive interaction parameter value, L_m, was used to express the phase separation. A two-phase region appears over the range of $0.79 < x < 0.97$ in Li_xCoO_2 [4, 57]. As shown in Fig. 7.16a, when Li is deintercalated by the electrochemical reaction

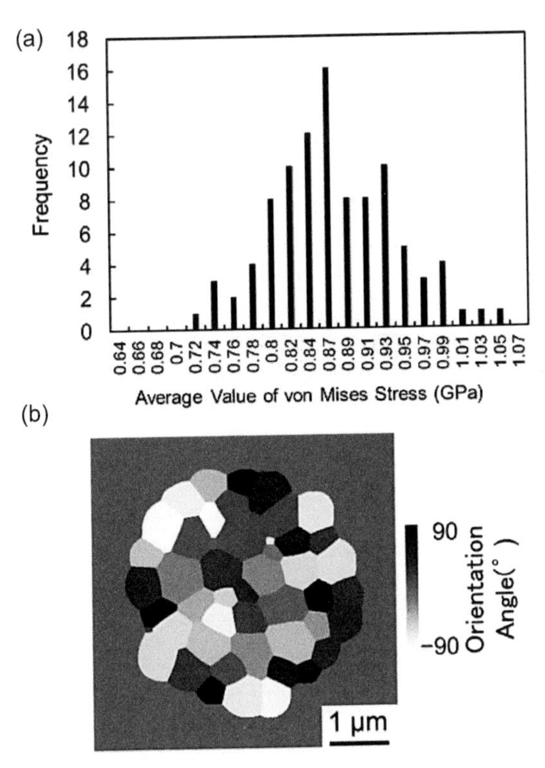

Figure 7.15 The microstructural characteristics of a Li_xCoO_2 particle surrounded by an electrolyte. (a) The variation in the average von Mises stress depending on the assignment of the orientation angle and (b) a structural image showing the mode of the stress value. Reprinted from Ref. [75], Copyright (2018), with permission from Elsevier.

during the charging process, a Li-poor region is generated in the Li-rich matrix because of a fluctuation in the Li concentration. In the present study, stress distribution was an important factor affecting the concentration fluctuation. Since the stress generated in secondary particles depends on the relative orientation angle between neighboring primary particles, non-uniform stress is generated in the secondary particles. When stress fluctuation occurs, as described in Section 7.4.1, Li diffusion begins to reduce the elastic strain energy, E_{str}, and induces fluctuations in Li concentration. However, because the fluctuation of Li increases the chemical free energy, G_{chem}, large fluctuations are suppressed and the fluctuations are 2 orders of magnitude smaller than the absolute average value of the concentration. Because initial precipitation nuclei are not necessarily required for the spinodal decomposition, phase separation takes place once a fluctuation occurs. Therefore, a concentration fluctuation caused by a stress fluctuation can be specified as the starting point of the phase separation.

Figure 7.17 shows crack propagation with regard to Li concentration. If the elastic constraint is weakened by a GB fracture, the average stress in the particle should decrease. The variation in the stress level between particles due to the presence or absence of cracks has a non-negligible effect on the stress value. When the σ_f value is 0.5 GPa, the fracture rate of GBs is about 20%. Conversely, the fracture ratio of the GBs is approximately 10% at a σ_f value of 1 GPa. Therefore, a significant GB fracture, where the primary particles are almost separated, does not occur. The changes in Young's modulus of a sintered pellet of $LiCoO_2$ have been experimentally determined as a function of the initial charging process, using nanoindentation [55]. Young's modulus was found to decrease by about 40% even when the charging interval was within the range of Li_xCoO_2 ($0.6 < x < 1$). It was also confirmed that this decrease was due to GB cracking. Therefore, it appears that the presupposition of crack propagation preferring the GB is reasonable. If the cracked GB ratio and the variation in Young's modulus are assumed to have a negative linear correlation, the simulated result of less than 30% is slightly smaller than the experimental value of approximately 40%. This discrepancy is thought to be due to the grain size mismatch (~1 μm vs. ~100 μm) and to differences in morphology (particulate vs. pellet) between studies. The distributions of the Li concentration, stress, and GB cracks are presented in Fig. 7.16c. GB cracks with lengths approximately equal to the grain size are distributed throughout the secondary particles.

As discussed above, the morphological anisotropy in the secondary particles due to the stress distribution has an important influence on the change of the state of the particles. The structural factors affecting the stress distribution are the primary particle size, the relative orientation angle, and constraint by the electrolyte.

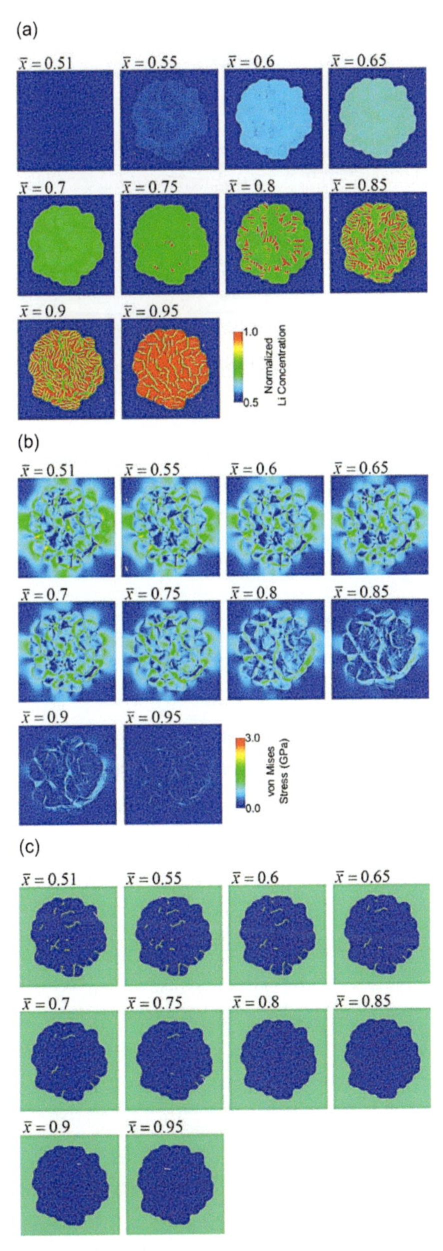

Figure 7.16 Variations in the morphological aspects of a Li_xCoO_2 particle with changes in the average Li concentration. The local distributions of (a) Li concentration, (b) stress value, and (c) GB cracks. The value of σ_f was set to 0.5 GPa. Reprinted from Ref. [75], Copyright (2018), with permission from Elsevier.

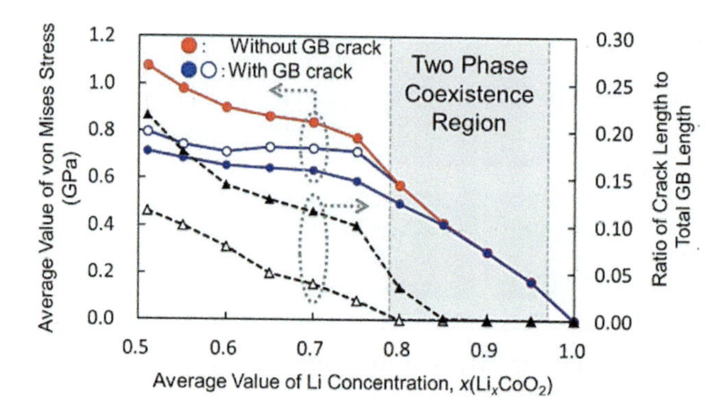

Figure 7.17 Changes in the GB crack ratio and stress value averaged over a Li_xCoO_2 particle. Filled (•) and open (○) circles denote the average values of the von Mises stress for σ_f of 0.5 and 1 GPa, respectively. Filled (▲) and open (Δ) triangles denote the GB crack ratios for σ_f of 0.5 and 1 GPa, respectively. Reprinted from Ref. [75], Copyright (2018), with permission from Elsevier.

7.5 Final Remarks

As a means of identifying ways of improving the performance of PEFCs and Li ion secondary batteries, the PFM was applied to the analysis of electrode materials with microstructures in the scale range of nanometers to micrometers. The results demonstrate that PFM studies can elucidate the driving forces behind nanostructure formation and permit a quantitative understanding of important aspects of the atomic diffusion properties of anisotropic materials. However, further detailed studies are still needed to assess a number of issues prior to the realization of high-performance batteries with fully designed microstructures. Specifically, the design of multicomponent alloys having robust nanostructures for practical applications will require the simulation of multiparticle systems, along with assessments of concentrations and particle size distributions.

In addition, the study of Li diffusivity in practical electrodes for use in LIBs will necessitate multiscale simulations that describe anisotropic Li diffusion on both the scale of one particle and the scale of the electrode level. Simultaneously, it will be necessary to reduce the simulation time by developing more efficient numerical algorithms.

Finally, present-day PFM studies are based on incorporating CALPHAD data and numerous material parameters. Therefore, parameter estimation using first-principles calculations should be a subject of future work so as to allow broad surveys of alloy phases in multi-element alloy systems.

References

1. Japan Institute of Metals (ed.) (1993). *Metals Data Book*, 3rd Ed.
2. Abe, T., Sundman, B., and Onodera, H. (2006). Thermodynamic assessment of the Cu–Pt system, *J. Phase Equilib. Diffus.*, **27**, pp. 5–13.
3. Abe, T. (2011). *Computational Materials Design: Computational Thermodynamics*, pp. 157–160.
4. Abe, T. and Koyama, T. (2011). Thermodynamic modeling of the $LiCoO_2$–CoO_2 pseudo-binary system, *Calphad*, **35**, pp. 209–218.
5. Akita, T., Taniguchi, A., Maekawa, J., Siroma, Z., Tanaka, K., Kohyama, M., and Yasuda, K. (2006). Analytical TEM study of Pt particle deposition in the proton-exchange membrane of a membrane-electrode-assembly, *J. Power Sources*, **159**, pp. 461–467.
6. Allen, S. M., Cahn, J. W. (1979). A microscopic theory for antiphase boundary motion and its application to antiphase domain coarsening, *Acta Metall.*, **27**, pp. 1085–1095.
7. Arblaster, J. W. (2006). Crystallographic properties of platinum, *Platinum Met. Rev.*, **50**, pp. 118–119.
8. Bard, A. J. and Faulkner, L. R. (2001). *Electrochemical Methods–Fundamentals and Applications*, 2nd Ed.
9. Barybin, A. and Shapovalov, V. (2011). Modification of Pawlow's thermodynamical model for the melting of small single-component particles, *J. Appl. Phys.*, **109**, pp. 034303.
10. Binder, K., Lebowitz, J. L., Phani, M. K. and Kalos, M. H. (1981). Monte carlo study of the phase diagrams of binary alloys with face centered cubic lattice structure, *Acta Metall.*, **29**, pp. 1655–1665.
11. Cahn, J. W. and Hilliard, J. E. (1958). Free energy of a nonuniform system. I. Interfacial free energy, *J. Chem. Phys.*, **28**, pp. 258–267.
12. Cahn, J. W. (1965). Phase separation by spinodal decomposition in isotropic systems, *J. Chem. Phys.*, **42**, pp. 93–99.
13. Chen, L.-Q. (2002). Phase-field models for microstructure evolution, *Annu. Rev. Mater. Res.*, **32**, pp. 113–140.

14. Choi, Y.-M., Pyun, S.-I., Bae, J.-S. and Moon, S.-I. (1995). Effects of lithium content on the electrochemical lithium intercalation reaction into $LiNiO_2$ and $LiCoO_2$ electrodes, *J. Power Sources*, **56**, pp. 25–30.

15. Dannenberg, A., Gruner, M. E., Hucht, A., and Entel, P. (2009). Surface energies of stoichiometric FePt and CoPt alloys and their implications for nanoparticle morphologies, *Phys. Rev. B*, **80**, p. 245438.

16. Debe, M. K. (2012). Electrocatalyst approaches and challenges for automotive fuel cells, *Nature*, **486**, pp. 43–51.

17. Deng, H., Hu, W., Shu, X., Zhao, L. and Zhang, B. (2002). Monte Carlo simulation of the surface segregation of Pt–Pd and Pt–Ir alloys with an analytic embedded-atom method, *Surf. Sci.*, **517**, pp. 177–185.

18. Dinsdale, A. T. (1991). SGTE data for pure elements, *Calphad*, **15**, pp. 317–425.

19. Fisher, C. A., Huang, R., Hitosugi, T., Moriwake, H., Kuwabara, A., Ikuhara, Y. H., Oki, H., and Ikuhara, Y. (2012). A high-coincidence twin boundary in lithium battery material $LiCoO_2$, *Nanosci. Nanotechnol. Lett.*, **4**, pp. 165–168.

20. Fredriksson, P. and Sundman, B. (2001). A thermodynamic assessment of the Fe-Pt system, *Calphad*, **25**, pp. 535–548.

21. Fu, B. Q., Liu, W. and Li, Z.L. (2009). Calculation of the surface energy of bcc-metals with the empirical electron theory, *Appl. Surf. Sci.*, **255**, pp. 8511–8519.

22. Greeley, J., Stephens, I. E. L., Bondarenko, A. S., Johansson, T. P., Hansen, H. A., Jaramillo, T. F., Rossmeisl, J., Chorkendorff, I., and Nørskov, J. K. (2009). Alloys of platinum and early transition metals as oxygen reduction electrocatalysts, *Nat. Chem.*, **1**, pp. 552–556.

23. Grolier, V. and Schmid-Fetzer, R. (2007). Experimental study of Au-Pt-Sn phase equilibria and thermodynamic assessment of the Au-Pt and Au-Pt-Sn systems, *J. Electron. Mater.*, **37**, pp. 264–278.

24. Hillert, M. (1998). *Phase Equilibria, Phase Diagrams and Phase Transformations* (Cambridge University Press).

25. Hilliard, J. E. (1970). *Phase Transformation*, Aaronson, H. I., ed. (American Society for Metals, Metals Park, OH), p. 497.

26. Hu, S. and Chen, L. (2001). A phase-field model for evolving microstructures with strong elastic inhomogeneity, *Acta Mater.*, **49**, pp. 1879–1890.

27. Ito, Y., Yamakawa, S., Hayashi, A., and Tatsumisago, M. (2017). Effects of the microstructure of solid-electrolyte-coated $LiCoO_2$ on its discharge

properties in all-solid-state lithium batteries, *J. Mater. Chem. A*, **5**, pp. 10658–10668.

28. Jang, Y.-I., Neudecker, B. J., and Dudney, N. J. (2001). Lithium diffusion in Li_xCoO_2 ($0.45 < x < 0.7$) intercalation cathodes, *Electrochem. Solid-State Lett.*, **4**, pp. A74–A77.
29. Khoutami, A., Legrand, B. and Tréglia G. (1993). On a surprising anisotropy of surface segregation in CuPt alloys, *Surf. Sci.*, **287–288**, pp. 851–856.
30. Kim, D., Saal, J. E., Zhou, L., Shang, S., Du, Y., and Liu, Z.-K. (2011). Thermodynamic modeling of fcc order/disorder transformations in the Co–Pt system, *Calphad*, **35**, pp. 323–330.
31. Kim, S., Wee, J., Peters, K., and Huang, H.-Y. S. (2018). Multiphysics coupling in lithium-ion batteries with reconstructed porous microstructures, *J. Phys. Chem. C*, **122**, pp. 5280–5290.
32. Kim, S. G., Kim, D. I., Kim, W. T., and Park, Y. B. (2006). Computer simulations of two-dimensional and three-dimensional ideal grain growth, *Phys. Rev. E*, **74**, pp. 061605.
33. Koyama, T. (2011). *Computational Materials Design: Simulation of Microstructures* (Uchida Rokakuho Publishing, Tokyo).
34. Koyama, T. and Tsukada, Y. (2012). *Elasticity of Microstructure and Phase Transformation: Introduction to Phase-Field Microelasticity Theory*.
35. Lu, G. and Zangari, G. (2006). Electrodeposition of platinum nanoparticles on highly oriented pyrolitic graphite, *Electrochim. Acta*, **51**, pp. 2531–2538.
36. Lu, X.-G., Sundman, B., and Ågren, J. (2009). Thermodynamic assessments of the Ni–Pt and Al–Ni–Pt systems, *Calphad*, **33**, pp. 450–456.
37. Lukas, H. L., Fries, S. G., and Sundman, B. (2007). *Computational Thermodynamics: The Calphad Method* (Cambridge University Press).
38. Mehl, M. J. and Papaconstantopoulos, D. A. (1996). Applications of a tight-binding total-energy method for transition and noble metals: elastic constants, vacancies, and surfaces of monatomic metals, *Phys. Rev. B*, **54**, pp. 4519–4530.
39. Miyazaki, T., Kitakami, O., Okamoto, S., Shimada, Y., Akase, Z., Murakami, Y., Shindo, D., Takahashi, Y. K., and Hono, K. (2005). Size effect on the ordering of $L1_0$ FePt nanoparticles, *Phys. Rev. B*, **72**, pp. 144419.
40. Moskovkin, P. and Hou, M. (2007). Metropolis Monte Carlo predictions of free Co–Pt nanoclusters, *J. Alloys Compd.*, **434–435**, pp. 550–554.

41. Nagpure, S. C., Bhushan, B., and Babu, S. S. (2013). Multi-scale characterization studies of aged Li-ion large format cells for improved performance: an overview, *J. Electrochem. Soc.*, **160**, pp. A2111–A2154.

42. Okazaki-Maeda, K., Morikawa, Y., Tanaka, S., and Kohyama, M. (2010). Structures of Pt clusters on graphene by first-principles calculations, *Surf. Sci.*, **604**, pp. 144–154.

43. Ouyang, Y., Chen H. and Zhong, X. (2003). Enthalpies of Formation of noble metal binary alloys bearing Rh or Ir, *J. Mater. Sci. Technol.*, **19**, pp. 243–246.

44. Park, M., Zhang, X., Chung, M., Less, G. B., and Sastry, A. M. (2010). A review of conduction phenomena in Li-ion batteries, *J. Power Sources*, **195**, pp. 7904–7929.

45. Preußner, J., Prins, S., Völkl, R., Liu, Z.-K., and Glatzel, U. (2009). Determination of phases in the system chromium–platinum (Cr–Pt) and thermodynamic calculations, *Mater. Sci. Eng. A*, **510–511**, pp. 322–327.

46. Ramadesigan, V., Northrop, P. W. C., De, S., Santhanagopalan, S., Braatz, R. D., and Subramanian, V. R. (2012). Modeling and simulation of lithium-ion batteries from a systems engineering perspective, *J. Electrochem. Soc.*, **159**, pp. R31–R45.

47. Reimers, J. N. and Dahn, J. R. (1992). Electrochemical and in situ X-ray diffraction studies of lithium intercalation in Li_xCoO_2, *J. Electrochem. Soc.*, **139**, pp. 2091–2097.

48. Rong, C.-b., Poudyal, N., Chaubey, G. S., Nandwana, V., Skomski, R., Wu, Y. Q., Kramer, M. J., and Liu, J. P. (2007). Structural phase transition and ferromagnetism in monodisperse 3 nm FePt particles, *J. Appl. Phys.*, **102**, pp. 043913.

49. Rong, C. b., Li, D., Nandwana, V., Poudyal, N., Ding, Y., Wang, Z. L., Zeng, H., and Liu, J. P. (2006). Size-dependent chemical and magnetic ordering in $L1_0$-FePt nanoparticles, *Adv. Mater.*, **18**, pp. 2984–2988.

50. Saito, Y. (2000). *An Introduction to the Kinetics of Diffusion Controlled Microstructural Evolutions in Materials* (Corona Publishing).

51. Schaefer, H. E. (1987). Investigation of thermal equilibrium vacancies in metals by positron annihilation, *Phys. Status Solidi A*, **102**, pp. 47–65.

52. Schröder, A., Klüppel, M., Schuster, R.H. and Heidberg, J. (2002). Surface energy distribution of carbon black measured by static gas adsorption, *Carbon*, **40**, pp. 207–210.

53. Schurmans, M., Luyten, J., Creemers, C., Declerck, R., and Waroquier, M. (2007). Surface segregation in CuPt alloys by means of an improved

modified embedded atom method, *Phys. Rev. B*, **76**.

54. Scrosati, B. and Garche, J. (2010). Lithium batteries: status, prospects and future, *J. Power Sources*, **195**, pp. 2419–2430.
55. Swallow, J. G., Woodford, W. H., McGrogan, F. P., Ferralis, N., Chiang, Y.-M., and Van Vliet, K. J. (2014). Effect of electrochemical charging on elastoplastic properties and fracture toughness of LiXCoO2, *J. Electrochem. Soc.*, **161**, pp. F3084–F3090.
56. Takeshita, T., Murata, H., Hatanaka T. and Morimoto Y. (2008). Analysis of Pt catalyst degradation of a PEFC cathode by TEM observation and macro model simulation, *ECS Trans.*, **16**, pp. 367–373.
57. Tatsukawa, E. and Tamura, K. (2014). Activity correction on electrochemical reaction and diffusion in lithium intercalation electrodes for discharge/charge simulation by single particle model, *Electrochim. Acta*, **115**, pp. 75–85.
58. Turchi, P. E. A., Drchal, V., and Kudrnovský, J. (2006). Stability and ordering properties of fcc alloys based on Rh, Ir, Pd, and Pt, *Phys. Rev. B*, **74**, pp. 064202.
59. Tyson, W. R. and W.A., M. (1977). Surface free energies of solid metals: estimation from liquid surface tension measurements, *Surf. Sci.*, **62**, pp. 267–276.
60. Ungár, T., Gubicza, J., Ribárik, G., Pantea, C. and Zerda, T.W. (2002). Microstructure of carbon blacks determined by X-ray diffraction profile analysis, *Carbon*, **40**, pp. 929–937.
61. Van der Ven, A., Aydinol, M., Ceder, G., Kresse, G., and Hafner, J. (1998). First-principles investigation of phase stability in Li_xCoO_2, *Phys. Rev. B*, **58**, pp. 2975.
62. Van der Ven, A., Ceder, G., Asta, M., and Tepesch, P. (2001). First-principles theory of ionic diffusion with nondilute carriers, *Phys. Rev. B*, **64**, pp. 184307.
63. Wang, G., Van Hove, M. A., Ross, P. N., and Baskes, M. I. (2005). Monte Carlo simulations of segregation in Pt-Ni catalyst nanoparticles, *J. Chem. Phys.*, **122**, pp. 024706.
64. Wang, Z. L., Petroski, J.M., Green, T.C. and El-Sayed, M.A. (1998). Shape transformation and surface melting of cubic and tetrahedral platinum nanocrystals, *J. Phys. Chem. B*, **102**, pp. 6145–6151.
65. Watanabe, M., Masumoto, T., Ping, D. H., and Hono, K. (2000). Microstructure and magnetic properties of FePt–Al–O granular thin films, *Appl. Phys. Lett.*, **76**, pp. 3971–3973.

66. Wen, C. J., Boukamp, B., Huggins, R., and Weppner, W. (1979). Thermodynamic and mass transport properties of "LiAl", *J. Electrochem. Soc.*, **126**, pp. 2258–2266.
67. Wen, Y.-N. and Zhang, J.-M. (2007). Surface energy calculation of the fcc metals by using the MAEAM, *Solid State Commun.*, **144**, pp. 163–167.
68. Wolverton, C. and Zunger, A. (1998). Cation and vacancy ordering in Li_xCoO_2, *Phys. Rev. B*, **57**, pp. 2242–2252.
69. Xie, J., Imanishi, N., Matsumura, T., Hirano, A., Takeda, Y., and Yamamoto, O. (2008). Orientation dependence of Li–ion diffusion kinetics in $LiCoO_2$ thin films prepared by RF magnetron sputtering, *Solid State Ionics*, **179**, pp. 362–370.
70. Yamakawa, S., Okazaki-Maeda, K., Kohyama, M., and Hyodo, S. (2008). Phase-field model for deposition process of platinum nanoparticles on carbon substrate, *J. Phys.: Conf. Ser.*, **100**, pp. 072042.
71. Yamakawa, S., Asahi, R., and Koyama, T. (2013). Phase-field modeling of phase transformations in platinum-based alloy nanoparticles, *Mater. Trans.*, **54**, pp. 1242–1249.
72. Yamakawa, S., Yamasaki, H., Koyama, T., and Asahi, R. (2013). Numerical study of Li diffusion in polycrystalline $LiCoO_2$, *J. Power Sources*, **223**, pp. 199–205.
73. Yamakawa, S., Asahi, R., and Koyama, T. (2014). Surface segregations in platinum-based alloy nanoparticles, *Surf. Sci.*, **622**, pp. 65–70.
74. Yamakawa, S., Yamasaki, H., Koyama, T., and Asahi, R. (2014). Effect of microstructure on discharge properties of polycrystalline $LiCoO_2$, *Solid State Ionics*, **262**, pp. 56–60.
75. Yamakawa, S., Nagasako, N., Yamasaki, H., Koyama, T., and Asahi, R. (2018). Phase-field modeling of stress generation in polycrystalline $LiCoO_2$, *Solid State Ionics*, **319**, pp. 209–217.
76. Zhang, Q., Guo, Q., and White, R. E. (2006). A new kinetic equation for intercalation electrodes, *J. Electrochem. Soc.*, **153**, pp. A301–A309.
77. Zhao, S., Wang, S. and Ye, H. (2001). Size-dependent melting properties of free silver nanoclusters, *J. Phys. Soc. Jpn.*, **70**, pp. 2953–2957.
78. Zoval, J. V., Lee, J., Gorer, S. and Penner, R.M. (1998). Electrochemical preparation of platinum nanocrystallites with size selectivity on basal plane oriented graphite surfaces, *J. Phys. Chem. B*, **102**, pp. 1166–1175.

Chapter 8

Device Simulation for Li-Ion Batteries

Naoki Baba
Toyota Central R&D Laboratories, Inc., Nagakute,
Aichi 480-1192, Japan
baba-n@mosk.tytlabs.co.jp

8.1 Introduction

Lithium-ion batteries (LiBs) are state-of-the-art power sources and have thus been one of the central targets of research and development for many years. In practice, LiBs are now employed as power sources for hybrid electric vehicles (HEVs) and electric vehicles (EVs) because they have more energy per unit weight and a relatively higher power density than conventional batteries.

It is expected that the demand for LiBs will increase continuously in the future. However, the earth's resources of battery materials are limited. Accordingly, battery manufacturers and researchers should pursue creating LiBs with the ultimate performance and maximizing the use of limited resources. For example, optimizing the mesoscale porous electrode structure and large-scale cell stack arrangement and ensuring the sophisticated state-of-charge management

Multiscale Simulations for Electrochemical Devices
Edited by Ryoji Asahi

ISBN 978-981-4800-71-6 (Hardcover), 978-0-429-29545-4 (eBook)
www.jennystanford.com

required under dynamic load conditions are challenges to be addressed.

Large-scale device simulation techniques would be quite helpful to ensure utilization of the high energy content of LiBs. When designing cell specifications and battery pack configurations, this would provide useful information that is difficult or impossible to obtain experimentally.

Device simulation techniques typically consist of simplified methods and submodels due to calculation cost limitations. Moreover, submodels are often required to be consistent with the molecular scale and mesoscale simulations discussed in previous chapters.

In this chapter, device simulation techniques for LiBs are described. The next section focuses on a multidimensional simulation method capable of evaluating both electrochemical and thermal behavior of LiBs under typical operating conditions. It is necessary to include a simplified and low-cost lumped battery model that has the ability to estimate heat generation rates as accurately as possible. A new lumped battery model that satisfies these requirements will be proposed. An electrochemical-thermal coupled simulation method using a code coupling interface (CCI) will then be presented. Section 8.3 attempts to propose a thermal abuse model for atypical conditions. The conventional thermal abuse models simulate exothermic phenomena by calculating the time variation in the degree of progress of the reaction. The drawback of this procedure is the lack of a reaction mechanism and scheme. Another thermal abuse modeling approach will thus be proposed.

8.2 Electrochemical-Thermal Coupled Device Simulation

LiB performance is strongly influenced by temperature because thermophysical properties of battery materials, such as the conductivity of the electrolyte and the diffusion coefficient of lithium ions, are dependent on the temperature. Device simulation techniques that can be used to evaluate the thermal performance of batteries are thus required to simultaneously consider electrochemical reactions and thermal behavior.

8.2.1 Conventional Device Simulation Approach

There have been numerous previous reports on thermal modeling approaches for LiBs. Pals and Newman performed 1D thermal modeling to calculate the temperature profiles in a cell stack [25, 26]. This work was based on the 1D macroscopic model developed by Doyle et al. [10], with the addition of a lumped heat generation term presented by Bernardi et al. [6]. Chen and Evans [7–9] presented a multidimensional thermal model that focused on heat transport inside the cell stack without considering the electrochemistry of the cell. In their work, the heat generation rate was estimated using experimental discharge curves based on the lumped heat generation formulas given by Bernardi et al. [6]. Kim et al. [17, 18] applied a 2D model to parallel plate battery electrodes to simulate not only the potential and current density distribution but also the temperature distribution. However, their modeling approach focused on ensuring current continuity on the electrodes, and lumped heat generation terms were applied.

The issues with conventional device simulation approaches are that the mesh generation of the actual cell design is inadequate and that the procedure to solve the electrochemical model is simplified due to calculation restrictions. Moreover, it is of significant concern that the electrochemical-thermal interaction procedure is a weak coupling method. In addition, temperature information for the electrode domain, which includes time variation and distribution, is not considered in the electrochemical battery model.

8.2.2 LiB Model Suitable for Device Simulation

A prime requisite for device simulation is to achieve multidimensional analysis of the actual device geometry. Multidimensional calculations generally require a lot of time and are costly. When performing an electrochemical-thermal coupled simulation, it is also obvious that the time required to solve the governing electrochemical equations is much higher than that to solve the heat conduction equation. Accordingly, from the standpoint of computer-aided engineering in the battery design phase, a battery model suitable for device simulation should be both accurate and efficient to be practical and productive.

8.2.2.1 Conventional LiB models

Conventional mathematical models can be categorized into two broad types.

The first type was proposed by Doyle et al. [10]. A schematic of this model, which consists of two composite electrodes and a separator, is given in Fig. 8.1 and is referred to as the Newman model hereafter. A mathematical description includes equations that describe (i) mass transport of lithium in the solid phases, (ii) mass transport of lithium ions in the solution phase, (iii) charge transport in the solid phases, and (iv) charge transport in the solution phase. The Newman model is a 1D macroscopic model across the thickness of the electrode at a local point on the electrode plane. From the viewpoint of application to the 3D electrochemical-thermal coupled simulation, the advantage of the Newman model is its ability to accurately estimate heat generation rates, although it has a disadvantage in terms of its high computational costs. The local heat generation rate is calculated from the detailed theoretical formula described below [12, 13, 27].

$$q=\sigma^{\mathrm{eff}}\nabla\phi_{\mathrm{s}}\cdot\nabla\phi_{\mathrm{s}}+(\kappa^{\mathrm{eff}}\nabla\phi_{\mathrm{e}}\cdot\nabla\phi_{\mathrm{e}}+\kappa_{\mathrm{D}}^{\mathrm{eff}}\nabla\ln c_{\mathrm{e}}\cdot\nabla\phi_{\mathrm{e}})$$

$$+a_{\mathrm{s}}\bar{i}_{\mathrm{n,j}}\left(\phi_{\mathrm{s}}-\phi_{\mathrm{e}}-U_j\right)+a_{\mathrm{s}}\bar{i}_{\mathrm{n,j}}T\frac{\partial U_j}{\partial T} \qquad (8.1)$$

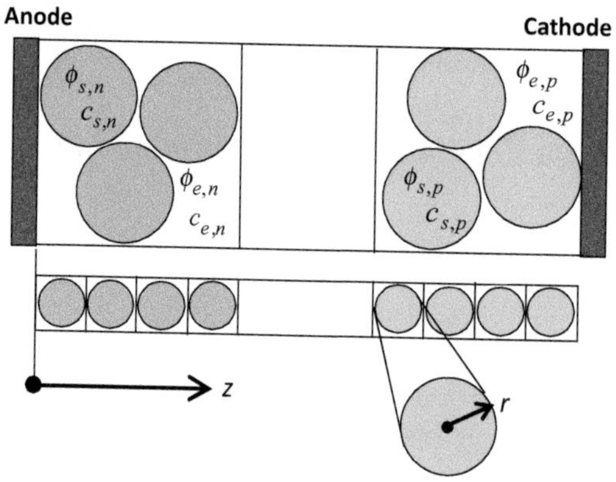

Figure 8.1 Schematic of the lithium-ion battery model proposed by Doyle and Newman. Reprinted from Ref. [5], Copyright (2014), with permission from Elsevier.

The notations of equations are summarized at the end of chapter. Equation 8.1 requires the local point values and gradients of potentials and concentrations in the solid and electrolyte phases. The Newman model is capable of estimating these values at each local point; therefore, the heat generation rate can be accurately estimated using Eq. 8.1. On the other hand, the Newman model requires the use of a micrometer-order mesh size to discretize a calculation domain across the thickness of each electrode in one dimension. As a result, if the entire cell geometry is discretized with the micrometer-order size, then the number of computational meshes will be far in excess of 10 million, which makes this too expensive and time consuming.

The second type of modeling is the single-particle (SP) model [14, 24, 30, 32, 35, 36]. A schematic of this model is shown in Fig. 8.2. In this model, each electrode is lumped into a single spherical particle. The advantages of the SP model are its simplicity and reduced computational time requirements. This model is orders of magnitude faster than the Newman model; however, a decisive drawback of the SP model is that the electrolyte phase diffusion limitations are ignored. Therefore, potentials in the electrolyte phase are set to zero and concentrations in the electrolyte phase are assumed to be constant, which causes inaccuracies in the estimations of the heat generation rate via Eq. 8.1.

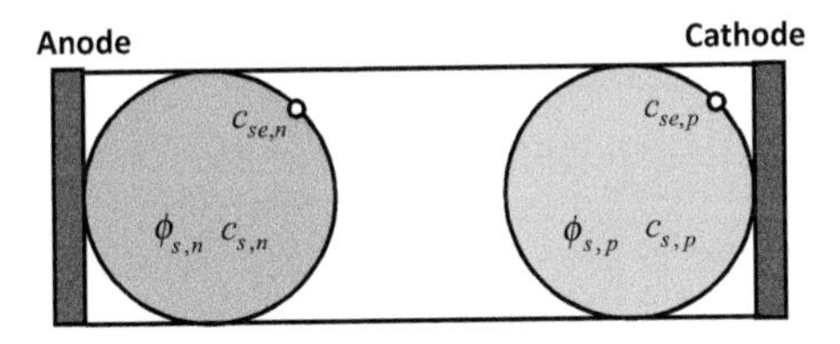

Figure 8.2 Schematic of the single-particle (SP) model. This model ignores the electrolyte phase diffusion limitations. Reprinted from Ref. [5], Copyright (2014), with permission from Elsevier.

8.2.2.2 Enhanced single-particle model

The basic concept of a new battery model is that it is a lumped model similar to the SP model, which is due to calculation cost limitations. Moreover, as a prime requirement, it should be able to accurately estimate the heat generation rate. Figure 8.3 shows a schematic of the new model, which is referred to hereafter as the enhanced single-particle (ESP) model. In the ESP model, each electrode is lumped into a single spherical particle within its electrolyte phase in order to take into account the electrolyte phase diffusion limitations. The potential and lithium concentration in the solid phase are also estimated in the same way as in the conventional SP model. The potential and lithium-ion concentration in the electrolyte phase are also estimated at representative positions in each electrode. A feature of the ESP model is that depth profiles for these physical quantities within each electrode are approximated by parabolic functions. As represented in Eq. 8.1, accurate estimation of the heat release rate requires point values and gradients for these physical quantities. These are calculated from approximated parabolic functions. The potentials and lithium-ion concentrations at interfaces between the negative electrode and the separator or between the positive electrode and the separator are implicitly calculated in the ESP model. Mass and charge transport between both electrodes in the electrolyte phase, which satisfies the conservation of mass and charge, are estimated using these boundary values.

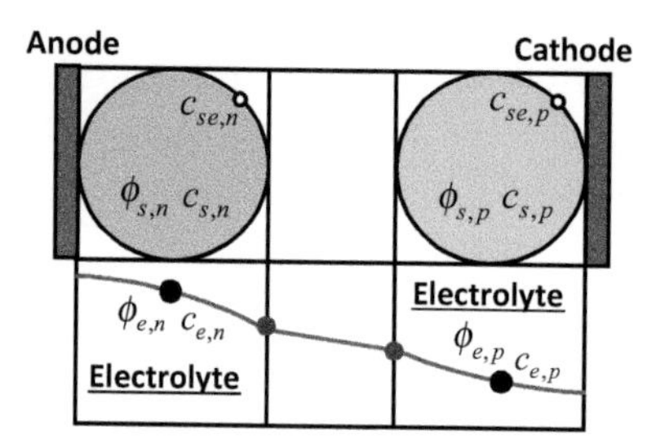

Figure 8.3 Schematic of the enhanced single-particle (ESP) model. In this model, the negative electrode and the positive electrode each is represented by a single spherical particle with the electrolyte phase. Reprinted from Ref. [5], Copyright (2014), with permission from Elsevier.

The procedure used to calculate concentrations in the ESP model is shown in Fig. 8.4. Fick's second law describes lithium transport in the solid phase, with the following boundary conditions for a spherical particle:

$$\frac{\partial c_s}{\partial t} = D_s\left[\frac{\partial^2 c_s}{\partial r^2} + \frac{2}{r}\frac{\partial c_s}{\partial r}\right] \tag{8.2}$$

$$-D_s \frac{\partial c_s}{\partial r} = 0 \quad \text{at} \quad r = 0 \tag{8.3}$$

$$-D_s \frac{\partial c_s}{\partial r} = \frac{\bar{i}_n}{F} \quad \text{at} \quad r = r_s \tag{8.4}$$

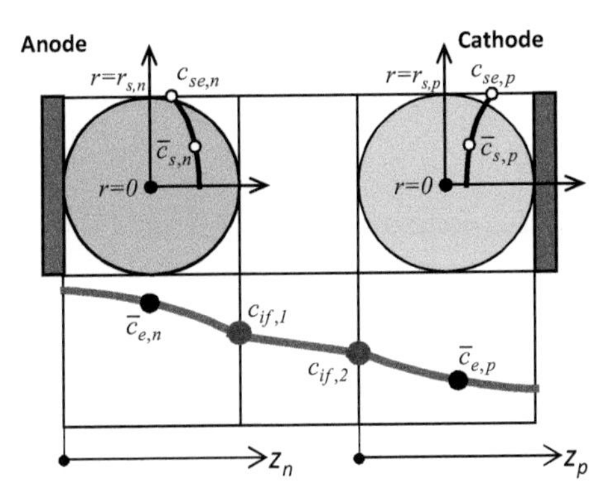

Figure 8.4 Calculation procedure of the concentration field of the ESP model. Reprinted from Ref. [5], Copyright (2014), with permission from Elsevier.

In the conventional SP model, the diffusion length [33] is introduced to simplify the diffusion equation. Assuming a parabolic concentration profile in the diffusion layer and using the volume average technique, the solutions for Eqs. 8.2–8.4 are as follows:

$$\frac{d\bar{c}_{s,j}}{dt} + \frac{15D_{s,j}}{r_{s,j}^{\ 2}}(\bar{c}_{s,j} - c_{se,j}) = 0 \tag{8.5}$$

and

$$-D_{s,j}\frac{(c_{se,j} - \bar{c}_{s,j})}{l_{se,j}} = \frac{j_j^{Li}}{a_{s,j}F}, \tag{8.6}$$

where $l_{se,j}$ is the diffusion length, which takes a value of $r_{s,j}/5$ for spherical particles. To reduce the computational costs, the diffusion length method is implemented in the ESP model in the same way.

The diffusion limitations in the electrolyte phase are also considered in the ESP model. The representative lithium-ion concentrations in the electrolyte phase for each electrode are defined using the volume average technique as follows:

$$\bar{c}_{e,n} = \frac{1}{L_n}\int_0^{L_n} c_{e,n}\, dz_n \tag{8.7}$$

$$\bar{c}_{\mathrm{e,p}} = \frac{1}{L_{\mathrm{p}}} \int_0^{L_{\mathrm{p}}} c_{\mathrm{e,p}} \mathrm{d}z_{\mathrm{p}} \tag{8.8}$$

In the ESP model, a parabolic concentration profile is assumed for each electrode. The defined depth positions of the representative concentrations are

$$z_{\mathrm{n,e}}^{*} = \frac{1}{\sqrt{3}} \cdot L_{\mathrm{n}} \tag{8.9}$$

and

$$z_{\mathrm{p,e}}^{*} = \left(1 - \frac{1}{\sqrt{3}}\right) \cdot L_{\mathrm{p}}, \tag{8.10}$$

where the extra pseudo coordinates z_{n} and z_{p} are defined as indicated in Fig. 8.4.

In addition, the concentrations at the negative electrode/separator interface and the separator/positive electrode interface are implicitly defined as $c_{\mathrm{if,1}}$ and $c_{\mathrm{if,2}}$, respectively. The interfacial balances of lithium-ion flux are

$$-D_{\mathrm{e,n}}^{\mathrm{eff}} \frac{c_{\mathrm{if,1}} - \bar{c}_{\mathrm{e,n}}}{\delta_{\mathrm{n}}} = -D_{\mathrm{e,sep}}^{\mathrm{eff}} \frac{c_{\mathrm{if,2}} - c_{\mathrm{if,1}}}{L_{\mathrm{sep}}} \tag{8.11}$$

and

$$-D_{\mathrm{e,sep}}^{\mathrm{eff}} \frac{c_{\mathrm{if,2}} - c_{\mathrm{if,1}}}{L_{\mathrm{sep}}} = -D_{\mathrm{e,p}}^{\mathrm{eff}} \frac{\bar{c}_{\mathrm{e,p}} - c_{\mathrm{if,2}}}{\delta_{\mathrm{p}}}, \tag{8.12}$$

where δ_{n} and δ_{p} represent the diffusion lengths in each electrode, which are expressed as [33]

$$\delta_{\mathrm{n}} = \frac{L_{\mathrm{n}}}{3} \text{ and } \delta_{p} = \frac{L_{\mathrm{p}}}{3}. \tag{8.13}$$

Solving Eqs. 8.11 and 8.12 gives the interfacial concentrations as

$$c_{\mathrm{if,1}} = \frac{\alpha_{\mathrm{n}} + \omega}{\alpha_{\mathrm{n}} + \omega + \gamma_{\mathrm{p}}} \cdot \bar{c}_{\mathrm{e,n}} + \frac{\gamma_{\mathrm{p}}}{\alpha_{\mathrm{n}} + \omega + \gamma_{\mathrm{p}}} \cdot \bar{c}_{\mathrm{e,p}}, \tag{8.14}$$

$$c_{\mathrm{if,2}} = \frac{\alpha_{\mathrm{n}}}{\alpha_{\mathrm{n}} + \omega + \gamma_{\mathrm{p}}} \cdot \bar{c}_{\mathrm{e,n}} + \frac{\omega + \gamma_{\mathrm{p}}}{\alpha_{\mathrm{n}} + \omega + \gamma_{\mathrm{p}}} \cdot \bar{c}_{\mathrm{e,p}}, \tag{8.15}$$

and

$$\alpha_{\mathrm{n}} = \frac{D_{\mathrm{e,n}}^{\mathrm{eff}}}{\delta_{\mathrm{n}}}, \quad \beta_{\mathrm{sep}} = \frac{D_{\mathrm{e,sep}}^{\mathrm{eff}}}{L_{\mathrm{sep}}}, \quad \gamma_{\mathrm{p}} = \frac{D_{\mathrm{e,p}}^{\mathrm{eff}}}{\delta_{\mathrm{p}}} \text{ and } \omega = \frac{\gamma_{p} \cdot \alpha_{\mathrm{n}}}{\beta_{\mathrm{sep}}}. \tag{8.16}$$

The equations used to calculate the time evolution of the volume-averaged lithium-ion concentration in the electrolyte phase are derived as described below. As for the negative electrode, the conservation of the lithium ion [10] in the electrolyte yields

$$\varepsilon_{\mathrm{e,n}} \frac{\partial c_{\mathrm{e,n}}}{\partial t} = D_{\mathrm{e,n}}^{\mathrm{eff}} \frac{\partial^2 c_{\mathrm{e,n}}}{\partial z_{\mathrm{n}}^{\ 2}} + \frac{1-t_+^{\circ}}{F} j_{\mathrm{n}}^{\mathrm{Li}}. \tag{8.17}$$

As shown in Fig. 8.4, the electrolyte phase concentration is represented by a parabolic profile

$$c_{\mathrm{e,n}} = a \cdot z_{\mathrm{n}}^{\ 2} + b \cdot z_{\mathrm{n}} + c, \tag{8.18}$$

where the three coefficients, a, b, and c can be determined using the following boundary conditions and Eq. 8.7:

$$\frac{\partial c_{\mathrm{e,n}}}{\partial z_{\mathrm{n}}} = 0 \quad \text{at} \quad z_{\mathrm{n}} = C \tag{8.19}$$

$$c_{\mathrm{e,n}} = c_{\mathrm{if,1}} \quad \text{at} \quad z_{\mathrm{n}} = L_{\mathrm{n}} \tag{8.20}$$

Application of volume averaging to Eq. 8.17 and substitution of Eq. 8.18 leads to

$$\varepsilon_{\mathrm{e,n}} \frac{\mathrm{d}\bar{c}_{\mathrm{e,n}}}{\mathrm{d}t} = D_{\mathrm{e,n}}^{\mathrm{eff}} \cdot (2a) + \frac{1-t_+^{\circ}}{F} j_{\mathrm{n}}^{\mathrm{Li}}. \tag{8.21}$$

Substituting the coefficient a into Eq. 8.21 yields

$$\varepsilon_{\mathrm{e,n}} \frac{\mathrm{d}\bar{c}_{\mathrm{e,n}}}{\mathrm{d}t} = A_{\mathrm{n}} \cdot \bar{c}_{\mathrm{e,n}} + B_{\mathrm{n}} \cdot \bar{c}_{\mathrm{e,p}} + \frac{1-t_+^{\circ}}{F} j_{\mathrm{n}}^{\mathrm{Li}}, \tag{8.22}$$

where the coefficients A_{n} and B_{n} are given by

$$A_{\mathrm{n}} = -\frac{1}{L_{\mathrm{n}}} \cdot \frac{\gamma_{\mathrm{p}} \cdot \alpha_{\mathrm{n}}}{\alpha_{\mathrm{n}} + \omega + \gamma_{\mathrm{p}}} \tag{8.23}$$

and

$$B_{\mathrm{n}} = \frac{1}{L_{\mathrm{n}}} \cdot \frac{\gamma_{\mathrm{p}} \cdot \alpha_{\mathrm{n}}}{\alpha_{\mathrm{n}} + \omega + \gamma_{\mathrm{p}}}. \tag{8.24}$$

For the positive electrode, the same procedure also yields the equation to calculate the time evolution of the volume-averaged lithium-ion concentration in the electrolyte phase:

$$\varepsilon_{\mathrm{e,p}} \frac{\mathrm{d}\bar{c}_{\mathrm{e,p}}}{\mathrm{d}t} = A_{\mathrm{p}} \cdot \bar{c}_{\mathrm{e,n}} + B_{\mathrm{p}} \cdot \bar{c}_{\mathrm{e,p}} + \frac{1-t_+^{\circ}}{F} j_{\mathrm{n}}^{\mathrm{Li}}, \tag{8.25}$$

where the coefficients A_p and B_p are given by

$$A_\mathrm{p} = \frac{1}{L_\mathrm{p}} \cdot \frac{\gamma_\mathrm{p} \cdot \alpha_\mathrm{n}}{\alpha_\mathrm{n} + \omega + \gamma_\mathrm{p}} \tag{8.26}$$

and

$$B_\mathrm{p} = -\frac{1}{L_\mathrm{p}} \cdot \frac{\gamma_\mathrm{p} \cdot \alpha_\mathrm{n}}{\alpha_\mathrm{n} + \omega + \gamma_\mathrm{p}}. \tag{8.27}$$

In the ESP model, Eqs. 8.5, 8.6, 8.22, and 8.25 are specified to solve mass transport in the system.

The procedure for the calculation of the potentials with the ESP model is shown in Fig. 8.5. Analogous to the concentrations, the potentials on the negative electrode/separator interface and the separator/positive electrode interface are implicitly defined as $\phi_{\mathrm{if},1}$ and $\phi_{\mathrm{if},2}$, respectively. The boundary conditions for the solid phase potential $\phi_{\mathrm{s,n}}$ in the negative electrode are described below. $\phi_{\mathrm{s,n}}$ at the copper current collector/composite negative electrode interface is arbitrarily set to zero. The solid phase current density is equated to the applied current density at the same interface.

$$\phi_{\mathrm{s,n}} = 0 \quad \text{at} \quad z_\mathrm{n} = 0 \tag{8.28}$$

$$-\sigma_\mathrm{n}^{\mathrm{eff}} \frac{\partial \phi_{\mathrm{s,n}}}{\partial z_\mathrm{n}} = I_{\mathrm{app}} \quad \text{at} \quad z_\mathrm{n} = 0 \tag{8.29}$$

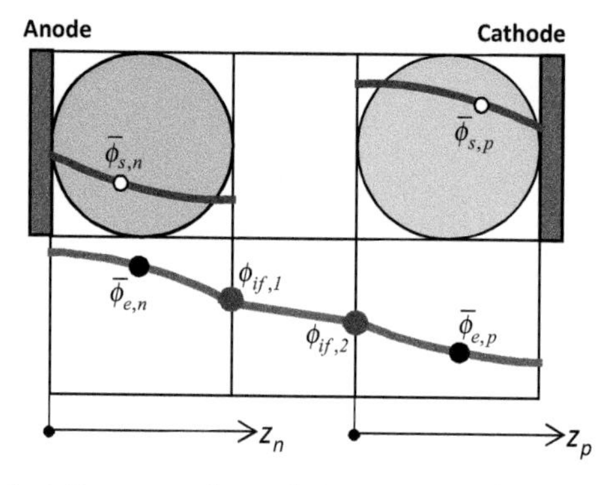

Figure 8.5 Calculation procedure of the potential field of the ESP model. Reprinted from Ref. [5], Copyright (2014), with permission from Elsevier.

At the negative electrode/separator interface, the charge flux in the solid phase is equated to zero:

$$\frac{\partial \phi_{\mathrm{s,n}}}{\partial z_\mathrm{n}} = 0 \quad \text{at} \quad z_\mathrm{n} = L_\mathrm{n} \tag{8.30}$$

Applying a parabolic profile for $\phi_{s,n}$ and satisfaction of the above boundary conditions yield

$$\phi_{s,n} = \frac{1}{2L_n} \cdot \frac{I_{app}}{\sigma_n^{eff}} \cdot z_n^2 - \frac{I_{app}}{\sigma_n^{eff}} \cdot z_n. \tag{8.31}$$

The representative potential in the negative electrode solid phase is defined using the volume average technique

$$\bar{\phi}_{s,n} = \frac{1}{L_n} \int_0^{L_n} \phi_{s,n} dz_n. \tag{8.32}$$

Substituting Eq. 8.31 into Eq. 8.32 gives

$$\bar{\phi}_{s,n} = -\frac{1}{3} \cdot \frac{L_n}{\sigma_n^{eff}} \cdot I_{app}. \tag{8.33}$$

This is equal to the point value that is calculated in the profile, Eq. 8.31, at

$$z_{n,s}^* = \left(1 - \frac{1}{\sqrt{3}}\right) \cdot L_n. \tag{8.34}$$

Equation 8.33 gives the solid phase potential, $\bar{\phi}_{s,n}$; therefore, the representative potential in the negative electrolyte phase, $\bar{\phi}_{e,n}$, is determined by the following Butler–Volmer equation:

$$\frac{I_{app}}{L_n} = a_{s,n} i_{0,n} \left[\exp\left(\frac{\alpha_{a,n} F}{RT} \eta_n \right) - \exp\left(\frac{\alpha_{c,n} F}{RT} \eta_n \right) \right], \tag{8.35}$$

where the surface overpotential η_n is defined to be

$$\eta_n = \bar{\phi}_{s,n} - \bar{\phi}_{e,n} - U_n - \bar{i}_{n,n} \cdot R_{f,n}. \tag{8.36}$$

Conservation of charge in the electrolyte phase results in [10]

$$\nabla \cdot (\kappa_j^{eff} \nabla \phi_{e,j}) + \nabla \cdot (\kappa_{D,j}^{eff} \nabla \ln c_{e,j}) + j_j^{Li} = 0. \tag{8.37}$$

From Eq. 8.37, the charge flux at the negative electrode/separator interface is equated to the applied current density

$$-\kappa_n^{eff} \left. \frac{\partial \phi_{e,n}}{\partial z_n} \right|_{z_n = L_n} - \kappa_{D,n}^{eff} \frac{\left. \frac{\partial c_{e,n}}{\partial z_n} \right|_{z_n = L_n}}{c_{if,1}} = I_{app}. \tag{8.38}$$

The second term on the left-hand side is estimated using Eqs. 8.14 and 8.18 so that the potential gradient at the negative electrode/separator interface (the first term) is obtained from Eq. 8.38. Next, a parabolic profile is assumed for the electrolyte potential in the depth direction of the negative electrode. This parabolic profile must satisfy the following three conditions:

$$\frac{\partial \phi_{\mathrm{e,n}}}{\partial z_{\mathrm{n}}} = 0 \quad \text{at} \quad z_{\mathrm{n}} = 0, \tag{8.39}$$

$$\frac{\partial \phi_{\mathrm{e,n}}}{\partial z_{\mathrm{n}}} = s_{\mathrm{if,1}} \quad \text{at} \quad z_{\mathrm{n}} = L_{\mathrm{n}}, \tag{8.40}$$

and

$$\bar{\phi}_{\mathrm{e,n}} = \frac{1}{L_{\mathrm{n}}} \int_0^{L_{\mathrm{n}}} \phi_{\mathrm{e,n}} \mathrm{d}z_{\mathrm{n}}, \tag{8.41}$$

where $s_{\mathrm{if,1}}$ is the electrolyte potential gradient at the negative electrode/separator interface, which is estimated from Eq. 8.38. Using Eqs. 8.39–8.41, the potential profile in the negative electrolyte phase is determined by

$$\phi_{\mathrm{e,n}} = \frac{s_{\mathrm{if,1}}}{2L_{\mathrm{n}}} \cdot z_{\mathrm{n}}^2 + \bar{\phi}_{\mathrm{e,n}} - \frac{s_{\mathrm{if,1}}}{6} \cdot L_{\mathrm{n}}. \tag{8.42}$$

According to Eq. 8.42, the interfacial potential $\phi_{\mathrm{if,1}}$ is expressed by

$$\phi_{\mathrm{if,1}} = \frac{s_{\mathrm{if,1}}}{3} \cdot L_{\mathrm{n}} + \bar{\phi}_{\mathrm{e,n}}. \tag{8.43}$$

Taking into account the continuity of the charge flux in the electrolyte phase, the charge transport in the separator results in:

$$\phi_{\mathrm{if,2}} = \phi_{\mathrm{if,1}} - \frac{1}{\kappa_{\mathrm{sep}}^{\mathrm{eff}}} \left(I_{\mathrm{app}} \cdot L_{\mathrm{sep}} + \kappa_{\mathrm{D,sep}}^{\mathrm{eff}} \cdot \ln \frac{c_{if,2}}{c_{if,1}} \right), \tag{8.44}$$

where the interfacial concentrations $c_{\mathrm{if,1}}$ and $c_{\mathrm{if,2}}$ can be estimated using Eqs. 8.14 and 8.15.

A similar calculation procedure is used to derive the parabolic profile for the positive electrolyte potential. The charge flux at the positive electrode/separator interface is equated to the applied current density

$$-\kappa_{\mathrm{p}}^{\mathrm{eff}} \left.\frac{\partial \phi_{\mathrm{e,p}}}{\partial z_{\mathrm{p}}}\right|_{z_{\mathrm{p}}=0} - \kappa_{\mathrm{D,p}}^{\mathrm{eff}} \frac{\left.\frac{\partial c_{\mathrm{e,p}}}{\partial z_{\mathrm{p}}}\right|_{z_{\mathrm{p}}=0}}{c_{\mathrm{if,2}}} = I_{\mathrm{app}}. \tag{8.45}$$

The second term on the left is estimated using Eq. 8.15 and the parabolic profile of $c_{e,p}$. This allows the potential gradient at the separator/positive electrode interface in the first term to be estimated. A parabolic profile is assumed for the electrolyte potential in the depth direction of the positive electrode. This parabolic profile must meet the following conditions:

$$\phi_{e,p} = \phi_{if,2} \quad \text{at} \quad z_p = 0, \tag{8.46}$$

$$\frac{\partial \phi_{e,p}}{\partial z_p} = s_{if,2} \quad \text{at} \quad z_p = 0, \tag{8.47}$$

and

$$\frac{\partial \phi_{e,p}}{\partial z_p} = 0 \quad \text{at} \quad z_p = L_p, \tag{8.48}$$

where $s_{if,2}$ is the electrolyte potential gradient at the separator/positive electrode interface, which is estimated from Eq. 8.45. Using Eqs. 8.46–8.48, the potential profile in the positive electrolyte phase is expressed by

$$\phi_{e,p} = -\frac{s_{if,2}}{2L_p} \cdot z_p^{\ 2} + s_{if,2} \cdot z_p + \phi_{if,2}. \tag{8.49}$$

The representative potential in the positive electrolyte is defined as

$$\bar{\phi}_{e,p} = \frac{1}{L_p} \int_0^{L_p} \phi_{e,p} dz_p = \frac{s_{if,2}}{3} L_p + \phi_{if,2}. \tag{8.50}$$

This is equal to the point value calculated in the profile of Eq. 8.49 at $z^*_{p,e}$ (Eq. 8.10). Let us turn to the positive electrode solid phase potential. Equation 8.50 gives the electrolyte phase potential; therefore, the representative potential in the positive solid phase, $\bar{\phi}_{s,p}$, is estimated by the following Butler–Volmer equation:

$$-\frac{I_{app}}{L_p} = a_{s,p} i_{0,p} \left[\exp\left(\frac{\alpha_{a,p} F}{RT} \eta_p \right) - \exp\left(\frac{\alpha_{c,p} F}{RT} \eta_p \right) \right], \tag{8.51}$$

where the surface overpotential η_p is defined to be

$$\eta_p = \bar{\phi}_{s,p} - \bar{\phi}_{e,p} - U_p - \bar{i}_{n,p} \cdot R_{f,p}. \tag{8.52}$$

A parabolic profile of $\phi_{s,p}$ is assumed and must satisfy the following conditions to determine the three coefficients:

$$\frac{\partial \phi_{\mathrm{s,p}}}{\partial z_{\mathrm{p}}} = 0 \quad \text{at} \quad z_{\mathrm{p}} = 0 \tag{8.53}$$

$$-\sigma_{\mathrm{s,p}}^{\mathrm{eff}} \frac{\partial \phi_{\mathrm{s,p}}}{\partial z_{\mathrm{p}}} = I_{\mathrm{app}} \quad \text{at} \quad z_{\mathrm{p}} = L_{\mathrm{p}} \tag{8.54}$$

$$\bar{\phi}_{\mathrm{s,p}} = \frac{1}{L_{\mathrm{p}}} \int_0^{L_p} \phi_{\mathrm{s,p}} \mathrm{d}z_{\mathrm{p}} \tag{8.55}$$

Satisfaction of these conditions yields

$$\phi_{\mathrm{s,p}} = -\frac{1}{2L_{\mathrm{p}}} \cdot \frac{I_{\mathrm{app}}}{\sigma_{\mathrm{s,p}}^{\mathrm{eff}}} \cdot z_{\mathrm{p}}^{2} + \frac{L_{\mathrm{p}}}{6} \cdot \frac{I_{\mathrm{app}}}{\sigma_{\mathrm{s,p}}^{\mathrm{eff}}} + \bar{\phi}_{s,p}. \tag{8.56}$$

The representative value, $\bar{\phi}_{\mathrm{s,p}}$, is obtained in the profile, Eq. 8.56, at

$$z_{\mathrm{p,s}}^{*} = \frac{1}{\sqrt{3}} \cdot L_{\mathrm{p}}. \tag{8.57}$$

Model equations of the ESP model are now specified completely. The heat generation rate during the charge-discharge process, which is Eq. 8.1 as revised for the ESP model, is expressed by

$$\begin{aligned} q = & \sum_{\mathrm{j=n,p}} (\sigma_{\mathrm{j}}^{\mathrm{eff}} \nabla \phi_{s,j} \cdot \nabla \phi_{s,j}) \\ & + \sum_{\mathrm{j=n,sep,p}} (\kappa_{\mathrm{j}}^{\mathrm{eff}} \nabla \phi_{\mathrm{e,j}} \cdot \nabla \phi_{\mathrm{e,j}} + \kappa_{\mathrm{D,j}}^{\mathrm{eff}} \nabla \ln c_{\mathrm{e,j}} \cdot \nabla \phi_{\mathrm{e,j}}) \\ & + \sum_{\mathrm{j=n,p}} a_{\mathrm{s,j}} \bar{i}_{\mathrm{n,j}} \left(\bar{\phi}_{\mathrm{s,j}} - \bar{\phi}_{\mathrm{e,j}} - U_{j} \right) + \sum_{\mathrm{j=n,p}} a_{\mathrm{s,j}} \bar{i}_{\mathrm{n,j}} T \frac{\partial U_{\mathrm{j}}}{\partial T}. \end{aligned} \tag{8.58}$$

The first term arises from the ohmic heat in the solid phase, while the second term results from the ohmic heat in the electrolyte phase. The summation of the last two terms accounts for irreversible and reversible heat associated with charge transfer at the solid/electrolyte interphases (SEIs). The third term represents the potential deviation from the equilibrium potential (irreversible heat). The fourth term results from the entropic effect (reversible heat). From Eq. 8.58, it is clear that the estimation of ohmic heat in the solid and electrolyte phases requires information on the gradients of potentials. The ESP model includes the profile information in the depth direction of each electrode. The gradients of concentrations and potentials in the solid phase are estimated at the positions of $z^*_{n,s}$ (Eq. 8.34) for the negative electrode and $z^*_{p,s}$ (Eq. 8.57) for the positive electrode using the corresponding parabolic profile. The gradients in the electrolyte phase are estimated at the positions of $z^*_{n,e}$ (Eq. 8.9) for the negative electrode and $z^*_{p,e}$ (Eq. 8.10) for the positive electrode.

8.2.2.3 Model validation

Discharge curves at various current densities were compared among battery models to validate the ESP model. The target cell of the calculation consisted of a carbon negative electrode (Li_xC_6), a manganese oxide positive electrode ($Li_yMn_2O_4$), and a plasticized polymer electrolyte. The solution used in the cell was 2M $LiPF_6$ in a 1:2 ratio mixture of ethylene carbonate/dimethyl carbonate in the plasticized polymer matrix [11].

Figure 8.6 shows a comparison of the simulated discharge curves for the Newman and conventional SP models. The difference between the simulation results becomes increasingly large at rates of discharge beyond 1C. The SP model assumes the concentration and potential in the electrolyte phase to be constant. However, the concentration distribution in the electrolyte phase cannot be ignored as the discharge rate becomes higher. The Newman model is capable of simulating this via the electrolyte phase charge and species conservation equations.

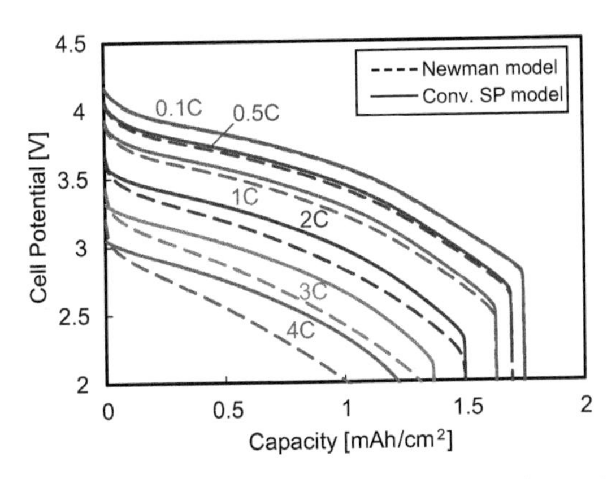

Figure 8.6 Comparison between discharge curves at various rates from the Newman model and those predicted by the conventional SP model. Reprinted from Ref. [5], Copyright (2014), with permission from Elsevier.

Figure 8.7 shows the validation of the ESP model. Good agreement is observed between the discharge curves for both Newman and ESP models for rates up to 2C. The ESP model includes the profile information for the concentration and potential in the solution phase. At higher rates, the discharge curve for the ESP model becomes closer to the Newman model in the later stage of the discharge process because the developed concentration field coincides approximately with the profile information. On the other hand, the discharge curves for the ESP model are below those for the Newman model in the early stage of discharge. Additional efforts have been made to reduce this error, as outlined in the following paragraphs.

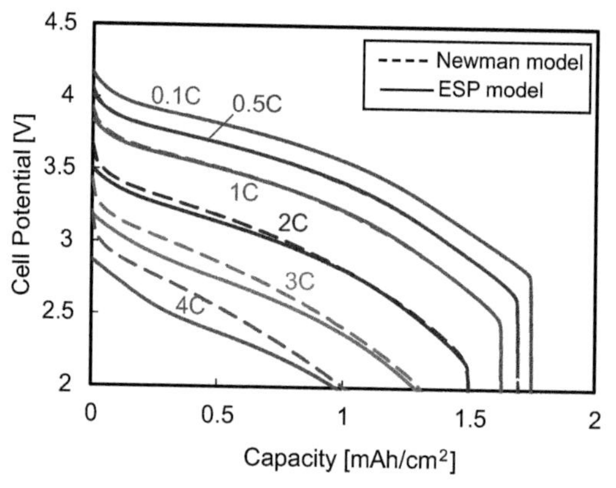

Figure 8.7 Comparison between discharge curves at various rates from the Newman model and those predicted by the ESP model. Reprinted from Ref. [5], Copyright (2014), with permission from Elsevier.

As with the conventional SP model, the ESP model employs the diffusion length method inside a spherical particle with a second-degree polynomial (Eqs. 8.5, 8.6). The method should be valid only after the diffusion layer builds up to a steady state. Therefore, under very high rates of discharge, the approximation that fixes the diffusion length at a constant value is inadequate [34, 35]. To make up for this shortcoming, the time-variant diffusion length concept was introduced by Wang and Srinivasan [34]:

$$l_{\mathrm{se,j}}(t) = l^*_{\mathrm{se,j}}\left(1 - \exp\left(-\chi_{\mathrm{s,j}} \cdot \frac{\sqrt{D_{\mathrm{s,j}} \cdot t}}{l^*_{\mathrm{se,j}}}\right)\right), \tag{8.59}$$

where $l^*_{se,j}$ takes the value of $r_{s,j}/5$ for spherical particles and $\chi_{s,j}$ is a tuning parameter. By developing a similar concept, the ESP model also employs the time-variant diffusion length method with respect to the electrolyte phase diffusion:

$$\delta_j(t) = \delta_j^* \left(1 - \exp\left(-\chi_{e,j} \cdot \frac{\sqrt{D_{e,j} \cdot t}}{\delta_j^*}\right)\right) \tag{8.60}$$

Here, δ_j^* represents the diffusion length in the electrolyte phase, given by Eq. 8.13. $\chi_{e,j}$ is a tuning parameter.

Figure 8.8 shows a comparison between the simulated discharge curves for the Newman and revised ESP models using Eqs. 8.59 and 8.60, which clearly indicates that the revised ESP model can provide results that are as accurate as those provided by the Newman model.

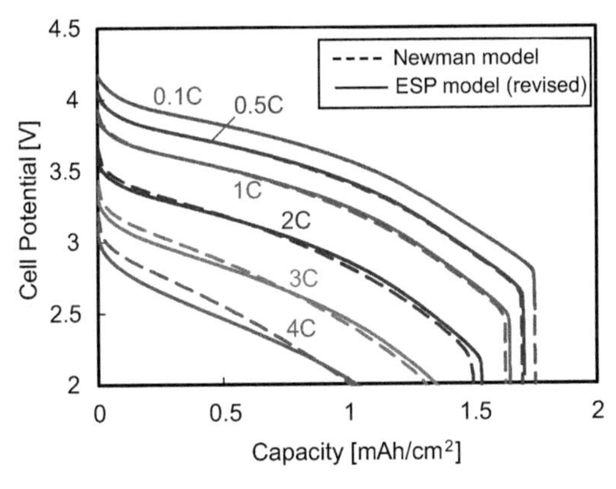

Figure 8.8 Comparison between discharge curves at various rates from the Newman model and those predicted by the revised ESP model. Reprinted from Ref. [5], Copyright (2014), with permission from Elsevier.

Figure 8.9 shows an experimental verification of the ESP model. The comparison of the simulated discharge curves for an 18650-type lithium-ion cell is shown. This cell was manufactured in-house and consists of a carbon negative electrode, a Ni-based oxide positive electrode, and a separator. The capacity of the cell is approximately 0.5 Ah. At a discharge rate of 2C, both conventional SP and ESP models show good agreement with the experimental discharge curve. At discharge rates of 5C and 10C, both Newman and

ESP models show good agreement with the experimental discharge curves. On the other hand, the discharge curves simulated with the conventional SP model show a significant error, which has the same tendency as shown in Fig. 8.6.

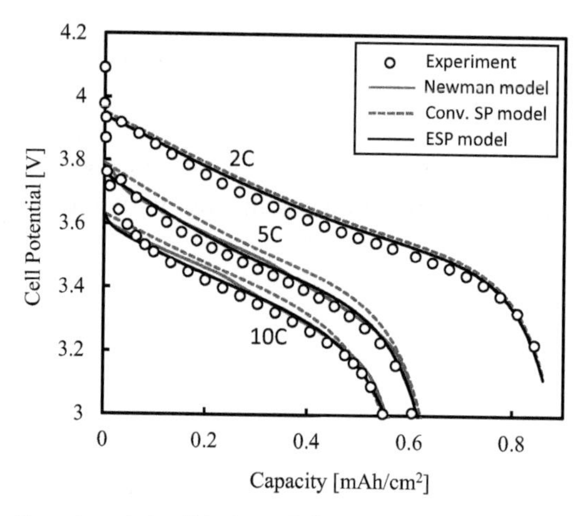

Figure 8.9 Experimental validation of the ESP model. This figure compares simulated discharged curves of the 18650-type lithium-ion cell (manufactured in-house) at various rates. The lines denote model predictions, while the circles denote the experimental data. Reprinted from Ref. [5], Copyright (2014), with permission from Elsevier.

8.2.3 Multidimensional Two-Way Coupling Device Simulation

The device simulation method for an LiB should be capable of simultaneously analyzing the electrochemical and thermal behavior. However, there are many difficulties in realizing this using the single code because the phenomena that occur in the cell are multiscale in both time and space and involve multiphysics. One possible solution to address these problems is the code-coupling method.

Figure 8.10 shows a schematic for the two-way electrochemical-thermal coupled simulation method. The term "two-way" means that physical quantities are exchanged at every computational time-step and physical properties are updated according to the exchanged physical quantities. A quasi-3D porous electrode solver is applied for the 2D unrolled plane of the spirally wound electrodes. The radius of curvature of a spirally wound object is sufficiently large (in relation to the thickness of the positive/negative pole electrode); therefore, the potential for error is small and a 2D deployment approximation is possible, even if it treats the electrochemical reaction between electrodes locally at the plane. Thus, the ESP model could be implemented into each 2D mesh. Conservation equations for charge and lithium-ions are solved in the 2D domain by estimation of their transport across the electrode thickness using the ESP model. This leads to potential and lithium-ion distributions across the 2D unrolled electrode plane. As a result, the distribution of the heat generation rate, which is estimated by Eq. 8.58, is obtained.

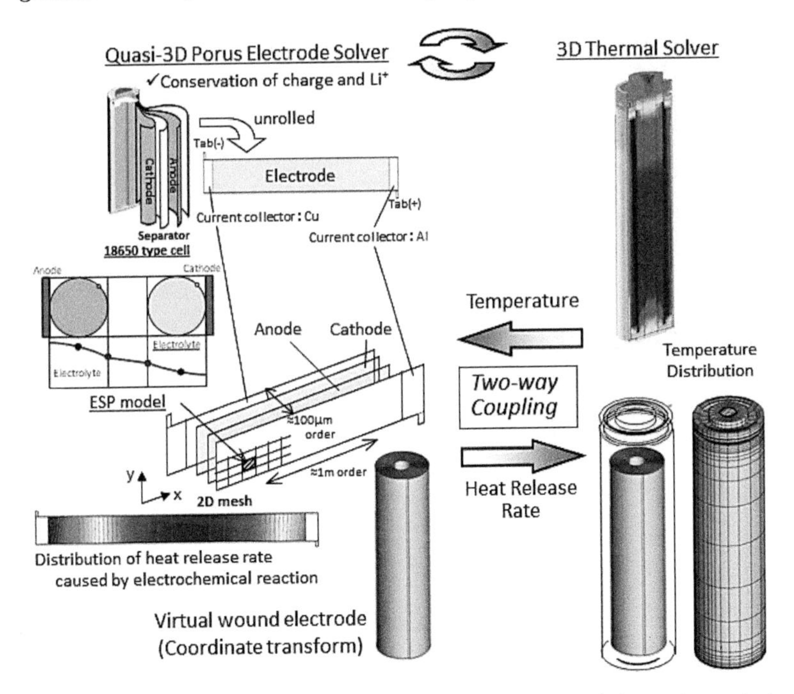

Figure 8.10 A schematic of the two-way electrochemical-thermal coupled simulation method of lithium-ion batteries. Reprinted from Ref. [5], Copyright (2014), with permission from Elsevier.

Next, a virtual spirally wound electrode is constructed using the coordinate transform of the unrolled electrode plane while maintaining the heat generation rate distribution information. Following this, estimated heat generation rate data are mapped to the 3D domain of the real electrode region and a 3D distribution of the mapped heat generation rate q_{mapped}, is reconstructed. This mapping procedure is performed using the CCI [1], by which both codes can easily exchange the specific data.

The 3D thermal solver simulates the temperature distribution of a real cell using the mapped heat release rate that results from the electrochemical reaction. The 3D mesh of the spirally wound electrode region in the actual cell geometry is then virtually unrolled onto the 2D electrode plane by the reverse coordinate transform. Temperature data are subsequently mapped to the 2D unrolled electrode plane using the CCI.

The two-way data exchange process between the two solvers is executed at every computational time-step. As a result, both electrochemical and thermal behavior can be simulated simultaneously. Physical properties are also updated, depending on the temperature at every time-step.

The formulation to implement the ESP model into the quasi-3D porous electrode solver is represented as follows. The ESP model is implemented into each 2D mesh of the unrolled electrode plane. The 3D governing equation of charge conservation in the negative electrolyte phase is given by

$$\kappa_{\text{n}}^{\text{eff}}\left(\frac{\partial^2\phi_{\text{e,n}}}{\partial x^2}+\frac{\partial^2\phi_{\text{e,n}}}{\partial y^2}+\frac{\partial^2\phi_{\text{e,n}}}{\partial z^2}\right)$$

$$+\kappa_{\text{D,n}}^{\text{eff}}\left(\frac{\partial^2\ln C_{\text{e,n}}}{\partial x^2}+\frac{\partial^2\ln C_{\text{e,n}}}{\partial y^2}+\frac{\partial^2\ln C_{\text{e,n}}}{\partial z^2}\right)+j_{\text{n}}^{\text{Li}}=0, \qquad (8.61)$$

where an x-y plane is the unrolled electrode plane. In the quasi-3D porous electrode solver, Eq. 8.61 is reduced to the 2D equation with an additional source term, as follows:

$$\kappa_{\text{n}}^{\text{eff}}\left(\frac{\partial^2\phi_{\text{e,n}}}{\partial x^2}+\frac{\partial^2\phi_{\text{e,n}}}{\partial y^2}\right)+\kappa_{\text{D,n}}^{\text{eff}}\left(\frac{\partial^2\ln C_{\text{e,n}}}{\partial x^2}+\frac{\partial^2\ln C_{\text{e,n}}}{\partial y^2}\right)+j_{\text{n}}^{\text{Li}}-\frac{i_{\text{sepa}}}{L_{\text{n}}}=0 \quad (8.62)$$

The last term on the left-hand side represents the charge transfer from the anode to the cathode via the separator and can be written using the ESP model, as follows:

$$i_{\text{sepa}} = -\kappa_{\text{n}}^{\text{eff}} \left.\frac{\partial \phi_{\text{e,n}}}{\partial z}\right|_{z=L_{\text{n}}} - \kappa_{\text{D,n}}^{\text{eff}} \left.\frac{\partial \ln C_{\text{e,n}}}{\partial z}\right|_{z=L_{\text{n}}}$$

$$= -\kappa_{\text{n}}^{\text{eff}} \frac{3}{L_{\text{n}}}(\phi_1 - \bar{\phi}_{\text{e,n}}) - \kappa_{\text{D,n}}^{\text{eff}} \frac{\frac{3}{L_{\text{n}}}(c_1 - \bar{c}_{\text{e,n}})}{c_1}. \tag{8.63}$$

Other 3D governing equations for charge and lithium-ion transfer can be formulated in an analogous procedure.

Figure 8.11 shows comparisons between the time evolution of temperature inside an 18650-type lithium-ion cell obtained via experiment and that calculated by the two-way coupled simulation method. The lower toothed lines compare time variations of the cell potential. The applied current is 2.5 A (5C). The ambient temperature and the initial temperature of the prepared cell were set at 0°C. It should be noted that the temperature in this figure is the average value of the measured temperatures because differences at each measured point were within 1 K, as shown in Fig. 8.12. However, the maximum temperature difference was approximately 5 K due to the large heat mass at the upper side. The two-way electrochemical-thermal coupled simulation method is thus capable of simulating the electrochemical and thermal behavior, simultaneously.

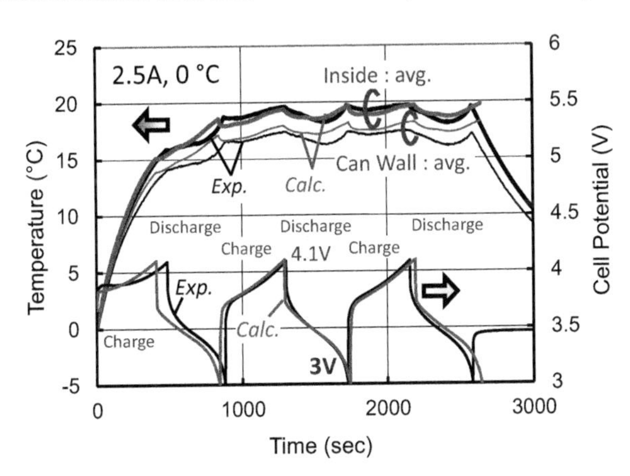

Figure 8.11 Comparison between time evolutions of temperatures for the 18650-type lithium-ion cell obtained from experiment and those predicted by the two-way coupled simulation method. The time evolutions of cell potential during the charge-discharge cycle are also shown in this figure. The applied current is 2.5 A (5C), and the ambient temperature is 0°C. Reprinted from Ref. [5], Copyright (2014), with permission from Elsevier.

Here, slightly more space will be devoted to discussing the thermal behavior based on each heat source term of the electrochemical heat release rate. The local electrochemical heat release rate consists of three heat source terms that are expressed in Eq. 8.1 and/or Eq. 8.58:

$$Q_{\text{elchem}} = Q_{\text{ohmic}} + Q_{\text{irrev}} + Q_{\text{rev}} \tag{8.64}$$

The first term on the right-hand side refers to the ohmic heat that results from ionic resistance. The second term expresses the irreversible heat that arises from overpotential and film resistance. The third term represents the reversible heat caused by the entropic effect of lithium intercalation or deintercalation [5]. These three terms are formulated as follows:

$$Q_{\text{ohmic}} = \sum_{\text{j=n,p}} \left(\sigma_{\text{j}}^{\text{eff}} \nabla \phi_{\text{s,j}} \cdot \nabla \phi_{\text{s,j}} \right) + \sum_{\text{j=n,sep,p}} \left(\kappa_{\text{j}}^{\text{eff}} \nabla \phi_{\text{e,j}} \cdot \nabla \phi_{\text{e,j}} + \kappa_{\text{D,j}}^{\text{eff}} \nabla \ln c_{\text{e,j}} \cdot \nabla \phi_{\text{e,j}} \right) \tag{8.65}$$

$$Q_{\text{irrev}} = \sum_{\text{j=n,p}} a_{\text{s,j}} \bar{i}_{\text{n,j}} \left(\bar{\phi}_{\text{s,j}} - \bar{\phi}_{\text{e,j}} - U_{\text{j}} \right) \tag{8.66}$$

$$Q_{\text{rev}} = \sum_{\text{j=n,p}} a_{\text{s,j}} \bar{i}_{\text{n,j}} T \frac{\partial U_{\text{j}}}{\partial T} \tag{8.67}$$

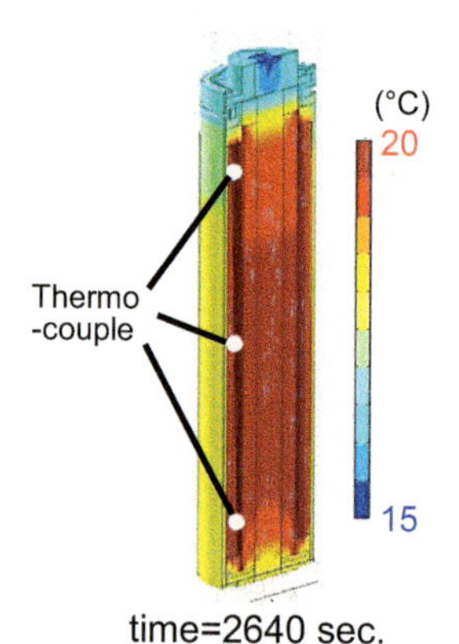

Figure 8.12 Comparison simulated temperature distribution inside the cell at the end of the charge-discharge cycle. The applied current is 2.5 A (5C), and the ambient temperature is 0°C. Reprinted from Ref. [5], Copyright (2014), with permission from Elsevier.

Figure 8.13 shows the effect of each of the three heat source terms with respect to time variations of temperature during one charge-discharge cycle at a rate of 5C (2.5 A) at 0°C. As shown in Fig. 8.13a, a small temperature fluctuation is observed inside the cell. Figures 8.13b and 8.13c indicate the time variations of the contribution ratios from each of the three heat source terms in relation to the total electrochemical heat release rate. From these figures, it is clear that the ohmic heat is similar for the discharge and charge processes. In addition, the irreversible heat is also similar for both processes and remains almost constant. A significant difference in the reversible heat is observed, which is expressed by Eq. 8.67. The time variation of the total electrochemical heat release rate is directly reflected in the thermal behavior of the cell. As a result, the small fluctuation in the temperature behavior observed is caused primarily by the reversible heat.

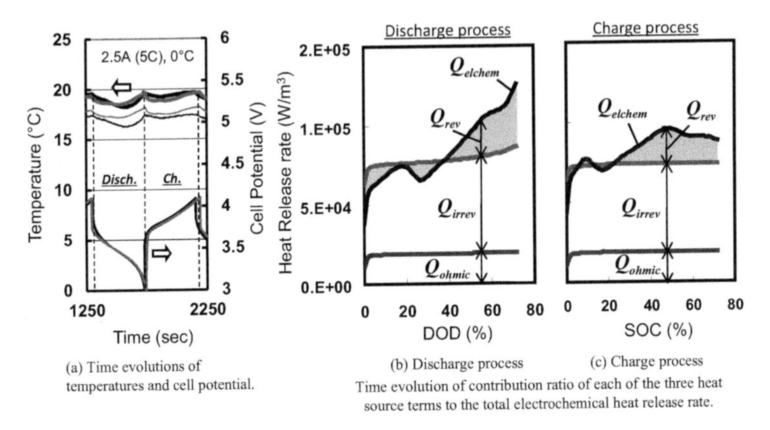

Figure 8.13 Effect of each heat source term on time variations of temperature of the cell during the charge-discharge cycle at the 5C rate (2.5 A), 0°C. The small fluctuation in the temperature behavior is mainly caused by the reversible heat resulting from an entropic effect due to lithium-ion intercalation or deintercalation. Reprinted from Ref. [5], Copyright (2014), with permission from Elsevier.

8.3 Thermal Abuse Simulation

LiBs can become more hazardous than other rechargeable batteries if mishandled. They may catch on fire and subsequently explode because chemical reactions between the electrode materials and the cell electrolyte occur at high temperatures under conditions of thermal heating, mechanical crushing, or electrical abuse. Larger cells to be carried aboard hybrid electric vehicles (HEVs) and electric vehicles (EVs) are intrinsically more vulnerable to thermal runaway due to the higher energy content. Therefore, there is a high possibility that human lives could be involved if such incidents occurred. According to statistics provided by a CPSC report [2], recalls of LiBs (18650 type) increased sharply after their capacity exceeded 2.4 Ah around 2004.

Many researchers have tried to model the reactions that occur in LiBs under abuse. Richard and Dahn proposed an abuse reaction model for lithium-intercalated graphite in an electrolyte based on accelerating rate calorimeter and differential scanning calorimeter (DSC) studies [28, 29]. In an analogous way, MacNeil and Dahn employed both calorimeters to analyze the reaction mechanism of cathode materials in an electrolyte at an elevated temperature and determined the reaction model [19–23]. Spotnitz and Franklin presented a review on the abuse behavior of LiBs and modeling work [31]. Some researchers have also presented simulation results on the thermal abuse behavior of commercial lithium-ion cells. Hatchard et al. simulated thermal abuse behavior by oven tests using a 1D model that targeted cylindrical and prismatic cells for the purposes of developing the model and testing accuracy [15]. Kim et al. presented the 3D thermal abuse model for large-format LiBs [16].

Any thermal abuse model presented in these reports uses the degree of reaction progress. However, the degree of reaction progress does not generally reflect the reaction mechanism. A thermal abuse model that uses the degree of reaction progress thus estimates the heat generation rate, regardless of the material amounts related to the reaction. In extreme cases, these models estimate a certain heat generation rate, even where no reactants exist. Accordingly, the conventional thermal abuse model could fail to estimate the heat generation rate when simulating the thermal abuse behavior of an actual device.

In this section, another approach to thermal abuse modeling, which concerns the reaction mechanism and formulates the heat generation rate on the basis of the reaction scheme, is proposed.

8.3.1 Thermal Abuse Modeling

The modeling approach of the thermal abuse behavior simulation is described in Fig. 8.14. First of all, DSC profiles of the anode and cathode materials in an electrolyte are measured individually. The thermal abuse reaction model is then established to ensure that it is available for simulation of the measured DSC profiles as accurately as possible. This model is implemented as a submodel in the quasi-3D porous electrode solver, which constitutes the two-way electrochemical-thermal coupled simulation method.

The basic concept of the modeling is as follows: A thermal abuse reaction model should be based on the reaction mechanism caused by cathode and anode active materials. The reaction mechanism is expressed by a simple overall reaction scheme. Key species that dominate the thermal abuse phenomena are identified from various analytical results, and the overall reaction rates that express the production and/or consumption rates of the key species are determined so that the DSC profiles can be simulated.

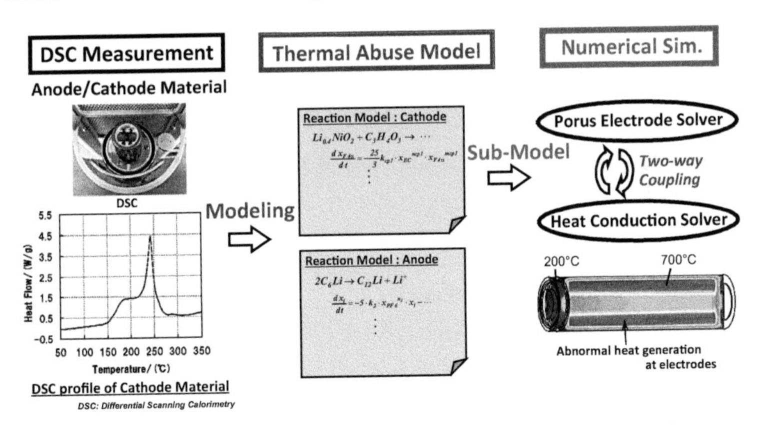

Figure 8.14 Modeling approach of the thermal abuse behavior simulation.

The following serves as one example of a thermal abuse reaction model for a charged cathode. Figure 8.15 shows the result for a DSC measurement of a sample of $Li_{0.4}NiO_2$ in a nonaqueous electrolyte. Two exothermic peaks are clearly observed in this profile, which suggests that two reactions are dominant for this thermal behavior. The work of Arai et al. [3] suggests that for delithiated nickelate, oxygen released by the decomposition of Li_xNiO_2 at an elevated temperature contributes to the combustion of the solvent. The following reaction mechanism is assumed to simulate the DSC profile. The first exothermic peak, around 180°C, is assumed to be due to the reaction whereby oxygen is released by the conformational change of $Li_{0.4}NiO_2$ to the rock-salt structure $Li_{2/7}Ni_{5/7}O$ and reacts with the solvent. The reaction equation is given by

$$\frac{25}{3}Li_{0.4}NiO_2 + C_3H_4O_3 \rightarrow \frac{25}{3}\left(\frac{7}{5}Li_{\frac{2}{7}}Ni_{\frac{5}{7}}O\right) + 3CO_2 + 2H_2O. \quad (8.68)$$

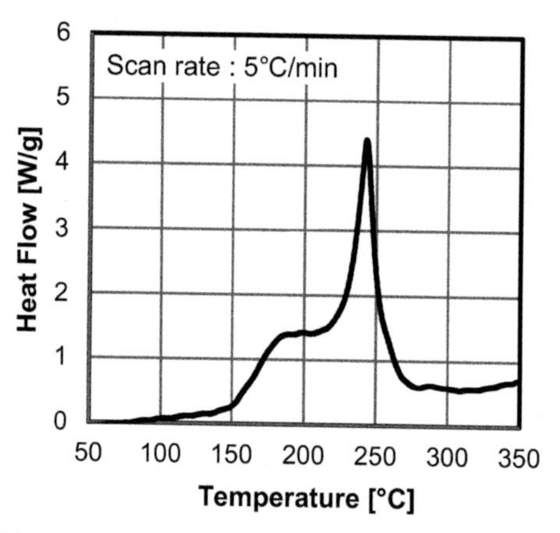

Figure 8.15 DSC measurement result on a sample of $Li_{0.4}NiO_2$ in a nonaqueous electrolyte.

The second exothermic peak, around 250°C, is assumed to be due to the reaction whereby more oxygen is released by the decomposition of rock salt $Li_{2/7}Ni_{5/7}O$ to NiO and reaction with the solvent. The reaction equation is assumed to be

$$25\left(\frac{7}{5}Li_{\frac{2}{7}}Ni_{\frac{5}{7}}O\right) + C_3H_4O_3 \rightarrow 5Li_2O + 25NiO + 3CO_2 + 2H_2O. \quad (8.69)$$

The production and consumption rates of the key species considered in these two reactions are given as follows:

$$\frac{dx_\alpha}{dt} = -\frac{25}{3}k_{cp1} \cdot x_{EC}{}^{ncp1} \cdot x_\alpha{}^{mcp1}, \quad (8.70)$$

$$\frac{dx_\beta}{dt} = \frac{25}{3}k_{cp1} \cdot x_{EC}{}^{ncp1} \cdot x_\alpha{}^{mcp1} - 25k_{cp2} \cdot x_{EC}{}^{ncp2} \cdot x_\beta{}^{mcp2}, \quad (8.71)$$

and

$$\frac{dx_\gamma}{dt} = 25k_{cp2} \cdot x_{EC}{}^{ncp2} \cdot x_\beta{}^{mcp2}, \tag{8.72}$$

where x_α, x_β, x_γ, and x_{EC} are the amounts of $Li_{0.4}NiO_2$, rock salt $Li_{2/7}Ni_{5/7}O$, NiO, and EC solvent in the DSC sample, respectively. k_{cp1} and k_{cp2} are rate constants for reactions 8.68 and 8.69, respectively, and follow an Arrhenius-type expression. Here, the rate of EC solvent consumption is expressed as

$$\frac{dx_{EC}}{dt} = -k_{cp1} \cdot x_{EC}{}^{ncp1} \cdot x_\alpha{}^{mcp1} - k_{cp2} \cdot x_{EC}{}^{ncp2} \cdot x_\beta{}^{mcp2}. \tag{8.73}$$

On the basis of the discussions of the reaction mechanism, the key species related to the thermal abuse reaction model of a charged cathode are determined to be oxygen released from the active material and the solvent molecules.

Next, a thermal abuse reaction model for Li-intercalated graphite carbon with an electrolyte is described. Figure 8.16 shows a measured DSC profile for a sample of C_6Li in an electrolyte, in which four exothermic regions are evident.

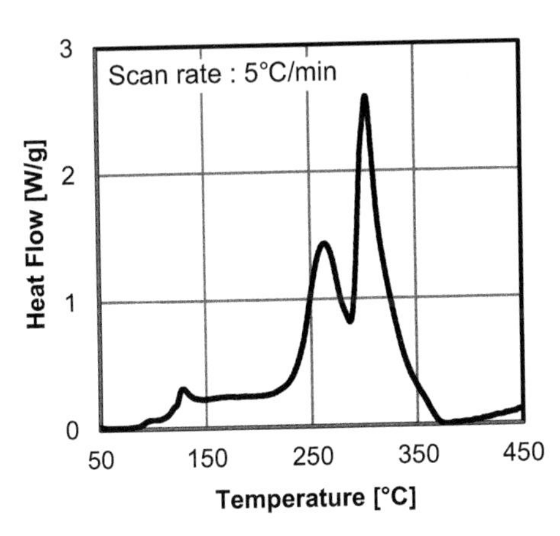

Figure 8.16 DSC measurement result on a sample of C_6Li in a nonaqueous electrolyte.

Richard and Dahn proposed a reaction model that results in the first exothermic peak, at around 120°C, and a subsequent prolonged exothermic phenomenon [28, 29]. They identified the first exothermic peak, at around 120°C as due to the decomposition of the SEI layer. The SEI layer is assumed to consist of stable (such as Li_2CO_3) and metastable components, such as $(CH_2OCO_2Li)_2$. The conversion of a metastable SEI to a stable SEI is responsible for the first peak. The depletion rate of the metastable SEI component is given by

$$\frac{dx_f}{dt} = -k_{ap1} \cdot x_f, \tag{8.74}$$

where x_f is the dimensionless amount of metastable SEI and k_{ap1} is the rate constant expressed as Arrhenius type.

The subsequent prolonged heat generation at 150°C–220°C is assumed to be due to the following mechanism: The prolonged heat must come from the reaction of intercalated Li with the solvent to form a new SEI. The lithium and solvent can react if Li ions diffuse and the corresponding electrons tunnel through the SEI. Accordingly, this is dependent on the transport of Li through the SEI layer. The consumption rate of intercalated Li caused by this reaction mechanism is limited by Li-ion diffusion and is expressed as:

$$\frac{dx_i}{dt} = -2\frac{k_{ap4} \cdot k_{dif}}{k_{ap4} + k_{dif}} \cdot x_{EC}^{\frac{1}{2}} \cdot x_i, \tag{8.75}$$

where x_i is the amount of Li in the active material, k_{ap4} is the rate constant, and k_{dif} is the diffusion rate constant.

At approximately 260°C in the DSC profile, the decomposition of PF_6^- and simultaneous reaction with the deintercalated Li from the active materials results in the second intensive peak. The reaction mechanism is assumed to be as follows:

$$2C_6Li \rightarrow C_{12}Li + Li^+ \tag{8.76}$$

$$PF_6^- \rightarrow PF_5 + F^- \tag{8.77}$$

$$5Li^+ + PF_5 \rightarrow P + 5LiF \tag{8.78}$$

The consumption rate of intercalated Li caused by this reaction mechanism is given by

$$\frac{dx_i}{dt} = -5k_{ap2} \cdot x_{PF6}^{\frac{1}{5}} \cdot x_i, \tag{8.79}$$

where x_{PF6} is the amount of PF_6^- in the electrolyte and k_{ap2} is the rate constant.

The third intensive heat generation, around 300°C, must be due to the reaction of the solvent with the deintercalated Li from the active materials. The reaction mechanism is assumed to be

$$C_{12}Li \rightarrow 2C_6 + Li^+ \tag{8.80}$$

and

$$2Li^+ + C_3H_4O_3 \rightarrow Li_2CO_3 + C_2H_4. \tag{8.81}$$

The consumption rate of intercalated Li caused by this reaction mechanism is expressed as

$$\frac{dx_i}{dt} = -2k_{ap3} \cdot x_{EC}^{\frac{1}{2}} \cdot x_i, \tag{8.82}$$

where k_{ap3} is the rate constant.

For the equations presented above, the total consumption rate of Li in the anode active materials during DSC measurement can be described by the following equation:

$$\frac{dx_i}{dt} = -2\frac{k_{ap4} \cdot k_{dif}}{k_{ap4} + k_{dif}} \cdot x_{EC}^{\frac{1}{2}} \cdot x_i - 5k_{ap2} \cdot x_{PF6}^{\frac{1}{5}} \cdot x_i - 2k_{ap3} \cdot x_{EC}^{\frac{1}{2}} \cdot x_i \tag{8.83}$$

The depletion rates of the PF_6^- lithium salt and the solvent are expressed, respectively, as follows:

$$\frac{dx_{PF6}}{dt} = -k_{ap2} \cdot x_{PF6}^{\frac{1}{5}} \cdot x_i \tag{8.84}$$

$$\frac{dx_{EC}}{dt} = -\frac{k_{ap4} \cdot k_{dif}}{k_{ap4} + k_{dif}} \cdot x_{EC}^{\frac{1}{2}} \cdot x_i - k_{ap3} \cdot x_{EC}^{\frac{1}{2}} \cdot x_i \tag{8.85}$$

Accordingly, the key species related to the thermal abuse reaction model of Li-intercalated graphite carbon are determined to be Li in the anode active materials, the PF_6^- lithium salt, and the solvent molecules.

Finally, the thermal abuse reaction model of a charged cell, which is simply modeled as a combination of the charged cathode model and the C_6Li model, is shown in Fig. 8.17. In this model, the thermal abuse anode–cathode interaction is determined by the depletion process of solvent molecules.

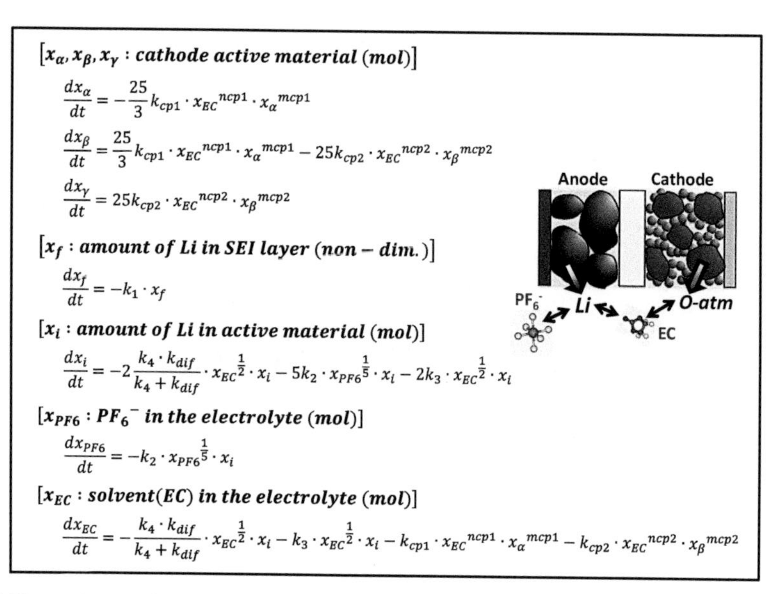

Figure 8.17 Thermal abuse reaction model for a charged cell.

8.3.2 Model Validation

Figure 8.18 (for a charged cathode) and Fig. 8.19 (for a C_6Li anode) show a comparison of the measured and simulated DSC profiles. Both figures suggest that the qualitative exothermic behavior can be simulated by the above models.

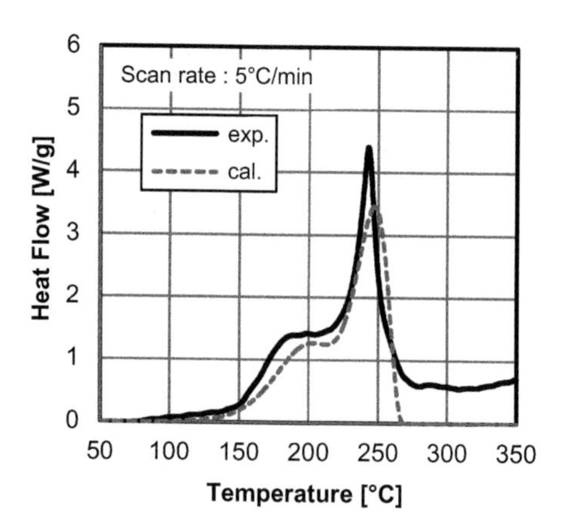

Figure 8.18 Comparison between measured and simulated DSC profiles on a sample of $Li_{0.4}NiO_2$ in a nonaqueous electrolyte.

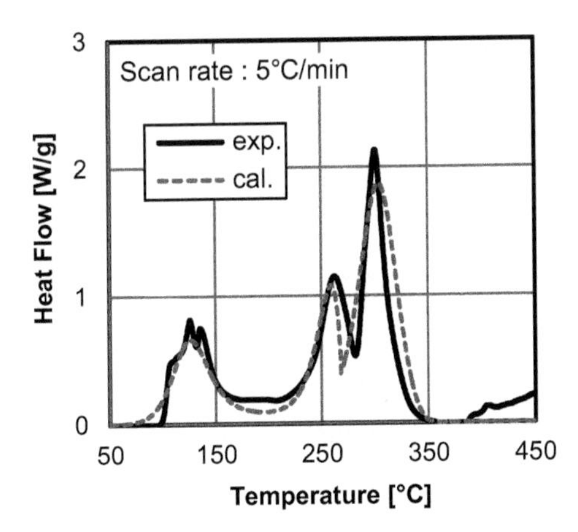

Figure 8.19 Comparison between measured and simulated DSC profiles on a sample of $Li_{0.4}NiO_2$ in a nonaqueous electrolyte.

Thermal abuse behavior was studied by a heating test of an in-house manufactured 18650-type lithium-ion cell with a capacity of 0.75 Ah. Figure 8.20 shows temperature histories of the can wall

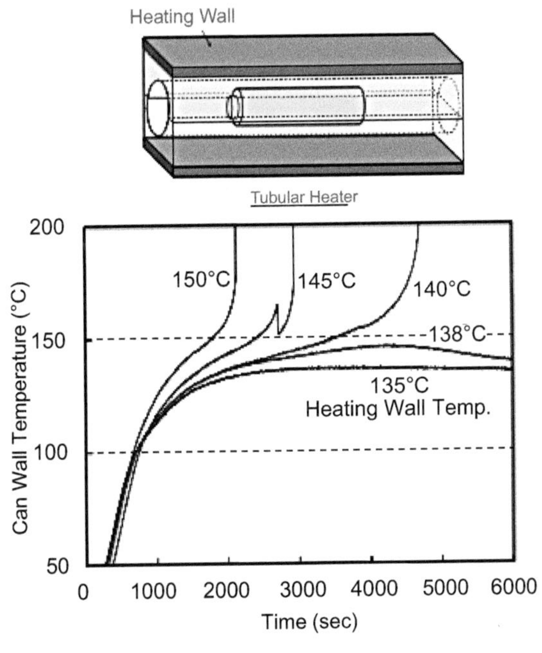

Figure 8.20 Thermal abuse behaviors during heating tests for a charged cell.

during heating tests as a function of the heating wall temperature setting. Thermal runaway here is caused by a 5°C temperature difference (135°C–140°C). Thermal abuse behavior that was simulated using the proposed thermal abuse reaction model is compared with experimental results in Fig. 8.21. The threshold

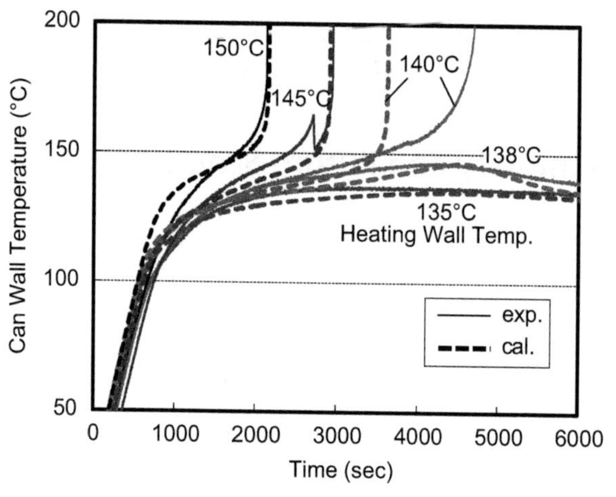

Figure 8.21 Comparison between measured and simulated thermal abuse behaviors during heating tests for a charged cell. The threshold temperature for thermal abuse is estimated with an accuracy of 5°C. Reprinted from Ref. [4], with permission from The Electrochemical Society of Japan.

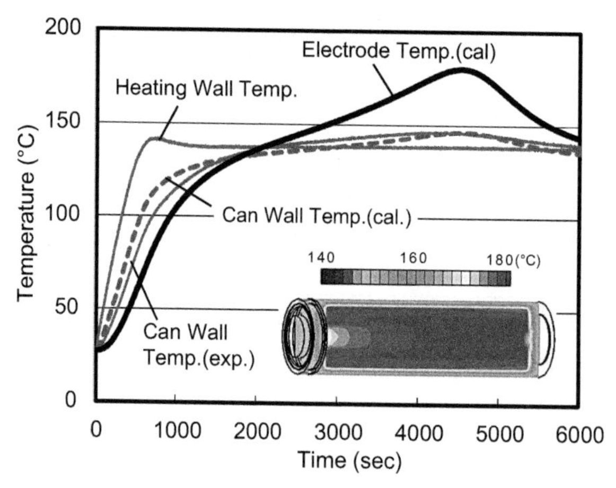

Figure 8.22 Simulated temperature history inside the cell during the heating test when the heating wall temperature is 138°C.

temperature for thermal abuse is estimated with an accuracy of 5°C. In the event of heating wall temperature 138°C, the thermal abuse phenomenon is suppressed because the heat loss from the can wall is greater than the heat generation caused by the early-stage thermal abuse reaction. Figure 8.22 shows the simulated temperature history inside the cell during the heating test with a wall temperature of 138°C. The maximum temperature inside the cell reached over 180°C at the electrode.

8.4 Summary and Future Scope

In this chapter, simulation techniques for LiB devices under typical use and thermal abuse conditions were presented. Particularly with regard to the thermal abuse model, the proposed model could be used to simulate the thermal behavior during heating. In addition, the proposed model could be used to simulate exothermic behavior after abnormal accidents, such as overcharging, short-circuiting, and nail penetration. Overcharge modeling is the most urgent requirement for extending the range of applications of thermal abuse model.

List of symbols

a_s	Specific interfacial area of an electrode (m^2/m^3)
c_e	Li^+ concentration in the electrolyte phase (mol/m^3)
c_s	Li concentration in the solid particles (mol/m^3)
c_{se}	Li concentration at the surface of the solid particles (mol/m^3)
D_e	Diffusion coefficient of Li^+ in the electrolyte phase (m^2/s)
D_s	Diffusion coefficient of Li^+ in the solid phase (m^2/s)
F	Faraday's constant (96,487 C/mol)
I_{app}	Applied current density (A/m^2)
i_0	Exchange current density (A/m^2)
$\bar{i}_n$	Superficial current density (A/m^2)
j	Local volumetric transfer current density due to charge transfer (A/m^3)
l_{se}	Diffusion length of Li^+ from solid/electrolyte interphase into solid phase (m)

L Thickness of n, sep, or p (m)
q Heat generation rate (W/m^3)
R Universal gas constant (8.3143 J/(mol·K))
R_f Film resistance on an electrode surface (Ω·m^2)
r Radial coordinate (m)
r_s Radius of the spherical particles (m)
T Absolute temperature (K)
t Time (s)
t_+° Transference number of Li^+ in solution
U Open-circuit potential of an electrode reaction (V)
x, y Coordinates on the unrolled electrode plane (m)
z Coordinate across the thickness of the electrode (m)

Greek symbols

α_a, α_c Anodic and cathodic transfer coefficients for an electrode reaction
ε Volume fraction of a phase
η Surface overpotential of an electrode reaction (V)
κ Conductivity of the electrolyte (S/m)
κ_D Diffusional conductivity of the electrolyte (A/m)
σ Conductivity of the electrode (S/m)
ϕ_s Electrical potential in the solid phase (V)
ϕ_e Electrical potential in the electrolyte phase (V)

Subscript and superscript

eff Effective
j = n or p
Li Lithium species
n Negative electrode
p Positive electrode
sep Separator

References

1. AVL FIRE™ user manual, AVL List GmbH. 2009.
2. Consumer Product Safety Commission (CPSC) Report. Dec. 2006.
3. Arai, H., Tsuda, M., Saito, K., Hayashi, M., and Sakurai, Y. (2002). Thermal reactions between delithiated lithium nickelate and electrolyte solutions, *J. Electrochem. Soc.*, **149**, pp. A401–A406.
4. Baba, N., Okuda, C., Inoue, T., Makimura, Y., Yoshida, H., and Nagaoka, M. (2012). Numerical simulation for thermal abuse behavior of lithium ion secondary batteries, *The 53rd Battery Symposium in Japan Proceedings*, **1A31**.
5. Baba, N., Yoshida, H., Nagaoka, M., Okuda, C., and Kawauchi, S. (2014). Numerical simulation of thermal behavior of lithium-ion secondary batteries using the enhanced single particle model, *J. Power Sources*, **252**, pp. 214–228.
6. Bernardi, D., Pawlikowski, E., and Newman, J. (1985). A general energy-balance for battery systems, *J. Electrochem. Soc.*, **132**, pp. 5–12.
7. Chen, Y. and Evans, J. W. (1993). Heat-transfer phenomena in lithium polymer-electrolyte batteries for electric vehicle application, *J. Electrochem. Soc.*, **140**, pp. 1833–1838.
8. Chen, Y. F. and Evans, J. W. (1994). 3-Dimensional thermal modeling of lithium-polymer batteries under galvanostatic discharge and dynamic power profile, *J. Electrochem. Soc.*, **141**, pp. 2947–2955.
9. Chen, Y. F. and Evans, J. W. (1996). Thermal analysis of lithium-ion batteries, *J. Electrochem. Soc.*, **143**, pp. 2708–2712.
10. Doyle, M., Fuller, T. F., and Newman, J. (1993). Modeling of galvanostatic charge and discharge of the lithium polymer insertion cell, *J. Electrochem. Soc.*, **140**, pp. 1526–1533.
11. Doyle, M., Newman, J., Gozdz, A. S., Schmutz, C. N., and Tarascon, J. M. (1996). Comparison of modeling predictions with experimental data from plastic lithium ion cells, *J. Electrochem. Soc.*, **143**, pp. 1890–1903.
12. Gu, W. B. and Wang, C. Y. (2000). Thermal and electrochemical coupled modeling of a lithium-ion cell in lithium batteries, *ECS Proc.*, **99-25**, pp. 748–762.
13. Gu, W. B. and Wang, C. Y. (2000). Thermal-electrochemical modeling of battery systems, *J. Electrochem. Soc.*, **147**, pp. 2910–2922.
14. Haran, B. S., Popov, B. N., and White, R. E. (1998). Determination of the hydrogen diffusion coefficient in metal hydrides by impedance spectroscopy, *J. Power Sources*, **75**, pp. 56–63.

15. Hatchard, T. D., MacNeil, D. D., Basu, A., and Dahn, J. R. (2001). Thermal model of cylindrical and prismatic lithium-ion cells, *J. Electrochem. Soc.*, **148**, pp. A755–A761.
16. Kim, G. H., Pesaran, A., and Spotnitz, R. (2007). A three-dimensional thermal abuse model for lithium-ion cells, *J. Power Sources*, **170**, pp. 476–489.
17. Kim, U. S., Shin, C. B., and Kim, C. S. (2008). Effect of electrode configuration on the thermal behavior of a lithium-polymer battery, *J. Power Sources*, **180**, pp. 909–916.
18. Kim, U. S., Shin, C. B., and Kim, C. S. (2009). Modeling for the scale-up of a lithium-ion polymer battery, *J. Power Sources*, **189**, pp. 841–846.
19. MacNeil, D. D., Christensen, L., Landucci, J., Paulsen, J. M., and Dahn, J. R. (2000). An autocatalytic mechanism for the reaction of Li_xCoO_2 in electrolyte at elevated temperature, *J. Electrochem. Soc.*, **147**, pp. 970–979.
20. MacNeil, D. D. and Dahn, J. R. (2001). The reaction of charged cathodes with nonaqueous solvents and electrolytes: II. $LiMn_2O_4$ charged to 4.2 V, *J. Electrochem. Soc.*, **148**, pp. A1211–A1215.
21. MacNeil, D. D. and Dahn, J. R. (2001). The reaction of charged cathodes with nonaqueous solvents and electrolytes: I. $Li_{0.5}CoO_2$, *J. Electrochem. Soc.*, **148**, pp. A1205–A1210.
22. MacNeil, D. D. and Dahn, J. R. (2001). Test of reaction kinetics using both differential scanning and accelerating rate calorimetries as applied to the reaction of Li_xCoO_2 in non-aqueous electrolyte, *J. Phys. Chem. A*, **105**, pp. 4430–4439.
23. MacNeil, D. D. and Dahn, J. R. (2002). The reactions of $Li_{0.5}CoO_2$ with nonaqueous solvents at elevated temperatures, *J. Electrochem. Soc.*, **149**, pp. A912–A919.
24. Ning, G. and Popov, B. N. (2004). Cycle life modeling of lithium-ion batteries, *J. Electrochem. Soc.*, **151**, pp. A1584–A1591.
25. Pals, C. R. and Newman, J. (1995). Thermal modeling of the lithium/polymer battery. 1. Discharge behavior of a single-cell, *J. Electrochem. Soc.*, **142**, pp. 3274–3281.
26. Pals, C. R. and Newman, J. (1995). Thermal modeling of the lithium/polymer battery. 2. Temperature profiles in a cell stack, *J. Electrochem. Soc.*, **142**, pp. 3282–3288.
27. Rao, L. and Newman, J. (1997). Heat-generation rate and general energy balance for insertion battery systems, *J. Electrochem. Soc.*, **144**, pp. 2697–2704.

28. Richard, M. N. and Dahn, J. R. (1999). Accelerating rate calorimetry study on the thermal stability of lithium intercalated graphite in electrolyte I. Experimental, *J. Electrochem. Soc.*, **146**, pp. 2068–2077.

29. Richard, M. N. and Dahn, J. R. (1999). Accelerating rate calorimetry study on the thermal stability of lithium intercalated graphite in electrolyte II. Modeling the results and predicting differential scanning calorimeter curves, *J. Electrochem. Soc.*, **146**, pp. 2078–2084.

30. Santhanagopalan, S., Guo, Q. Z., Ramadass, P., and White, R. E. (2006). Review of models for predicting the cycling performance of lithium ion batteries, *J. Power Sources*, **156**, pp. 620–628.

31. Spotnitz, R. and Franklin, J. (2003). Abuse behavior of high-power, lithium-ion cells, *J. Power Sources*, **113**, pp. 81–100.

32. Subramanian, V. R., Ritter, J. A., and White, R. E. (2001). Approximate solutions for galvanostatic discharge of spherical particles: I. Constant diffusion coefficient, *J. Electrochem. Soc.*, **148**, pp. E444–E449.

33. Wang, C. Y., Gu, W. B., and Liaw, B. Y. (1998). Micro-macroscopic coupled modeling of batteries and fuel cells: I. Model development, *J. Electrochem. Soc.*, **145**, pp. 3407–3417.

34. Wang, C. Y. and Srinivasan, V. (2002). Computational battery dynamics (CBD): electrochemical/thermal coupled modeling and multi-scale modeling, *J. Power Sources*, **110**, pp. 364–376.

35. Zhang, Q. and White, R. E. (2007). Comparison of approximate solution methods for the solid phase diffusion equation in a porous electrode model, *J. Power Sources*, **165**, pp. 880–886.

36. Zhang, Q. and White, R. E. (2008). Capacity fade analysis of a lithium ion cell, *J. Power Sources*, **179**, pp. 793–798.

Chapter 9

Device Simulations in Fuel Cells

Takahisa Suzuki, Kenji Kudo, Ryosuke Jinnouchi, and Yu Morimoto
Toyota Central R&D Laboratories, Inc., Nagakute, Aichi 480-1192, Japan
takahisa@mosk.tytlabs.co.jp

9.1 Introduction

Polymer electrolyte fuel cells (PEFCs) are promising energy-conversion devices for automotive applications and stationary power sources. The largest obstacle to their commercialization is the high cost. The most costly component is the electrode catalyst, which usually consists of platinum group metals (PGMs). Reducing the amount of PGMs used in the devices is one of the main issues in the development of PEFC systems. The United States Department of Energy set a technical total loading target of 0.125 mg PGM/cm^2 electrode area by 2020 for use in light-duty transportation applications.[a]

A reduction in the amount of catalyst required, especially in the cathode, can cause performance loss. In 1992, Wilson and Gottesfeld

[a]https://energy.gov/eere/fuelcells/doe-technical-targets-polymer-electrolyte-membrane-fuel-cell-components

Multiscale Simulations for Electrochemical Devices
Edited by Ryoji Asahi

ISBN 978-981-4800-71-6 (Hardcover), 978-0-429-29545-4 (eBook)
www.jennystanford.com

fabricated membrane electrode assemblies (MEAs) with cathode platinum loadings of 0.07, 0.12, and 0.17 mg·cm^{-2} and compared the resulting cell performances [39]. They reported that the MEAs with 0.12 and 0.17 mg·cm^{-2} loadings yielded nearly equivalent performances but the device with 0.07 mg·cm^{-2} Pt loading showed a significantly lower performance. A systematic analysis revealed that the additional performance loss resulting from the loading reduction of the catalyst layer did not stem from the molecular diffusion through the pores in the electrode [24, 25] or activation loss due to the Pt surface oxides [34]. This performance loss was repeatedly verified, but the underlying physical mechanisms were not clarified [9, 26].

Experimental studies of PEFCs have been aided by modeling studies and vice versa. Performance models usually incorporate electrochemical reactions in the electrodes and species transport in the cell. These phenomena are interrelated, and understanding the effect of each phenomenon on the overall performance of the devices is not a trivial issue. Experiments must be properly designed to identify the origin of the performance loss, which can then be analyzed by a tuned model that is customized to the individual experiments, as follows.

Identifying the various origins of the cell potential loss is the first step in enhancing PEFC performance. Activation overpotential can be described by the Tafel [36] or Butler–Volmer equations [3], and corresponding parameters can be measured using microelectrodes [27–29]. The specific activity factor can be modified by the coverage of Pt oxides, which inhibits the oxygen reduction reaction (ORR) [8, 34, 37]. Ohmic loss originates from the resistance of the polymer electrolyte, which increases with decreasing water content [32, 43]. The water content at equilibrium increases with increasing relative humidity [11, 44]. The water content in the membrane depends on the operating current density and is significantly influenced by electroosmotic drag and diffusion. The transport parameters were measured as a function of water content [21, 43, 45]. A 1D transmission line model was used to evaluate the effective ionic conductivity of the catalyst layer [17–20]. The mass transport loss likely originated from diffusion through the gas diffusion layer (GDL) [3, 36] and is significant when the GDL is flooded with liquid water [6, 23, 30].

Mathematical MEA models have been used to quantify the overall performance as a function of the parameters governing each loss. The water balance in a cell was initially described using a 1D through-plane model [2] as well as 2D through-plane and along-the-channel MEA models [6]. The water content distribution in the through-plane direction of the membrane was calculated using the measured transport parameters [1]. Transport resistance in the GDL was investigated using models that calculated mass transport in the through-plane and across-channel directions [22, 41, 42]. Full 3D models have been used for comprehensive analyses of the species, reaction, and temperature distributions in the devices [4, 38]. Among these models, through-plane macrohomogeneous models have been successfully applied as elementary tools to investigate transport resistance. These models are supported by first-principles simulation because they include phenomenological parameters.

This chapter discusses the contribution of the modeling activities to understanding the origin of the performance loss accompanying catalyst loading reduction. The discussion is devoted to oxygen transport resistance at the interface between the catalyst surface and the covering ionomer in the catalyst layer using both macrohomogeneous models and molecular dynamics (MD) simulations.

9.2 The Conventional Model

9.2.1 Reactions and Structure of PEFCs

Basically, a PEFC consists of two electrodes separated by a polymer electrolyte membrane. Proton exchange membranes are widely used as membranes, and the discussion is focused on acidic membranes in this chapter. On the anode side, the hydrogen oxidation reaction (HOR) occurs and protons and electrons are produced:

$$2H_2 \rightarrow 4H^+ + 4e^- \quad (9.1)$$

Protons move through the membrane, and electrons move through the external circuit to the cathode. On the cathode side, the ORR occurs and water is generated:

$$O_2 + 4H^+ + 4e^- \rightarrow 2H_2O \quad (9.2)$$

Figure 9.1 shows a schematic of a PEFC. The MEA (Fig. 9.1h) consists of a polymer electrolyte membrane (Fig. 9.1e), catalyst layers (Fig. 9.1d,f) on both sides of the membrane, and GDLs (Fig. 9.1g). The MEA is placed between the current collectors (Fig. 9.1a) on which flow fields are formed. A GDL is a sheet composed of carbon microfibers hydrophobized with polytetrafluoroethylene (PTFE). It often features another layer on one side, called a microporous layer (MPL, Fig. 9.1c) composed of carbon black and PTFE. The MPL side of the GDL is laminated to the catalyst layer, and when a GDL features an MPL, the rest of the MPL is called a GDL substrate (Fig. 9.1b).

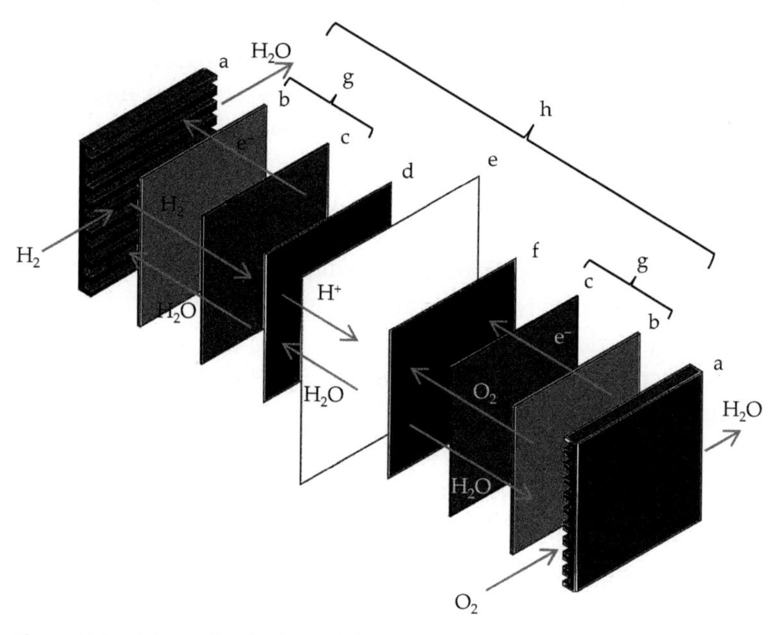

Figure 9.1 Schematic of a PEFC. (a) Current collector with flow fields; (b) GDL substrate; (c) microporous layer; (d) anode catalyst layer; (e) polymer electrolyte membrane; (f) cathode catalyst layer; (g) GDL; (h) MEA.

The membrane is typically composed of a fluorinated sulfonic acid ionomer that can conduct protons. The catalyst layer is a porous electrode usually composed of catalyst-loaded conductive carbon black covered with an ionomer. The ionomer in the catalyst layer forms continuous paths for protons, and the carbon black nanoparticles form continuous paths for electrons. Micrographs of the MEA components are shown in Fig. 9.2. Figure 9.2a shows a close-up view of the catalyst particles in the catalyst layer. The ionomer covering the catalyst particle is approximately 10 nm thick. Figure 9.2b shows a cross section of the catalyst layer where the size of the primary carbon black particles is approximately 50 nm. The pore size in the catalyst layer is of the same order as the primary particle size or larger. Figures 9.2c and 9.2d show the surface of an MPL and a GDL substrate, respectively. The pore size of the MPL is of the same order as the primary carbon black particle size, and that of the GDL substrate is several tens of micrometers or larger.

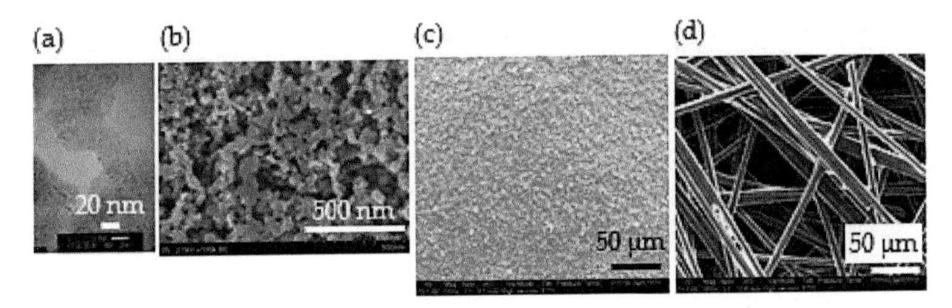

Figure 9.2 Micrographs of MEA components. (a) A transmission electron micrograph of a catalyst layer (Akimoto, Y., unpublished). (b) A scanning electron micrograph cross section of a catalyst layer fabricated by focused ion beam milling. Reprinted from Ref. [35], Copyright (2013), with permission of Elsevier. (c) A scanning electron micrograph of the surface of an MPL (Tsusaka, K., unpublished). (d) A scanning electron micrograph of the surface of a GDL substrate (Tsusaka, K., unpublished).

9.2.2 Mass Transport Resistance on the Cathode Side

A decreased oxygen concentration on the cathode catalyst surface is a more serious issue than a decreased hydrogen concentration on the anode catalyst surface because the ORR rate constant is much smaller than the HOR rate constant. Systems that use hydrogen and air as reactants often show oxygen transport limitation because the oxygen is diluted by nitrogen in the atmosphere to a final concentration of 21% to begin with. The oxygen concentration decreases along the path from the flow field to the catalyst surface because the driving force for oxygen transport is mainly diffusion—the concentration decreases in the direction of the diffusive flux. This scenario is schematically shown in Fig. 9.3a. In the GDL and catalyst layer pores, molecular and/or Knudsen diffusion controls the flux. In the catalyst layer, oxygen dissolves into the ionomer and diffuses toward the catalyst surface.

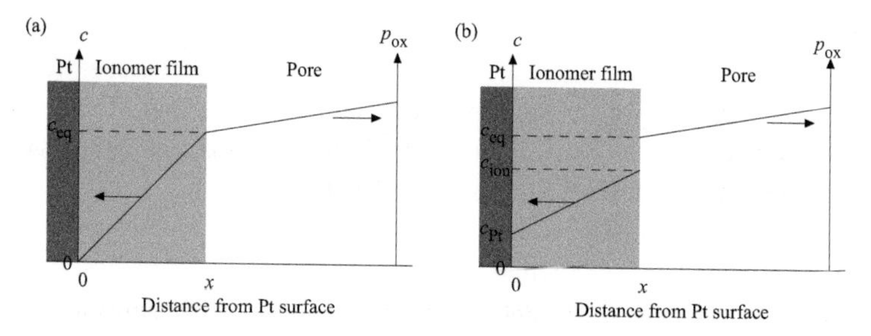

Figure 9.3 Schematic illustration of the oxygen concentration profile when limiting current density is observed. The partial pressure of O_2, p_{ox}, is a converted value that is equivalent to the concentration in the ionomer, *c*. The oxygen diffuses from right (flow field/GDL interface) to left (Pt surface). (a) The conventional model in which equilibrium is attained at the Pt/ionomer as well as the ionomer/pore interface. (b) The interface resistance model in which a concentration jump takes place at the Pt/ionomer and ionomer/pore interfaces.

When this simple picture accurately describes the oxygen transport, a constant slope should be observed in a Tafel plot (cathode overpotential vs. log(current density)) in the relatively low current density region before the influence of mass transport limitation is observed. In reality, the slope (referred to as the Tafel slope) increases with current density and doubles at larger current densities before the influence of limiting current density is observed. This discrepancy was first explained by the activation loss caused by surface oxides on the Pt catalyst [37] and later explained by mass transport loss caused by agglomeration, as observed by scanning electron microscopy [5].

9.2.3 The Agglomerate Model

The conventional agglomerate model is schematically illustrated in Fig. 9.4a and was developed from micrographs of the catalyst layer [10, 12, 13]. The conventional model assumes that the catalyst layer comprises spherical agglomerates of radius R, composed of catalyst-loaded carbon black particles and an ionomer, as shown in Fig. 9.4b. The ionomer fills the interspace among the carbon black particles, and the model assumes that this region is homogeneous, although it is actually microscopically inhomogeneous. The agglomerate may be covered with an ionomer film of thickness δ. Oxygen must diffuse from the outside of the agglomerate to be consumed within the structure.

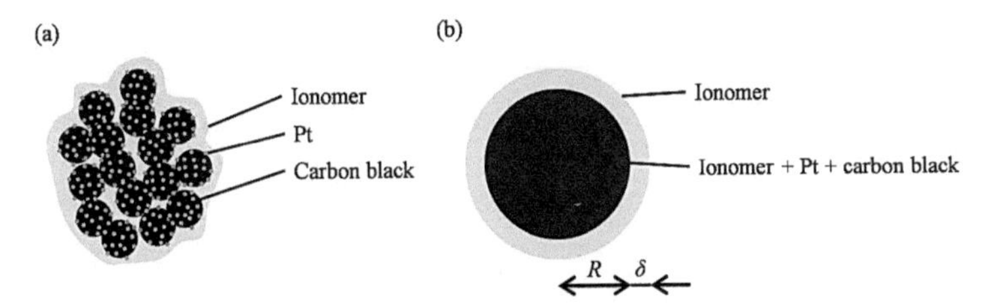

Figure 9.4 Agglomerate model. (a) Schematic illustration of a structure in the catalyst layer. (b) The agglomerate assumed in the model.

If oxygen diffusion obeys Fick's law and the ORR rate is described by the Tafel equation, then, at steady state, the mass conservation equation for oxygen is as follows:

$$-D\frac{\mathrm{d}^2c}{\mathrm{d}r^2}-D\frac{2}{r}\frac{\mathrm{d}c}{\mathrm{d}r}=-\frac{1}{4F}ai_0\frac{c}{c_0}\exp\left(-\frac{\alpha F}{\mathfrak{R}T}\eta\right) \tag{9.3}$$

for $0\le r\le R$ and

$$\frac{\mathrm{d}^2c}{\mathrm{d}r^2}+\frac{2}{r}\frac{\mathrm{d}c}{\mathrm{d}r}=0 \tag{9.4}$$

for $R \le r \le R + \delta$, where r is the distance from the center of the agglomerate to the edge, D is the diffusion coefficient of oxygen at $0 \le r \le R$, c is the oxygen concentration, F is Faraday's constant, a is the surface area of the catalyst in the unit volume of the agglomerate where $r \le R$, i_0 is the reference current density, c_0 is the reference concentration, α is the apparent transfer coefficient, $\mathfrak{R}$ is the universal gas constant, T is the temperature, and η is the overpotential.[b]

Assuming that the overpotential η is independent of the radius r, the solutions of Eqs. 9.3 and 9.4 are

$$c=\frac{\sinh\frac{r}{r_0}\Big/\sinh\frac{R}{r_0}}{r/r_0}c\Big|_{r=R} \tag{9.5}$$

and

$$c=c\Big|_{r=R}+\left[\frac{1}{1-\frac{R}{R+\delta}}-\frac{1}{\left(\frac{1}{R}-\frac{1}{R+\delta}\right)r}\right]\left(c\Big|_{r=R+\delta}-c\Big|_{r=R}\right), \tag{9.6}$$

[b]The diffusion coefficient is not included in the conservation equation for $R\le r\le R+\delta$ because there is no source term.

where

$$\frac{1}{r_0}=\sqrt{\frac{1}{c_0 D}\frac{1}{4F}ai_0\exp\left(-\frac{\alpha F}{\mathfrak{R}T}\eta\right)}. \tag{9.7}$$

The reaction rate in an agglomerate, Φ_1, can be calculated from the flux at $x = R$:

$$\Phi_1=-\frac{4\pi R^2 c|_{r=R+\delta}}{D_0\left(1+\frac{\delta}{R}\right)+D\left(\frac{R}{r_0}\coth\frac{R}{r_0}-1\right)}, \tag{9.8}$$

where D_0 is the diffusion coefficient at $R \le r \le R + \delta$. The parameters R and δ can be determined from direct observation of the agglomerates or by fitting the model performance prediction to the experimental results. When the number density of the agglomerate is equal to n, the generation of oxygen in the cathode catalyst layer is as follows:

$$\Phi_0 = n\Phi_1. \tag{9.9}$$

The number density of the agglomerate can be calculated from the porosity, ε, as

$$1-\varepsilon_0=\frac{4}{3}\pi(R+\delta)^3 n. \tag{9.10}$$

9.2.4 Macrohomogeneous 1D Through-Plane Model of an MEA

A useful experiment for determining the parameters R and δ is to operate the cell in differential mode, where the gases are supplied in large excess of stoichiometry. In this mode, the reaction distribution from the gas inlet to the outlet can be neglected, and only mass transport in the through-plane direction is relevant to cell performance. The performance at steady state can then be modeled by the continuity equation

$$\frac{\partial N}{\partial x}=\Phi, \tag{9.11}$$

where x is the coordinate in the through-plane direction, N is the flux in the x direction, and Φ is the source term.

The model is generated from the species continuity equations. The pressure change in the pore and heat conduction in the various layers can be neglected. In addition, the change in the electrode potential in the x direction can be neglected in the catalyst layers because the electron conductivity is much larger than the proton conductivity.

The flux of species i in the pore is represented by the sum of the advective and diffusive fluxes:

$$N_i = c_{\text{tot}} X_i v + J_i, \tag{9.12}$$

where c_{tot} is the total gas concentration in the pore, X_i is the mole fraction of species i, v is the average gas velocity, and J_i is the diffusion flux of species i. The diffusive flux can be described by Stefan–Maxwell equation

$$\frac{\partial X_i}{\partial x} = -\sum_{j=1}^{N} \frac{X_j J_i - X_i J_j}{c_{\text{tot}} D_{ij}}, \tag{9.13}$$

where N is the number of species and D_{ij} is the binary diffusion coefficient for the i–j gas pair.

The water flux in the ionomer, N_{w}, is the sum of the diffusive and electroosmotic fluxes

$$N_{\text{w}} = -\frac{\rho_{\text{I}}}{M_{\text{I}}} D_{\text{w}} \frac{\partial \lambda}{\partial x} + \frac{\beta}{F} i, \tag{9.14}$$

where ρ_{I} and M_{I} are the density and equivalent weight of the dry ionomer, respectively; λ is the water content (moles of water per mole of sulfonic acid group) of the ionomer, β is the electro-osmotic drag coefficient, and i is the proton current density.

The ionomer potential, φ, can be calculated using Ohm's law

$$i = -\kappa \frac{\partial \varphi}{\partial x}, \tag{9.15}$$

where κ is the proton conductivity.

The HOR rate in the anode catalyst layer, Φ_{H}, is described as a linearized form of the Butler–Volmer equation,

$$\Phi_{\text{H}} = -\frac{a^{\text{ACL}}}{2F} i_0^{\text{HOR}} \frac{c_{\text{H}}}{c_0^{\text{H}_2}} \frac{2F}{\mathcal{R}T} \eta, \tag{9.16}$$

where a^{ACL} is the catalyst area per unit volume of the catalyst layer, i_0^{HOR} is the reference current density of the HOR, and $c_0^{\text{H}_2}$ is the reference hydrogen concentration.

The generation rate of water vapor in the catalyst layer, Φ_w, is proportional to the difference in water activity between the ionomer (a_I) and pore (a_{gas}):

$$\Phi_w = a^{pore} k_{evap} \frac{p_{sat}}{\mathfrak{R}T}\left(a_I - a_{gas}\right), \qquad (9.17)$$

where a^{pore} is the ionomer area per unit volume of the catalyst layer, k_{evap} is the rate constant of evaporation, and p_{sat} is the saturation vapor pressure.

The continuity equations in two adjacent layers are combined under the imposition of boundary conditions to form the performance model. The model conditions are summarized as follows:

- The gas concentration is set at the GDL substrate-flow field boundary.
- The species concentration in a phase is continuous at two adjacent layers.
- The molar flux of the species is continuous at two adjacent layers.
- The ionic potential in the ionomer is defined as zero at the anode MPL–anode catalyst layer interface.
- The ionic potential in the ionomer is continuous at two adjacent layers.
- The proton flux is continuous at two adjacent layers.

9.3 From the Agglomerate to Interface Resistance Models

9.3.1 Problems in the Conventional Agglomerate Model

The agglomerate radius, R, and film thickness, δ, were determined by fitting the performance predictions to the experimental results [35]. Two operating conditions with different oxygen partial pressures, p_{ox}, were selected for the fitting: low concentration (p_{ox} = 1.9 kPa) and near the concentration of humidified air (p_{ox} = 19.5 kPa). The best fit obtained under these conditions was then tested at p_{ox} = 9.3 kPa, as shown in Fig. 9.5. The predictions were in fairly good agreement with the experimental data at current densities of <0.5 $A \cdot cm^{-2}$. However, agreement over the entire range could not be achieved even at p_{ox} = 19.5 kPa, but this disagreement is not the main problem of the model.

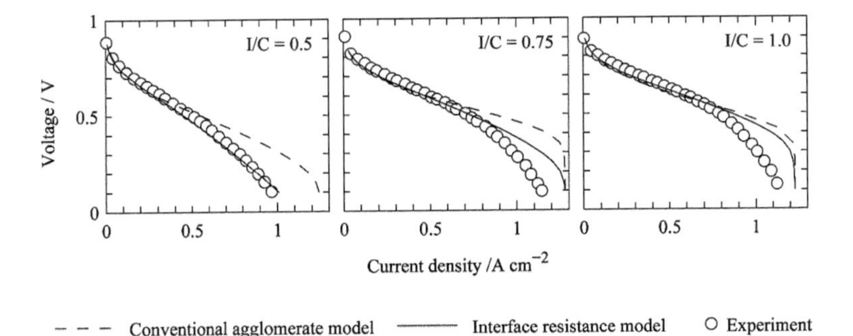

Figure 9.5 Comparison of performance between the model predictions and experimentally determined results at different ionomer-to-carbon ratios (I/C). Dashed line: conventional agglomerate model; solid line: interface resistance model; circles: experimental results. Reprinted from Ref. [35], Copyright (2013), with permission of Elsevier.

The major problem in the conventional model is that the agglomerate diameter $2(R + \delta)$ used is much larger than that typically observed in the micrographs of state-of-the-art catalyst layers. Table 9.1 shows the agglomerate diameter values, which are larger than 400 nm. In contrast, the typical sizes observed in the micrographs are similar to that of the primary particles, which is approximately 50 nm, as shown in Fig. 9.2. As a result, this model cannot be used to improve the structure of the catalyst layer as the obvious approach involving agglomerate size reduction is irrelevant to the improvement.

Table 9.1 Typical agglomerate sizes

	Agglomerate size (nm)	
Cathode I/C	Conventional agglomerate model	Interface resistance model
0.5	677	44
0.75	488	56
1.0	703	81

Source: Reprinted from Ref. [35], Copyright (2013), with permission of Elsevier.

9.3.2 Oxygen Transport Resistance in Thin Ionomer Films

We suspected that the difference between the agglomerate size in the model and the observed particle size in the micrograph arose from the overestimation of oxygen permeability of the thin ionomer film covering the catalyst particle because in the model, it was assumed to be equal to the value of films thicker than 20 mm. This assumption was supported by a study regarding another irregularity of thin ionomer films, which reported that the proton conductivity of Nafion films decreased with decreasing film thickness [31]. Thus, it is clear that thin ionomer films exhibit different properties from bulk ionomers because the interfaces inside the film may induce anisotropy. Accordingly, the permeability of thin Nafion films was experimentally estimated by measuring the mass-transport-limited current density of the ORR, i_d, using a Pt ultramicroelectrode [35]. As expected, thin Nafion films (thickness of 166 to 809 nm) exhibit lesser permeability than the thick (96 μm) film. However, the i_d did not approach infinity at the limit of zero thickness, as shown in Fig. 9.6.

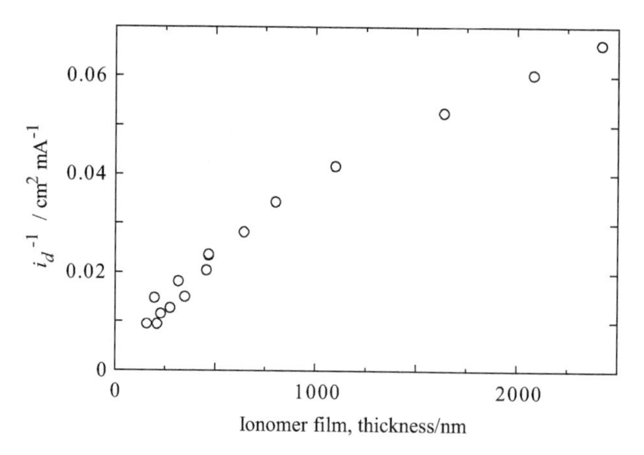

Figure 9.6 The inverse of the mass-transport-limited current density as a function of film thickness. Reprinted from Ref. [35], Copyright (2013), with permission of Elsevier.

The nonzero value of the intercept implies that the model shown in Fig. 9.3a is not applicable to the system but can be explained by assuming concentration (activity) jumps at the gas–ionomer and/or ionomer–Pt interface, as illustrated in Fig. 9.3b [16]. In this model, dissolution from the gas to the ionomer occurs at a rate proportional to the difference between the equilibrium (c_{eq}) and actual concentrations at the interface (c_{ion}). Transport from the ionomer to the Pt surface occurs at a rate proportional to the difference between the concentration near the Pt surface (c_{Pt}) and that on the Pt surface (assumed to be zero). When the mass-transport limitation is observed,

$$\frac{|i_d|}{4F} = \frac{1}{R_{ion}}\left(c_{eq} - c_{ion}\right) = D\frac{c_{ion} - c_{Pt}}{x} = \frac{1}{R_{Pt}}c_{Pt}, \tag{9.18}$$

where R_{ion} and R_{Pt} are the interface transport resistances and x is the film thickness. By eliminating c_{ion} and c_{Pt}, we obtain

$$\frac{c_{eq}}{|i_d|/4F} = \left(R_{ion} + R_{Pt}\right) + \frac{x}{D}. \tag{9.19}$$

The value of $R_{ion} + R_{Pt}$ (defined as $R_{interface}$) was derived from the intercept shown in Fig. 9.6. By assuming $R_{Pt} = 0$ and setting $c|_{r=R+\delta}$ in Eq. 9.8, the local current density, i_{local}, can be described as follows:

$$\frac{i_{local}}{4F} = \frac{1}{R_{ion}}\left(c_{eq} - c|_{r=R+\delta}\right). \tag{9.20}$$

When the largest $R_{interface}$ is used, the agglomerate size decreases and values close to the actual particle size observed in the micrographs (Table 9.1 and Fig. 9.2) are obtained without deteriorating prediction error (Fig. 9.5).[c] This result clearly shows that assuming interface resistance can reasonably explain the experimental results, as opposed to assuming that the agglomerate is much larger than that observed experimentally.[d,e]

9.3.3 Origin of the Interface Resistance

Using the limiting current density to estimate $R_{interface}$ is inadequate in that it cannot yield R_{ion} and R_{Pt} separately. Other experimental

---

[c]Use of the agglomerate model remains valid because the carbon support is Ketjenblack, where Pt is supported on the surface and in primary pores.

[d]The more accurate measurement of i_d revealed that the value of k^{-1} was similar to the smaller value obtained in Fig. 9.6 [16]. This may change the agglomerate size for the best fit but does not disprove the interface resistance hypothesis.

[e]The discrepancy between the model predictions and experiments at a high current density likely arises from the potential dependence or transport resistance, which is affected by the adsorbed species on the Pt surface.

methods have not succeeded in providing conclusive evidence as to which type of interface resistance more significantly affects oxygen transport [7, 15, 33, 40]. Performing numerical simulations is effective for tackling such problems. For this system, MD simulation can yield theoretical answers to the question of interface influence on interfacial transport resistance and oxygen transport hindrance [14].

A Nafion ionomer was modeled as a hexamer whose repeat unit is a PTFE-like main chain with a sidechain of a sulfonic acid group at the end of the chain. The Pt surface was modeled using a three-atomic layer Pt(111) slab with $8 \times 5\sqrt{3}$ 2D periodicity. Two Nafion ionomers were placed on the Pt surface with oxygen molecules, H_3O^+, and H_2O. The number of H_3O^+ ions and H_2O molecules were adjusted so that the system exhibited specific water content.

To obtain the mass-transport-limited current density by MD simulation, 100 O_2 molecules were placed above the ionomer, as shown in Fig. 9.7a. When an O_2 molecule reached the Pt surface, it was translated into the gas phase. Thus, the O_2 concentration was kept zero at the Pt surface and a diffusion-limited O_2 flux was obtained. The force field parameters were re-established in advance to overcome problems in using the conventional parameters.

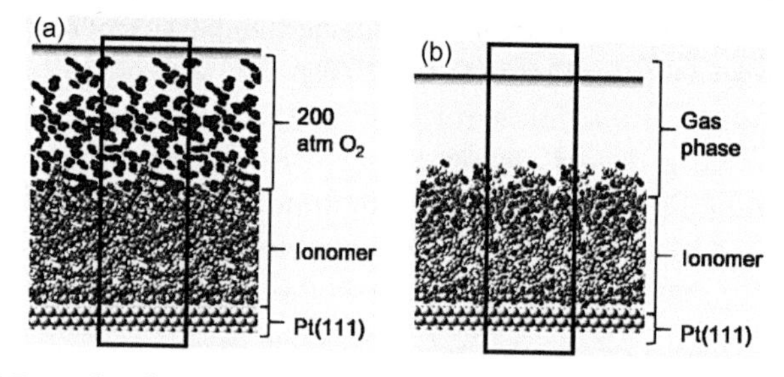

Figure 9.7 Nafion/Pt interfacial models used for (a) simulating O_2 permeation and (b) calculating the Helmholtz free energy profile. Reprinted from Ref. [14], Copyright (2016), with permission of Elsevier.

The density profile of O_2 at water content of 6 mol H_2O·mol $SO_3^{-\ -1}$ is shown in Fig. 9.8 (gray line) as a function of distance from the Pt surface. As O_2 approaches Pt from the gas phase, its density initially decreases from the gas phase to the ionomer phase at the gas–ionomer interface (between planes I and II) and then further decreases to zero at Z = 1 nm, where Z is the distance from the

Pt surface. Oxygen permeation from plane I to II as well as from plane II to I was 9.2×10^2 A·cm^{-2}·atm^{-1}, whereas the flux was only 0.94 A·cm^{-2}·atm^{-1} at the ionomer–Pt interface. This indicates that the O_2 profile equilibrates at the Nafion–gas interface and the rate of net O_2 permeation is determined by permeation at the ionomer–Pt interface.

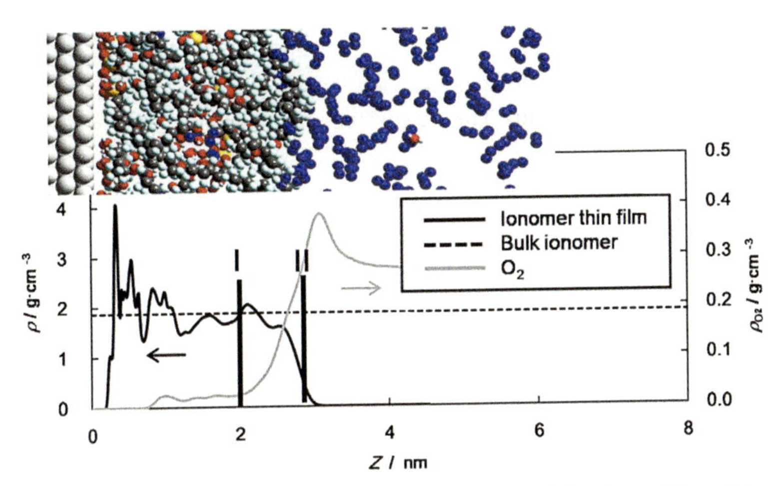

Figure 9.8 A snapshot of the ionomer/Pt interface and density profiles of the Nafion ionomer (solid black line) and oxygen molecules (solid gray line). The density of the bulk Nafion is shown as the dashed black line. Reprinted from Ref. [14], Copyright (2016), with permission of Elsevier.

The density profile can be used to determine the free energy profile of the O_2 molecules. The Helmholtz free energy profile of O_2 with respect to O_2 gas ΔA was calculated using the structure shown in Fig. 9.7b, and the resulting profile is shown in Fig. 9.9 for different water contents, λ. Two distinct peaks can be observed where the higher peak occurs at the ionomer–Pt interface (indicated by A) and the lower peak at the gas–ionomer interface (indicated by B). The diffusion-limited current density, i_L, can be determined using the higher peak (A) as follows:

$$i_L \propto c_{O_2(g)} \exp\left(-\frac{\Delta A_A}{k_B T}\right), \tag{9.21}$$

where $c_{O_2(g)}$ is the O_2 concentration in the gas, ΔA_A is the peak height in region A, and k_B is the Boltzmann constant. The position of peak

A is in the region where the ionomer density is larger than the bulk ionomer density, as observed in Fig. 9.8 (solid black line). Thus, the high-density ionomer phase near the ionomer–Pt interface is likely the origin of the oxygen transport resistance.

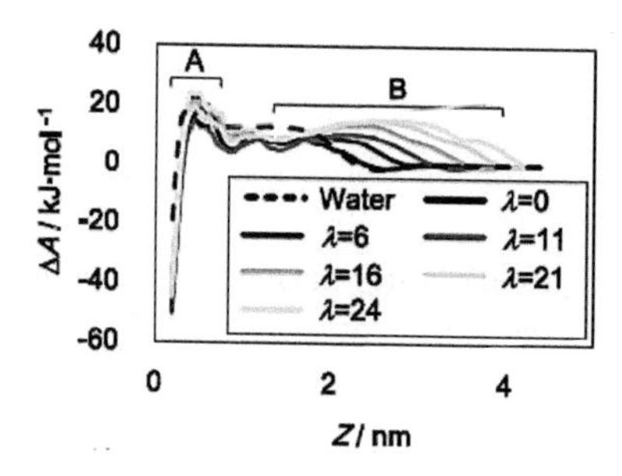

Figure 9.9 Helmholtz free energy profiles ΔA of O_2 molecules with respect to gas. Reprinted from Ref. [14], Copyright (2016), with permission of Elsevier.

9.4 Summary and Future Scope

The conventional agglomerate model for the catalyst layer of PEFCs was scrutinized to overcome performance loss when Pt loading is reduced. Although the performance curve predicted by the model can be fit to the experimental data to some degree, large agglomerates (several hundred nanometers), which are not observed in the state-of-the-art catalyst layers, must be assumed in the model. Measurement of the limiting current density of the ORR at the Pt microelectrode covered with thin Nafion films revealed interfacial transport resistance at the Pt–Nafion and/or Nafion–gas interface. Discrepancies in the structure of the catalyst layer were resolved by adding the interfacial resistance model that includes an oxygen concentration jump at the interfaces. MD simulations predicted that the rate-determining step of oxygen permeation was the interfacial permeation at the ionomer–Pt interface. The resistance at the interface was attributed to the dense ionomer layer formed in the immediate vicinity of the Pt surface.

Predictions of phenomenological models have provided novel insights into the process taking place in the cell by comparison with experimental results. New breakthrough will be achieved in this field through a combination of appropriately designed experiments, careful comparison of experimental results, and model predictions, allowing physical insight into the origin of the discrepancy.

References

1. Büchi, F. N. and Scherer, G. G. (2001). Investigation of the transversal water profile in nafion membranes in polymer electrolyte fuel cells, *J. Electrochem. Soc.*, **148**, pp. A183–A188.
2. Bernardi, D. M. (1990). Water-balance calculations for solid-polymer-electrolyte fuel cells, *J. Electrochem. Soc.*, **137**, pp. 3344–3350.
3. Bernardi, D. M. and Verbrugge, M. W. (1992). A mathematical model of the solid-polymer-electrolyte fuel cell, *J. Electrochem. Soc.*, **139**, pp. 2477–2491.
4. Berning, T. and Djilali, N. (2003). A 3D, multiphase, multicomponent model of the cathode and anode of a PEM fuel cell, *J. Electrochem. Soc.*, **150**, pp. A1589–A1598.
5. Broka, K. and Ekdunge, P. (1997). Modelling the PEM fuel cell cathode, *J. Appl. Electrochem.*, **27**, pp. 281–289.
6. Fuller, T. F. and Newman, J. (1993). Water and thermal management in solid-polymer-electrolyte fuel cells, *J. Electrochem. Soc.*, **140**, pp. 1218–1225.
7. Gómez-Marín, A. M., Berná, A., and Feliu, J. M. (2010). Spectroelectrochemical studies of the Pt(111)/Nafion interface cast electrode, *J. Phys. Chem. C*, **114**, pp. 20130–20140.
8. Gottesfeld, S. (2008). Some observations on the oxygen reduction reaction (ORR) at platinum catalysts based on post year 2000 reports, *ECS Trans.*, **6**, pp. 51–67.
9. Greszler, T. A., Caulk, D., and Sinha, P. (2012). The impact of platinum loading on oxygen transport resistance, *J. Electrochem. Soc.*, **159**, pp. F831–F840.
10. Guo, Q. and White, R. E. (2004). A steady-state impedance model for a PEMFC cathode, *J. Electrochem. Soc.*, **151**, pp. E133–E149.
11. Hinatsu, J. T., Mizuhata, M., and Takenaka, H. (1994). Water uptake of perfluorosulfonic acid membranes from liquid water and water vapor, *J. Electrochem. Soc.*, **141**, pp. 1493–1498.
12. Ihonen, J., Jaouen, F., Lindbergh, G., Lundblad, A., and Sundholm, G. (2002). Investigation of mass-transport limitations in the solid polymer fuel cell cathode: II. Experimental, *J. Electrochem. Soc.*, **149**, pp. A448–A454.

13. Jaouen, F., Lindbergh, G., and Sundholm, G. (2002). Investigation of mass-transport limitations in the solid polymer fuel cell cathode: I. Mathematical model, *J. Electrochem. Soc.*, **149**, pp. A437–A447.

14. Jinnouchi, R., Kudo, K., Kitano, N., and Morimoto, Y. (2016). Molecular dynamics simulations on O_2 permeation through Nafion ionomer on platinum surface, *Electrochim. Acta*, **188**, pp. 767–776.

15. Kodama, K., Jinnouchi, R., Suzuki, T., Murata, H., Hatanaka, T., and Morimoto, Y. (2013). Increase in adsorptivity of sulfonate anions on Pt (111) surface with drying of ionomer, *Electrochem. Commun.*, **36**, pp. 26–28.

16. Kudo, K., Jinnouchi, R., and Morimoto, Y. (2016). Humidity and temperature dependences of oxygen transport resistance of Nafion thin film on platinum electrode, *Electrochim. Acta*, **209**, pp. 682–690.

17. Lefebvre, M. C., Martin, R. B., and Pickup, P. G. (1999). Characterization of ionic conductivity profiles within proton exchange membrane fuel cell gas diffusion electrodes by impedance spectroscopy, *Electrochem. Solid-State Lett.*, **2**, pp. 259–261.

18. Liu, Y., Murphy, M. W., Baker, D. R., Gu, W., Ji, C., Jorne, J., and Gasteiger, H. A. (2009). Proton conduction and oxygen reduction kinetics in PEM fuel cell cathodes: effects of ionomer-to-carbon ratio and relative humidity, *J. Electrochem. Soc.*, **156**, pp. B970.

19. Liu, Y., Ji, C., Gu, W., Jorne, J., and Gasteiger, H. A. (2011). Effects of catalyst carbon support on proton conduction and cathode performance in PEM fuel cells, *J. Electrochem. Soc.*, **158**, pp. B614.

20. Liu, Y. X., Ji, C. X., Gu, W. B., Baker, D. R., Jorne, J., and Gasteiger, H. A. (2010). Proton conduction in PEM fuel cell cathodes: effects of electrode thickness and ionomer equivalent weight, *J. Electrochem. Soc.*, **157**, pp. B1154–B1162.

21. Motupally, S., Becker, A. J., and Weidner, J. W. (2000). Diffusion of water in Nafion 115 membranes, *J. Electrochem. Soc.*, **147**, pp. 3171–3177.

22. Natarajan, D. and Van Nguyen, T. (2001). A two-dimensional, two-phase, multicomponent, transient model for the cathode of a proton exchange membrane fuel cell using conventional gas distributors, *J. Electrochem. Soc.*, **148**, pp. A1324–A1335.

23. Nguyen, T. V. (1996). A gas distributor design for proton-exchange-membrane fuel cells, *J. Electrochem. Soc.*, **143**, pp. L103–L105.

24. Nonoyama, N., Okazaki, S., Weber, A. Z., Ikogi, Y., and Yoshida, T. (2011). Analysis of oxygen-transport diffusion resistance in proton-exchange-membrane fuel cells, *J. Electrochem. Soc.*, **158**, pp. B416.

25. Ohma, A., Mashio, T., Sato, K., Iden, H., Ono, Y., Sakai, K., Akizuki, K., Takaichi, S., and Shinohara, K. (2011). Analysis of proton exchange membrane fuel cell catalyst layers for reduction of platinum loading at Nissan, *Electrochim. Acta*, **56**, pp. 10832–10841.

26. Owejan, J. P., Owejan, J. E., and Gu, W. (2013). Impact of platinum loading and catalyst layer structure on PEMFC performance, *J. Electrochem. Soc.*, **160**, pp. F824–F833.

27. Parthasarathy, A., Martin, C. R., and Srinivasan, S. (1991). Investigations of the O_2 reduction reaction at the platinum/Nafion® interface using a solid-state electrochemical cell, *J. Electrochem. Soc.*, **138**, pp. 916–921.

28. Parthasarathy, A., Srinivasan, S., Appleby, A. J., and Martin, C. R. (1992). temperature dependence of the electrode kinetics of oxygen reduction at the platinum/Nafion® interface: a microelectrode investigation, *J. Electrochem. Soc.*, **139**, pp. 2530–2537.

29. Parthasarathy, A., Srinivasan, S., Appleby, A. J., and Martin, C. R. (1992). Pressure dependence of the oxygen reduction reaction at the platinum microelectrode/Nafion interface: electrode kinetics and mass transport, *J. Electrochem. Soc.*, **139**, pp. 2856–2862.

30. Pasaogullari, U. and Wang, C. Y. (2004). Liquid water transport in gas diffusion layer of polymer electrolyte fuel cells, *J. Electrochem. Soc.*, **151**, pp. A399–A406.

31. Siroma, Z., Kakitsubo, R., Fujiwara, N., Ioroi, T., Yamazaki, S.-i., and Yasuda, K. (2009). Depression of proton conductivity in recast Nafion® film measured on flat substrate, *J. Power Sources*, **189**, pp. 994–998.

32. Sone, Y., Ekdunge, P., and Simonsson, D. (1996). Proton conductivity of Nafion 117 as measured by a four-electrode AC impedance method, *J. Electrochem. Soc.*, **143**, pp. 1254–1259.

33. Subbaraman, R., Strmcnik, D., Paulikas, A. P., Stamenkovic, V. R., and Markovic, N. M. (2010). Oxygen reduction reaction at three-phase interfaces, *ChemPhysChem*, **11**, pp. 2825–2833.

34. Subramanian, N. P., Greszler, T. A., Zhang, J., Gu, W., and Makharia, R. (2012). Pt-oxide coverage-dependent oxygen reduction reaction (ORR) kinetics, *J. Electrochem. Soc.*, **159**, pp. B531.

35. Suzuki, T., Kudo, K., and Morimoto, Y. (2013). Model for investigation of oxygen transport limitation in a polymer electrolyte fuel cell, *J. Power Sources*, **222**, pp. 379–389.

36. Springer, T. E., Zawodzinski, T. A., and Gottesfeld, S. (1991). Polymer electrolyte fuel cell model, *J. Electrochem. Soc.*, **138**, pp. 2334–2342.

37. Uribe, F. A., Wilson, M. S., Springer, T. E., and Gottesfeld, S. (1992). Oxygen reduction (ORR) at the Pt/recast ionomer interface and some general comments on the ORR at Pt/aqueous electrolyte interfaces, *ECS Proc.*, **92–11**, pp. 494–509.
38. Wang, Y. and Wang, C.-Y. (2006). A nonisothermal, two-phase model for polymer electrolyte fuel cells, *J. Electrochem. Soc.*, **153**, pp. A1193–A1200.
39. Wilson, M. S. and Gottesfeld, S. (1992). High performance catalyzed membranes of ultra-low Pt loadings for polymer electrolyte fuel cells, *J. Electrochem. Soc.*, **139**, pp. L28–L30.
40. Wood, D. L., Chlistunoff, J., Majewski, J., and Borup, R. L. (2009). Nafion structural phenomena at platinum and carbon interfaces, *J. Am. Chem. Soc.*, **131**, pp. 18096–18104.
41. Yamada, H., Hatanaka, T., Murata, H., and Morimoto, Y. (2006). Measurement of flooding in gas diffusion layers of polymer electrolyte fuel cells with conventional flow field, *J. Electrochem. Soc.*, **153**, pp. A1748–A1754.
42. Yi, J. S. and Van Nguyen, T. (1999). Multicomponent transport in porous electrodes of proton exchange membrane fuel cells using the interdigitated gas distributors, *J. Electrochem. Soc.*, **146**, pp. 38–45.
43. Zawodzinski, T. A., Neeman, M., Sillerud, L. O., and Gottesfeld, S. (1991). Determination of water diffusion coefficients in perfluorosulfonate ionomeric membranes, *J. Phys. Chem.*, **95**, pp. 6040–6044.
44. Zawodzinski, T. A., Derouin, C., Radzinski, S., Sherman, R. J., Smith, V. T., Springer, T. E., and Gottesfeld, S. (1993). Water uptake by and transport through Nafion® 117 membranes, *J. Electrochem. Soc.*, **140**, pp. 1041–1047.
45. Zawodzinski, T. A., Davey, J., Valerio, J., and Gottesfeld, S. (1995). The water content dependence of electro-osmotic drag in proton-conducting polymer electrolytes, *Electrochim. Acta*, **40**, pp. 297–302.

Index